CITES Orchid Checklist

Volume 5

Bulbophyllum and allied taxa (Orchidaceae)
For the genera:

*Acrochaene, Bulbophyllum, Chaseella, Codonosiphon,
Drymoda, Monomeria, Monosepalum, Pedilochilus,
Saccoglossum, Sunipia* and *Trias*

T0141649

Compiled by:

Sieder A., Rainer H., Kiehn M.

Final editing by:

Alexandra Bell, Chris Brodie and Rosemary Simpson

Kew Publishing
Royal Botanic Gardens, Kew

First published in 2009 by Royal Botanic Gardens, Kew Richmond, Surrey, TW9 3AB, UK (electronic version 2007)

www.kew.org

ISBN 978 1 84246 225 6

British Library Cataloguing in Publication Data
A catalogue record for this book is available from the British Library.

Produced with the financial assistance of the University of Vienna (Faculty of Life Sciences), the Austrian Ministry of Agriculture and Forestry, Environment and Water Management (CITES Management Authority of Austria), and the Royal Botanic Gardens, Kew.

Cover design by Publishing, Design and Photography, Royal Botanic Gardens, Kew.

Printed in the United Kingdom by Henry Ling Limited

For information or to purchase all Kew titles please visit www.kewbooks.com or email publishing@kew.org

All proceeds go to support Kew's work in saving the world's plants for life

Mixed Sources
Product group from well-managed forests and other controlled sources
www.fsc.org Cert no. SA-COC-001860
© 1996 Forest Stewardship Council
FSC

The paper used in this book contains material sourced from responsibly managed and sustainable commercial forests, certified in accordance with the FSC (Forestry Stewardship Council).

Acknowledgements / Remerciements / Agradecimientos

For help and comments during the preparation and publication of this checklist / pour leur assistance et leurs commentaires pendant la préparation et la publication de cette Liste / por asistencia y comentarios en la preparación y publicación de esta lista./ The compilers would like to thank / Les auteurs tiennent à remercier / Los autores desean expresar su agradecemiento a:

Max Abensperg-Traun (Austrian Ministry of Agriculture and Forestry, Environment and Water Management, Vienna), Chris Brodie (Conventions and Policy Section, Royal Botanic Gardens, Kew), Wolfgang Brunnbauer (Library, Museum of Natural History, Vienna), Gunter Fischer (Department of Biogeography, University of Vienna), Irawati (Botanic Gardens of Indonesia, Bogor), Noel McGough, Rosemary Simpson, Alexandra Bell and Lucy Garrett (Conventions and Policy Section, Royal Botanic Gardens, Kew), Gabriele Palfinger (Library, Museum of Natural History, Vienna), Robert Stangl (Botany Library, University of Vienna), Albert-Dieter Stevens (Botanic Garden and Herbarium, Free University of Berlin), Walter Till (Herbarium, University of Vienna), the staff of the library and the Botanical Garden, University of Leiden (The Netherlands), and the members and observers of the CITES Plants Committee.

We would particularly like to thank / Nous souhaitons remercier plus particulièrement / Deseamos dar especialmente las gracias a / For logistic and financial support for the preparation and publication of the checklist – without their help this project could not have been successfully completed / pour leur aide logistique et financière pour la préparation et publication de la présente Liste. Sans leur assistance, l'accomplissement de ce projet n'aurait pas été possible. / por el apoyo logístico y financiero para preparar y publicar esta lista, ya que sin su asistencia esta publicación no habría visto la luz:

The Austrian Ministry of Agriculture and Forestry, Environment and Water Management (CITES Management Authority of Austria); the Royal Botanic Gardens, Kew; the CITES Secretariat, Geneva; the Austrian Science Foundation (FWF, P-17124-B03); and the University of Vienna (Department of Biogeography, Faculty of Life Sciences).

This book is dedicated to Deborah Rhoads Lyon, friend and CITES wordsmith, who will be greatly missed but always remembered

Cet ouvrage est dédié à la mémoire de Deborah Rhoads Lyon, amie et artisantraductrice de CITES, qui sera à jamais regrettée

Este libro está dedicado a Deborah Rhoads Lyon, amiga y maestra artesana de las palabras de la CITES, a la que siempre echaremos mucho de menos

CONTENTS

Preamble

Table des matières

TABLE DES MATIERES

Préambule

ÍNDICE

Preámbulo

CITES CHECKLIST – BULBOPHYLLUM

PREAMBLE

1. Background
This checklist for *Bulbophyllum* and allied taxa (Orchidaceae) was prepared as a result of recommendations endorsed by the CITES Conference of the Parties and has been approved by the CITES Nomenclature Committee. It was adopted in electronic version as a standard reference at the 14th meeting of the Confernce of the Parties in June 2007.

The preparation of this comprehensive checklist was made possible through the financial support of the Austrian Ministry of Agriculture and Forestry, Environment and Water Management - the CITES Management Authority of Austria.

It was prepared following a global consultation process with specialists on the taxonomy and systematics of this group. A draft version of this checklist was made available for comment on the CITES website and Parties were informed of this through Notification 2005/049. The comments received by Parties were incorporated in a revised text.

As this checklist is designed to be used by non–specialists the compliers have taken a pragmatic approach to dealing with the large number of species included in the Bulbophyllinae. Therefore, the genus *Bulbophyllum* is dealt with in a broad sense. This does not express any opinion or preference of the compilers as regards the taxonomic rank or circumscription of subdivisions or segregates of the Bulbophyllinae. Such statements will, if at all, only be possible after the finalizing of the ongoing molecular studies on the Bulbophyllinae (i.e., by B. Gravendeel, Leiden, and M. Kiehn & G. Fischer, Vienna/Salzburg).

If taxa now considered to be members of the Bulbophyllinae have synonyms assigned to genera outside of the Bulbophyllinae, these names are included in the alphabetical list. Such genera outside of the Bulbophyllinae containing erroneously attributed taxa are not included in the alphabetical list of taxa and only with the relevant taxa in the context of the Bulbophyllinae.

2. How to use the Checklist
It is intended that this Checklist be used as a quick reference for checking accepted names, synonymy and distribution. The reference is therefore divided into three main parts:

Part I: All names in current use
An alphabetical list of all accepted names and synonyms included in this checklist.

Part II: Accepted names in current use
Separate lists for each genus. Each list is ordered alphabetically by the accepted name and details are given on current synonyms and distribution.

Part III: Country checklist
Accepted names from all genera included in this Checklist are ordered alphabetically under country of distribution.

Preamble

3. Conventions employed in Parts I, II and III

a) Accepted names are presented in **bold roman** type.
Synonyms are presented in *italic* type.

b) Duplicate names

In Part I, the author's name appears after each taxon where the taxon name appears twice or more. For example:

All Names
Bulbophyllum curtisii (Hook. f.) J. J. Sm.
Bulbophyllum curtisii (Hook. f.) Ridl.

i) Where a synonym occurs more than once, but refers to different species, for example, *Bulbophyllum curtisii*, for both **Bulbophyllum corolliferum** and **Bulbophyllum purpurascens**, an asterisk indicates the species most likely to be encountered in trade - if this is known. For example:

All Names	**Accepted Name**
*Bulbophyllum curtisii (Hook. f.) J. J. Sm…....**Bulbophyllum corolliferum**	
Bulbophyllum curtisii (Hook. f.) Ridl………..**Bulbophyllum purpurascens**	

*Species most likely to be in trade (in this example, **Bulbophyllum corolliferum**).

ii) Where an accepted name and a synonym are the same, but refer to different species, for example, **Bulbophyllum acuminatum** (accepted name) and *Bulbophyllum acuminatum* (a synonym of **Bulbophyllum andreeae**), the name with an asterisk is the species most likely to be seen in trade - if this is known. For example:

All Names	**Accepted Name**
*Bulbophyllum acuminatum (Ridl.) Ridl.	
Bulbophyllum acuminatum Schltr. ……………. **Bulbophyllum andreeae**	

*Species most likely to be in trade (in this example, **Bulbophyllum acuminatum** (Ridl.) Ridl.).

NB: In examples b) i) and b) ii) it is necessary to double-check by reference to the distribution as detailed in Part II. For instance, in the example b) ii), if the name given was '*Bulbophyllum acuminatum*' and it was known that the plant in question came from Papua New Guinea this would indicate that the species was **Bulbophyllum andreeae,** being traded under the synonym *Bulbophyllum acuminatum*. **Bulbophyllum acuminatum** is only found in Indonesia, Malaysia, Singapore and Thailand.

c) New names, combinations and synonyms in *Bulbophyllum* Thouars (Orchidaceae) adopted by CITES

In preparations for this checklist, several new names and new combinations for treated taxa were necessary. In order to be effectively and validly published, these names are included in Annex I. These new names, combinations and synonyms are indicated in the main text with the symbol †.

For example:

All Names   **Accepted Name**
Bulbophyllum arminii †

Bulbophyllum arminii is a new name in *Bulbophyllum* Thouars (Orchidaceae). The name is published in Annex I.

Bulbophyllum arminii Sieder & Kiehn nom. nov.

New name for *Hapalochilus bandischii* Garay, Hamer & Siegerist

d)   Names not formally adopted by CITES

Since the adoption of the checklist by the 14th meeting of the Conference of the Parties (CoP) in The Hague, the Netherlands, June 2007, a number of new species and combinations of names and synonyms have been recognised in *Bulbophyllum* and its allies. These names are not formally included in this checklist as accepted names recognised by CITES. However, in order to make this checklist as user-friendly as possible, such changes are informally included in Part I of the checklist and are marked with the symbol "‡" to distinguish them from the adopted names. For example:

All Names   **Accepted Name**
‡ *Bulbophyllum calimanianum* ‡

The name *Bulbophyllum calimanianum* is in italics enclosed in the symbols ‡. *Bulbophyllum calimanianum* does not appear in bold in Part I, as the name was not included in the original checklist and thus is not formally accepted by CITES. *Bulbophyllum calimanianum* was accepted by the compilers after the adoption of the checklist by the CITES CoP.

All Names   **Accepted Name**
Bulbophyllum cornutum (Blume) Rchb. f.
‡*Bulbophyllum cornutum* (Blume) Rchb. f. .....*Hamularia cornuta*‡

The name **Bulbophyllum cornutum** (Blume) Rchb. f. is in bold and was adopted by the CITES CoP as the accepted name. However, the same name, *Bulbophyllum cornutum* (Blume) Rchb. f. and *Hamularia cornuta* are in italics and enclosed by symbols ‡. In this instance *Bulbophyllum cornutum* (Blume) Rchb. f. has been demoted to a synonym and *Hamularia cornuta* is promoted as the accepted name recognised by the compilers after the adoption of the checklist by the CITES CoP. **Bulbophyllum cornutum** (Blume) Rchb. f. is the offical CITES name.

All Names   **Accepted Name**
Bulbophyllum concavum...............................**Bulbophyllum cornutum**
                                                     (Blume) Rchb. f.
‡*Bulbophyllum concavum*.............................*Hamularia cornuta*‡

The name *Bulbophyllum concavum* is a synonym of **Bulbophyllum cornutum** (Blume) Rchb. f. seen in bold and was adopted by the CITES CoP as the accepted name. However, the same name, *Bulbophyllum concavum* and *Hamularia cornuta* are in italics and enclosed by symbols ‡. In this instance *Bulbophyllum concavum* is a synonym of *Hamularia cornuta* as recognised by the compilers after the adoption of the checklist by the CITES CoP. **Bulbophyllum cornutum** (Blume) Rchb. f. is the adopted CITES name and *Bulbophyllum concavum* is its synonym.

e)   For author abbreviations the following references have been used.

Brummitt, R.K. & Powell C.E. (1992). Authors of plant names. Royal Botanic Gardens, Kew. UK.

The International Plant Names Index (Accessed multiple times, latest access September, 2007). Published on the internet, http://www.ipni.org

f)   The CD-ROM contains the following files:

'CITESOrchidChecklistV.pdf', an Adobe Acrobat® file. This file contains CITES Orchid Checklist 5 and can be viewed using Adobe Reader®. You will need Adobe Acrobat Reader® installed on your computer to view this and the following files (can be downloaded from www.adobe.com).

'CITES_Orchid_Checklist4.pdf', an Adobe Acrobat® file. This file contains CITES Orchid Checklist 4.

'Introduction.pdf', an Adobe Acrobat® file. This files contains the introductory text to the CITES Orchid Checklists Volumes 1-3.

'PART_I_All_names.pdf', an Adobe Acrobat® file. This file contains an alphabetical list of all accepted names and synonyms for the genera as included in CITES Orchid Checklists Volumes 1-3.

'PART_II_Names_in_current_use.pdf', an Adobe Acrobat® file. This file contains an ordered alphabetical list of the accepted names and their current synonyms and distribution for the genera in CITES Orchid Checklists Volumes 1-3.

'PART_III_Country_distribution.pdf', an Adobe Acrobat® file. This file contains accepted names from all genera included in CITES Orchid Checklists Volumes 1-3, ordered alphabetically under country of distribution.

Navigation of the Adobe Acrobat® files is made easy using either the thumbnails on the left hand side of the document or active links within the document. All active links within the document are highlighted in blue.

4. Geographical areas
Country names follow the United Nations standard found in the UNTERM website (http://unterm.un.org/).

5. Orchidaceae controlled by CITES
The family Orchidaceae is listed on Appendix II of CITES. In addition the following taxa are listed on Appendix I at time of publication:

Aerangis ellisii
Dendrobium cruentum
Laelia jongheana

Laelia lobata
Paphiopedilum spp.
Peristeria elata
Phragmipedium spp.
Renanthera imschootiana

6. Abbreviations, botanical terms and Latin*

Not all the following abbreviations, botanical terms and Latin appear in this Checklist. However, they have been included as a useful reference. Note: words in *italics* are Latin.

ambiguous name a name which has been applied to different taxa by different authors, so that it has become a source of ambiguity
anon. anonymous; without author or author unknown
auct. *auctorum*: of authors
CITES Convention on International Trade in Endangered Species of Wild Fauna and Flora
cultivar an individual, or assemblage of plants maintaining the same distinguishing features, which has been produced or is maintained (propagated) in cultivation
cultivation the raising of plants by horticulture or gardening; not immediately taken from the wild
descr. *descriptio*: the description of a species or other taxonomic unit
distribution where plants are found (geographical)
ed. editor
edn. edition (book or journal)
eds. editors
epithet the last word of a species, subspecies, or variety (etc.), for example: *speciosum* is the species epithet for the species *Bulbophyllum speciosum*
escape a plant that has left the boundaries of cultivation (e.g. a garden) and is found occurring in natural vegetation
ex *ex*: after; may be used between the name of two authors, the second of whom validly published the name indicated or suggested by the first
excl. *exclusus*: excluded
forma *forma*: a taxonomic unit inferior to variety
hort. *hortorum*: of gardens (horticulture); raised or found in gardens; not a plant of the wild
ICBN International Code for Botanical Nomenclature
in prep. in preparation
in sched. *in scheda*: on a herbarium specimen or label
in syn. *in synonymia*: in synonymy
incl. including
ined. *ineditus*: unpublished
introduction a plant which occurs in a country, or any other locality, due to human influence (by purpose or chance); any plant which is not native
key a written system used for the identification of organisms (e.g. plants)
leg. *legit*: he gathered; the collector
misspelling a name that has been incorrectly spelt; not a new or different name
morphology the form and structure of an organism (e.g. a plant)
name causing confusion a name that is not used because it cannot be assigned unambiguously to a particular taxon (e.g. a species of plant)
native an organism (e.g. a plant) that occurs naturally in a country, or region, etc.
naturalized a plant which has either been introduced (see introduction) or has escaped (see escape) but which looks like a wild plant and is capable of reproduction in its new environment

Preamble

nec *neque*: and not, also not, neither

nom. *nomen*: name

nom. ambig. *nomen ambiguum*: ambiguous name

nom. cons. prop. *nomen conservandum propositum*: name proposed for conservation under the rules of the International Code for Botanical Nomenclature (ICBN)

nom. illeg *nomen illegitimum*: illegitimate name

nom. nud. *nomen nudum*: name published without description

nomenclature branch of science concerned with the naming of organisms (e.g. plants)

non *non*: not

only known from cultivation a plant which does not occur in the wild, only in cultivation

orthographic variant an alternative spelling for the same name

p.p. *pro parte*: partly, in part

provisional name name given in anticipation of a valid description

sens. *sensu*: in the sense of; the manner in which an author interpreted or used a name

sens. lat. *sensu lato*: in the broad sense; a taxon (usually a species) and all its subordinate taxa (e.g. subspecies) and/or other taxa sometimes considered as distinct

sensu *sensu*: in the sense of; the manner in which an author interpreted or used a name

sic *sic*, used after a word that looks wrong or absurd, to show that it has been quoted correctly

spp. species

ssp. subspecies

synonym a name that is applied to a taxon but which cannot be used because it is not the accepted name - the synonym or synonyms form the synonymy

taxa plural of taxon

taxon a named unit of classification, e.g. genus, species, subspecies

var. variety

*thanks to Dr Aaron Davis, RBG Kew, for the provision of this guide.

Preamble

LISTE CITES - BULBOPHYLLUM

PRÉAMBULE

1. Contexte

Cette Liste regroupant les *Bulbophyllum* et taxa affins (*Orchidacées*), établie à partir des recommandations adoptées par la Conférence des Parties de CITES, a reçu l'approbation du Comité de la Nomenclature de CITES. Elle a été adoptée dans sa version électronique en tant que référence normalisée lors de la 14e réunion de la Conférence des Parties, en juin 2007.

La préparation de la présente Liste détaillée a été rendue possible grâce au soutien financier du ministère autrichien de l'Agriculture, des Forêts, de l'Environnement et de Gestion de l'Eau – l'Organe de Gestion CITES pour l'Autriche.

Elle a été préparée à la suite d'un processus de consultation globale mené auprès de spécialistes de la taxonomie et de la systématique de ce groupe. Une version préliminaire de cette Liste a été publiée, pour commentaires, sur le site web de CITES, ce dont les Parties ont été informées à travers la Notification 2005/049. Les commentaires reçus par les Parties ont été incorporés à une version corrigée du texte.

Cette Liste étant destinée à être utilisée par des non–spécialistes, ses auteurs ont adopté une approche pragmatique pour traiter le grand nombre d'espèces appartenant à la sous-tribu Bulbophyllinae. Le genre *Bulbophyllum* doit par conséquent être pris dans un sens large. Ce fait ne représente toutefois nullement une quelconque opinion ou préférence de ses auteurs concernant le rang taxonomique ou la circonscription des subdivisions ou des isolés appartenant aux Bulbophyllinae. De telles affirmations ne seront possibles, en tout état de cause, qu'à l'issue des études moléculaires actuellement en cours concernant les Bulbophyllinae (i.e., par B. Gravendeel, de Leyde, et M. Kiehn & G. Fischer, de Vienne/Salzbourg).

Lorsque des taxa actuellement considérés comme appartenant aux Bulbophyllinae ont des synonymes affectés à des genres n'appartenant pas aux Bulbophyllinae, ces noms sont inclus dans la liste alphabétique. Cependant, les genres n'appartenant pas aux Bulbophyllinae et incluant des taxa erronément attribués ne sont pas inclus dans la liste alphabétique des taxa, mais uniquement parmi les taxa appropriés dans le contexte des Bulbophyllinae.

2. Comment utiliser la Liste?

Cette Liste vise à être utilisée en tant que liste de référence afin de permettre de vérifier rapidement les noms acceptés, les synonymes et la répartition géographique. Elle est divisée en trois parties principales:

Première partie: tous les noms actuellement en usage
Liste alphabétique de tous les noms acceptés et des synonymes inclus dans la Liste.

Deuxième partie: noms acceptés d'usage courant
Il existe une liste pour chaque genre. Chacune de ces listes contient les noms acceptés, par ordre alphabétique, et inclut des indications concernant les synonymes actuels et la répartition géographique.

Préambule

Troisième partie: liste par pays
Les noms acceptés de tous les genres inclus dans cette Liste apparaissent par ordre alphabétique pour chaque pays de l'aire de répartition.

3. Conventions utilisées dans la première, la deuxième et la troisième partie

a)   Les noms acceptés sont en **caractères gras**.
    Les synonymes sont en *italique*.

b)   Noms identiques pour des taxa différents:

    Dans la première partie, le nom de l'auteur apparaît après chaque taxon lorsque le taxon est mentionné deux fois ou plus. Exemple:

    Tous les noms
    Bulbophyllum curtisii (Hook. f.) J. J. Sm.
    Bulbophyllum curtisii (Hook. f.) Ridl.

i)   Lorsque le synonyme apparaît plus d'une fois mais fait référence à des noms acceptés différents – par exemple, *Bulbophyllum curtisii*, (synonyme à la fois de **Bulbophyllum corolliferum** et de **Bulbophyllum purpurascens**), le nom comportant un astérisque est celui de l'espèce la plus susceptible d'être trouvée dans le commerce, lorsqu'il est connu. Exemple:

    Tous les noms   **Noms acceptés**
    *Bulbophyllum curtisii (Hook. f.) J. J. Sm...... **Bulbophyllum corolliferum**
    Bulbophyllum curtisii (Hook. f.) Ridl............. **Bulbophyllum purpurascens**

    *Espèce la plus susceptible d'être trouvée dans le commerce (dans cet exemple, **Bulbophyllum corolliferum**)

ii)   Lorsque le nom accepté et un synonyme sont identiques, mais renvoient à des espèces différentes - par exemple, **Bulbophyllum acuminatum** (nom accepté) et *Bulbophyllum acuminatum* (synonyme de **Bulbophyllum andreeae**), le nom comportant un astérisque est celui de l'espèce la plus susceptible d'être trouvée dans le commerce, lorsqu'il est connu. Exemple:

    Tous les noms   **Noms acceptés**
    *Bulbophyllum acuminatum (Ridl.) Ridl.
    Bulbophyllum acuminatum Schltr. .... **Bulbophyllum andreeae**

    *Espèce la plus susceptible d'être trouvée dans le commerce (dans cet exemple, **Bulbophyllum acuminatum** (Ridl.) Ridl.).

    NB: Dans les exemples b i) et b ii), il faut procéder à une double vérification en se reportant à la répartition géographique indiquée dans la deuxième partie. Ainsi, dans l'exemple b ii), si le nom donné est *Bulbophyllum acuminatum* et que l'on sait que la plante vient de **Papouasie-Nouvelle-Guinée**, cela indique qu'il s'agit de **Bulbophyllum andreeae**, commercialisée sous le synonyme *Bulbophyllum acuminatum*. **Bulbophyllum acuminatum** ne pousse qu'en **Indonésie**, en **Malaisie**, à **Singapour** et en **Thaïlande**.

c)   Les nouveaux noms, combinaisons et synonymes appartenant à la sous-tribu *Bulbophyllum* Thouars (Orchidacées) adoptés par CITES.

Dans le cadre des travaux préparatoires de cette Liste, l'adoption d'un certain nombre de nouveaux noms et de nouvelles combinaisons concernant les taxa en question s'est révélé nécessaire. Afin de permettre leur publication valide et efficace, ces noms ont été regroupés dans l'Annexe I.

Pour permettre d'identifier ces nouveaux noms, combinaisons et synonymes dans le corps du texte, ces termes sont accompagnés du symbole †. Par exemple:

Tous les noms   **Noms acceptés**
Bulbophyllum arminii†

Bulbophyllum arminii est un nouveau nom parmi les *Bulbophyllum* Thouars (Orchidacées), et figure, à ce titre, dans l'Annexe I.

Bulbophyllum arminii Sieder & Kiehn nom. nov.
Nouveau nom pour *Hapalochilus bandischii* Garay, Hamer & Siegerist

d)   Noms non formellement acceptés par CITES

Depuis l'adoption de la présente Liste, à l'occasion de la 14ᵉ réunion de la Conférence des Parties à La Haye (Pays-Bas) en juin 2007, un certain nombre de nouvelles espèces, de combinaisons de noms et de synonymes ont été reconnus comme appartenant au genre *Bulbophyllum* et espèces affines. Ces noms ne sont pas formellement inclus dans la présente Liste en tant que noms acceptés et reconnus par CITES. Toutefois, afin de rendre ladite Liste le plus conviviale possible vis-à-vis de l'utilisateur, ces termes ont été inclus, bien que de manière informelle, dans la Partie n 1 de la Liste, et identifiés grâce au symbole ‡ afin de permettre leur distinction par rapport aux noms officiellement adoptés. Par exemple:

Tous les noms   **Nom accepté**
‡ *Bulbophyllum calimanianum* ‡

Le nom *Bulbophyllum calimanianum* est en italique et encadré par les symboles ‡. Le nom *Bulbophyllum calimanianum* n'ayant pas été inclus dans la Liste originelle ni été accepté par CITES, il n'apparaît pas en caractères gras dans la Partie n 1. *Bulbophyllum calimanianum* n'a en effet été accepté que lors de la compilation, postérieurement à l'adoption de la Liste par la Conférence des Parties de CITES.

Tous les noms   **Noms acceptés**
Bulbophyllum cornutum (Blume) Rchb. f.
‡*Bulbophyllum cornutum* (Blume) Rchb. f. .... *Hamularia cornuta*‡

Le nom **Bulbophyllum cornutum** (Blume) Rchb. f., en caractères gras, a été adopté par la Conférence des Parties de CITES en tant que nom accepté.

Cependant, ce même nom, *Bulbophyllum cornutum* (Blume) Rchb. f., ainsi que celui de *Hamularia cornuta,* sont en italiques et encadrés par les symboles ‡. En l'occurrence, *Bulbophyllum cornutum* (Blume) Rchb. f. a été dégradé au rang de synonyme, et *Hamularia cornuta* promu à la catégorie de nom accepté reconnu par les compilateurs ultérieurement à l'adoption de la Liste par la Conférence des Parties de

CITES. **Bulbophyllum cornutum** (Blume) Rchb. f. est le nom officiel de CITES.

Tous les noms	Noms acceptés
Bulbophyllum concavum	**Bulbophyllum cornutum** **(Blume) Rchb. f.**
‡*Bulbophyllum concavum*	*Hamularia cornuta* ‡

Le nom *Bulbophyllum concavum* est un synonyme de **Bulbophyllum cornutum** (Blume) Rchb. f., en caractères gras, qui a été adopté par la Conférence des Parties de CITES en tant que nom accepté. Toutefois, ce même nom, *Bulbophyllum concavum,* ainsi que *Hamularia cornuta,* figurent également, en italiques et encadrés par les symboles « ‡ ». En l'occurrence, *Bulbophyllum concavum* est un synonyme de *Hamularia cornuta,* et reconnu comme tel par les compilateurs ultérieurement à l'adoption de la Liste par la Conférence des Parties de CITES. **Bulbophyllum cornutum** (Blume) Rchb. f. est le nom qui a été adopté par CITES, et *Bulbophyllum concavum* est son synonyme.

e) En ce qui concerne les abréviations concernant les auteurs, les références suivantes ont été utilisées.

Brummitt, R.K. & Powell C.E. (1992). Authors of plant names. Royal Botanic Gardens, Kew. Royaume-Uni.

The International Plant Names Index (Consulté à de nombreuses reprises, la dernière fois en septembre 2007). Publication électronique disponible sur l'internet, depuis l'URL suivante: http://www.ipni.org

f) Le CD-ROM contient les fichiers suivants:

'CITESOrchidChecklistV.pdf', un fichier Adobe Acrobat® qui contient le livre CITES Orchid Checklist 5 et peut être visualisé en utilisant Adobe Reader®. Adobe Acrobat Reader® doit être installé sur le système de l'utilisateur pour pouvoir visualiser ce fichier. Pour télécharger ce logiciel, visitez le site www.adobe.com.

'CITES_Orchid_Checklist4.pdf', un fichier Adobe Acrobat® qui contient le livre CITES Orchid Checklist 4.

'Introduction.pdf', un fichier Adobe Acrobat® qui contient l'introduction aux Volumes 1-3 des Listes des Orchidées CITES.

'PART_I_All_names.pdf', un fichier Adobe Acrobat® qui contient une liste par ordre alphabétique de tous les noms acceptés et les synonymes pour les genres, comme celles qui apparaissent dans les Volumes 1-3 des Listes des Orchidées CITES.

'PART_II_Names_in_current_use.pdf', un fichier Adobe Acrobat® qui contient une liste par ordre alphabétique des noms acceptés et des synonymes actuels, ainsi que la distribution du genre, comme celles qui apparaissent dans les Volumes 1-3 des Listes des Orchidées CITES.

'PART_III_Country_distribution.pdf', un fichier Adobe Acrobat® qui contient les noms acceptés de tous les genres compris dans les Volumes 1-3 des Listes des Orchidées CITES, par ordre alphabétique, sous la rubrique 'pays de distribution'.

Les signets ('bookmarks') à gauche de l'écran ou les liens actifs qui se trouvent dans le document lui-même permettent de naviguer aisément en utilisant le logiciel Adobe Acrobat®. Tous les liens actifs du document sont affichés en bleu.

4. Régions géographiques

Les noms des pays sont conformes au standard de l'ONU tel qu'il figure sur le site web de l'UNTERM (http://unterm.un.org/).

5. Orchidées soumises aux contrôles CITES

La famille des Orchidacées est inscrite à l'Annexe II de la CITES. Par ailleurs, les taxa suivants étaient inscrits à l'Annexe I au moment de la publication de la Liste:

Aerangis ellisii
Dendrobium cruentum
Laelia jongheana
Laelia lobata
Paphiopedilum spp.
Peristeria elata
Phragmipedium spp.
Renanthera imschootiana

6. Abréviations, termes botaniques et mots latins*

Ces termes de botanique, noms latins et abréviations ne sont pas tous utilisés dans la Liste. Ils ne sont fournis ici qu'à titre de référence. Note : les mots *en italique* sont d'origine latine.

ambiguous name (nom ambigu) nom donné à différents taxa par différents auteurs, ce qui crée une ambiguïté
anon. anonyme; sans auteur
auct. *auctorum*: d'auteurs
CITES Convention sur le commerce international des espèces de faune et de flore sauvages menacées d'extinction
cultivar spécimen ou groupe de plantes conservant les mêmes caractéristiques distinctives, produites ou conservées (propagées) en culture
cultivation (culture) obtention de plantes par horticulture ou jardinage, par opposition au prélèvement dans la nature
descr. *descriptio* description d'une espèce ou d'une autre entité taxonomique
distribution (aire de répartition géographique) région(s) où se trouvent les plantes
ed. éditeur
edn. édition (d'un livre ou d'un périodique)
eds. éditeurs
epithet (épithète) dernier mot d'une espèce, sous-espèce ou variété (etc.). Exemple: *speciosum* est l'épithète de l'espèce *Bulbophyllum speciosum*.
escape (échappée) qualifie une plante qui a quitté l'enceinte de culture (un jardin, par exemple) et qu'on retrouve parmi la végétation naturelle
ex *ex* d'après; peut être utilisé entre deux noms d'auteurs, dont le second a validement publié le nom d'après les indications ou suggestions du premier
excl. *exclusus* exclu
hort. *hortorum* de jardins (horticole); plante cultivée ou se trouvant dans des jardins horticoles, par opposition à une plante d'origine sauvage
ICBN (CINB) Code international de la nomenclature botanique
in prep. en préparation
in sched. *in scheda* sur un spécimen d'herbier ou une étiquette
in syn. *in synonymia* en synonymie

incl. incluant

ined. *ineditus* non publié

introduction résultat d'une activité humaine (volontaire ou non) aboutissant à ce qu'une plante non indigène se retrouve dans un pays ou une région

key (clé) système écrit utilisé pour la détermination d'organismes (plantes, par exemple)

leg. *legit* 'il ramassa'; le collecteur

misspelling (faute d'orthographe) nom mal orthographié, par opposition à un nom nouveau ou différent

morphology (morphologie) forme et structure d'un organisme (d'une plante, par exemple)

name causing confusion (nom causant une confusion) nom qui n'est pas utilisé parce qu'il ne peut être assigné sans ambiguïté à un taxon particulier (à une espèce de plante, par exemple)

native (indigène) qualifie un organisme (une plante, par exemple) prospérant naturellement dans un pays, une région, etc.

naturalized (naturalisée) qualifie une plante introduite (voir introduction) ou échappée (voir échappée) qui ressemble à une plante sauvage et qui se propage dans son nouvel environnement

nec *neque* : et non, et ... ne ... pas, ni, non plus, et non

nom. *nomen* nom

nom. ambig. *nomen ambiguum* nom ambigu

nom. cons. prop. *nomen conservandum propositum* nom dont le maintien a été proposé d'après les règles du Code international de la nomenclature botanique *(International Code of Botanical Nomenclature)*

nom. illeg *nomen illegitimum*: nom illégitime

nom. nud. *nomen nudum*: nom publié sans description

nomenclature branche de la science qui nomme les organismes (les plantes, par exemple)

non *non* pas

only known from cultivation (connue seulement en culture) qualifie une plante que l'on ne trouve pas à l'état sauvage

orthographic variant (variante orthographique) même nom orthographié différemment

pro parte *pro parte* partiellement, en partie

provisional name (nom provisoire) nom donné en attente d'une description.

sens. *sensu* au sens de; manière dont un auteur interprète ou utilise un nom

sens. lat. *sensu lato* au sens large; un taxon (habituellement une espèce) et tous ses taxa inférieurs (sous-espèces, etc.) et/ou d'autres taxa parfois considérés comme distincts

sic *sic*, utilisé après un mot qui semble faux ou absurde; indique que ce mot est cité textuellement

synonym (synonyme) nom donné à un taxon mais qui ne peut être utilisé parce que ce n'est pas le nom accepté ; le ou les synonymes forment la synonymie

taxa pluriel de taxon

taxon unité taxonomique à laquelle on a attribué un nom - genre, espèce, sous espèce, etc.

var. variété

* Nous remercions M. Aaron Davis, des Jardins Botaniques Royaux de Kew, de nous avoir fourni ce guide.

LISTA CITES - BULBOPHYLLUM

PREÁMBULO

1. Antecedentes

Esta Lista que reúne las plantas del género *Bulbophyllum* y taxa afines (Orchidaceae) ha sido preparada como resultado de las recomendaciones apoyadas por la Conferencia de las Partes de la CITES y ha sido aprobada por el Comité de Nomenclatura de la CITES. La versión eletronica de la Lista fue adoptada como referencia normalizada en la 14 reunión de la Conferencia de las Partes en junio de 2007.

La preparación de esta Lista exhaustiva ha sido posible gracias al apoyo financiero del Ministerio de Agricultura, Silvicultura, Medio Ambiente y Gestión del Agua de Austria - la Autoridad Administrativa CITES de Austria.

En la preparación de la Lista se siguió un proceso de consulta global con especialistas en la taxonomía y sistemática de este grupo taxonómico. Se incluyó una versión preliminar de la Lista en la página Web de la CITES y se solicitaron comentarios al respecto informando a las partes mediante la Notificación 2005/049. Los comentarios recibidos por las Partes se han incorporado en la versión revisada del texto.

Dado que esta Lista está dirigida a personas no especializadas, los autores han adoptado un enfoque pragmático a la hora de tratar el gran numero de especies incluidas en el grupo Bulbophyllinae. Por lo tanto, el género Bulbophyllum se trata en un sentido amplio. Esto no refleja ninguna opinión o preferencia de los autores respecto de la categoría taxonómica o la circunscripción de las subdivisiones o de los taxa segregados en el grupo Bulbophyllinae. Para que sea posible realizar este tipo de afirmaciones, habrá que esperar al menos a que concluyan los estudios moleculares en curso sobre Bulbophyllinae (a cargo de B. Gravendeel, Leiden, y M. Kiehn y G. Fischer, Viena/Salzburgo).

Cuando los taxa que actualmente se consideran incluidos en Bulbophyllinae tienen sinónimos asignados a géneros no incluidos en Bulbophyllinae, dichos nombres están incluidos en la lista alfabética. Sin embargo, los géneros no incluidos en Bulbophyllinae que contienen taxa atribuidos de forma errónea no están incluidos en la lista alfabética de taxa sino solamente con los taxa pertinentes en el contexto del grupo Bulbophyllinae.

2. Cómo emplear esta Lista

La idea es que esta Lista se utilice como referencia rápida para comprobar los nombres aceptados, los sinónimos y la distribución. Así, pues, la referencia se divide en tres partes principales:

Parte I: Todos los nombres de uso actual
Una lista de todos los nombres y sinónimos aceptados, en orden alfabético.

Parte II: Nombres aceptados de uso actual
Listas independientes para cada género. En cada lista se presentan por orden alfabético los nombres aceptados, con información sobre sinónimos y distribución.

Parte III: Lista por países
Los nombres aceptados para todos los géneros incluidos en esta Lista se presentan por orden alfabético según el país de distribución.

3. Sistema de presentación utilizado en las Partes I, II y III

a)   Los nombres aceptados se presentan en tipo de letra negrita y romano. Los sinónimos se presentan en letra cursiva.

b)   Nombres duplicados

En la Parte I, el nombre del autor aparece después de cada taxón, cuando dicho taxón se cita en más de una ocasión. Por ejemplo:

Todos los nombres
Bulbophyllum curtisii (Hook. f.) J. J. Sm.
Bulbophyllum curtisii (Hook. f.) Ridl.

i)   Donde un sinónimo aparece dos veces, pero se refiere a diferentes nombres aceptados, a saber, *Bulbophyllum curtisii*, (un sinónimo de ambas **Bulbophyllum corolliferum** y **Bulbophyllum purpurascens**), el nombre acompañado de un asterisco se refiere a la especie es más probable encontrar en el comercio, cuándo se sepa. Por ejemplo:

Todos los nombres	**Nombre aceptado**
*Bulbophyllum curtisii (Hook. f.) J. J. Sm...........	**Bulbophyllum corolliferum**
Bulbophyllum curtisii (Hook. f.) Ridl..............	**Bulbophyllum purpurascens**

*La especie que con mayor probabilidad se encontrará en el comercio (en este ejemplo, **Bulbophyllum corolliferum**).

ii)   Si un nombre aceptado es igual al sinónimo, pero se refiere a especies diferentes, a saber, **Bulbophyllum acuminatum** (nombre aceptado) y *Bulbophyllum acuminatum* (un sinónimo de **Bulbophyllum andreeae**), el nombre acompañado de un asterisco se refiere a la especie que es más probable encontrar en el comercio, cuando se sepa. Por ejemplo:

Todos los nombres	**Nombre aceptado**
*Bulbophyllum acuminatum (Ridl.) Ridl.	
Bulbophyllum acuminatum Schltr.	**Bulbophyllum andreeae**

*La especie que es más probable encontrar en el comercio (en este **Bulbophyllum acuminatum** (Ridl.) Ridl.).

NB: En los ejemplos b) i) y b) ii) es preciso efectuar una doble verificación en lo que concierne a la distribución, como se indica en la Parte II. Por ejemplo, en el caso b)ii), si se dio el nombre 'Bulbophyllum acuminatum' y se sabe que la planta en cuestión procede de Papua Nueva Guinea, querrá decir que la especie era **Bulbophyllum andreeae** comercializada bajo el sinónimo *Bulbophyllum acuminatum*. ya que **Bulbophyllum acuminatum** se encuentra únicamente en Indonesia, Malasia, Singapur y Tailandia.

c)   Nombres, combinaciones y sinónimos nuevos en *Bulbophyllum* Thouars (Orchidaceae) adoptados por la CITES

En la preparación de esta Lista han sido necesarios varios nombres nuevos y varias combinaciones nuevas para los taxa que contiene. Dichos nombres están incluidos en el Anexo I con el fin de que su publicación sea válida y efectiva.

Dichos nombres, combinaciones y sinónimos nuevos se han indicado en el texto principal con el símbolo †. Véase el siguiente ejemplo:

Todos los nombres   **Nombre aceptado**
Bulbophyllum arminii †
Bulbophyllum arminii es un nombre nuevo en *Bulbophyllum* Thouars (Orchidaceae). Dicho nombre está publicado en el Anexo I.

Bulbophyllum arminii Sieder & Kiehn nom. nov.
Nombre nuevo para *Hapalochilus bandischii* Garay, Hamer y Siegerist

d)   Nombres no adoptados formalmente por la CITES

    Desde la adopción de la Lista de referencia por la 14 reunión de la Conferencia de las Partes en La Haya (Países Bajos) en junio de 2007, se han reconocido varias especies nuevas y combinaciones de nombres y sinónimos en el género *Bulbophyllum* y taxa afines. Dichos nombres no están incluidos formalmente en esta Lista como nombres aceptados reconocidos por la CITES. No obstante, con el fin de que esta Lista sea lo mas práctica posible para los usuarios, dichos cambios se han incluido de manera informal en la Parte I de la Lista y están marcados con el símbolo ‡ para distinguirlos de los nombres adoptados. Por ejemplo:

Todos los nombres   **Nombre aceptado**
‡ *Bulbophyllum calimanianum* ‡

El nombre *Bulbophyllum calimanianum* está en cursiva, entre los símbolos ‡. *Bulbophyllum calimanianum* no aparece en negrita en la Parte I, dado que el nombre no estaba incluido en la Lista original y no ha sido aceptado formalmente por la CITES. *Bulbophyllum calimanianum* fue aceptado por los autores después de que la Lista fuera adoptada por la CdP de la CITES.

Todos los nombres   **Nombre aceptado**
Bulbophyllum cornutum (Blume) Rchb. f.
‡*Bulbophyllum cornutum* (Blume) Rchb. f. ....*Hamularia cornuta*‡

El nombre **Bulbophyllum cornutum** (Blume) Rchb. f. está en negrita y fue adoptado por la CdP de la CITES como el nombre aceptado. Sin embargo, el mismo nombre, *Bulbophyllum cornutum*, (Blume) Rchb. f. y *Hamularia cornuta* están en cursiva, entre los símbolos ‡. En este caso, *Bulbophyllum cornutum* (Blume) Rchb. f. ha pasado a la categoría de sinónimo y *Hamularia cornuta* ha pasado a ser el nombre aceptado, reconocido por los autores después de que la Lista fuera adoptada por la CdP de la CITES. **Bulbophyllum cornutum** (Blume) Rchb. f. es el nombre oficial para la CITES.

Todos los nombres   **Nombre aceptado**
Bulbophyllum concavum .... **Bulbophyllum cornutum**
                                         (Blume) Rchb. f.
‡*Bulbophyllum concavum* .... *Hamularia cornuta* ‡

El nombre *Bulbophyllum concavum* es un sinónimo de **Bulbophyllum cornutum** (Blume) Rchb. f., que está en negrita y fue adoptado por la CdP de la CITES como el nombre aceptado.

Sin embargo, el mismo nombre, *Bulbophyllum concavum*, y *Hamularia cornuta* están en cursiva, entre los simbolos ‡. En este caso, *Bulbophyllum concavum* es un sinónimo de *Hamularia cornuta*, reconocido por los autores después de que la Lista fuera adoptada por la CdP de la CITES. **Bulbophyllum cornutum** (Blume) Rchb. f. es el nombre adoptado por la CITES, y *Bulbophyllum concavum* es su sinónimo.

e) Las siguientes referencias se han utilizado para las abreviaciones correspondientes a los autores:
Brummitt, R.K. y Powell C.E. (1992). Authors of plant names. Royal Botanic Gardens, Kew. Reino Unido.

El Indice Internacional de Nombres de las Plantas, en inglés: International Plant Names Index (consultado en multiples ocasiones, la última en septiembre de 2007). Publicado en Internet en la dirección http://www.ipni.org

f) El CD-ROM contiene los siguientes archivos:
"CITESOrchidChecklistV.pdf", en formato de Adobe Acrobat®. Este archivo contiene el libro de la *CITES Orchid Checklist 5* y se puede visualizar con el programa Acrobat Reader®. Vd. deberá tener Adobe Acrobat Reader® instalado en su ordenador para ver este archivo (puede descargarse de www.adobe.com).

"CITES_Orchid_Checklist4.pdf", en formato de Adobe Acrobat®. Este archivo contiene el libro de la *CITES Orchid Checklist 4.*

"Introduction.pdf", en formato de Adobe Acrobat®. Este archivo contiene el texto de introducción a los volúmenes 1-3 de las *CITES Orchid Checklists.*

"PART_I_All_names.pdf", en formato de Adobe Acrobat®. Este archivo contiene una lista alfabética de todos los nombres y sinónimos aceptados para los géneros, igual que en los volúmenes 1-3 de las *CITES Orchid Checklists.*

"PART_II_Names_in_current_use.pdf", en formato de Adobe Acrobat®. Este archivo contiene una lista, en orden alfabético, de los nombres aceptados, sus sinónimos de uso actual, y la distribución de los géneros, igual que en los volúmenes 1-3 de las *CITES Orchid Checklists*

"PART_III_Country_distribution.pdf", en formato de Adobe Acrobat®. Este archivo contiene los nombres aceptados de todos los géneros incluidos en los volúmenes 1-3 de las *CITES Orchid Checklists*, en orden alfabético por países de distribución.

Se navega fácilmente por el archivo de Adobe Acrobat® con el uso de los marcadores dinámicos que se encuentran al margen izquierdo del documento, o con los hipervínculos incluidos dentro del mismo. Todos los hipervínculos dentro del documento están resaltados en azul.

4. Áreas geográficas
Los nombres de países corresponden a la terminología utilizada en las Naciones Unidas, disponible en el sitio Web de UNTERM (http://unterm.un.org/).

5. Orchidaceae controladas por la CITES

La familia Orchidaceae está incluida en el Apéndice II de CITES. Además, en el momento de esta publicación, están incluidos en el Apéndice I los siguientes taxa:
Aerangis ellisii
Dendrobium cruentum
Laelia jongheana
Laelia lobata
Paphiopedilum spp.
Peristeria elata
Phragmipedium spp.
Renanthera imschootiana

6. Abreviaturas, términos botánicos y expresiones latinas*

Estas abreviaturas, términos botánicos y expresiones latinas se han incluido como referencia útil, aunque no todas aparecen en la Lista. Nota: las palabras en *cursiva* son latinas.

ambiguous name (nombre ambiguo) un nombre utilizado por distintos autores para diferentes taxa, convirtiéndose en una fuente de ambigüedad
anon. anonymous (anónimo) sin autor o autor desconocido
auct. *auctorum* de autores
CITES Convención sobre el Comercio Internacional de Especies Amenazadas de Fauna y Flora Silvestres
cultivar un ejemplar, o una agrupación de plantas, que tiene los mismos rasgos distintivos, y que se ha producido o se mantiene (se reproduce) en condiciones de cultivo
cultivation (cultivo) el cultivo de plantas mediante horticultura o jardinería; no extraída directamente del medio silvestre
descr. *descriptio*: la descripción de una especie o de otra unidad taxonómica
distribution (distribución) donde se encuentran las plantas (geográfica)
ed. editor
edn. edición (libro o revista)
eds. editores
epithet (epíteto) la última palabra de una especie, subespecie o variedad (etc.), por ejemplo: *speciosum* es el epíteto de la especie *Bulbophyllum speciosum*.
escape (asilvestrada) una planta que ha traspasado los límites de su lugar de cultivo (p.e.: un jardín) y prospera entre la vegetación natural
ex *ex*: después; puede utilizarse entre los nombres de dos autores, el segundo de los cuales publicó el nombre indicado o sugerido por el primero
excl. *exclusus*: excluida
forma *forma*: una unidad taxonómica inferior al nivel de variedad
hort. *hortorum*: de jardines (horticultura); se cultiva o se encuentra en jardines; no se trata de una planta silvestre
ICNB (CINB) Código Internacional de Nomenclatura Botánica
in prep. en preparación
in sched. *in scheda*: en un espécimen de herbario o etiqueta
in syn. *in synonymia*: en sinonimia
incl. inclusive
in ed. *ineditus*: inédito
introduction (introducción) una planta que se da en un país, o en cualquier otra localidad, por influencia antropogénica (intencionadamente o por casualidad); cualquier planta alóctona; que no sea nativa
key (clave) un sistema escrito utilizado para la identificación de organismos (p.e.: plantas)

leg. *legit*: él recolectó; el recolector o coleccionista
misspelling (error de ortografía) un nombre que se ha escrito incorrectamente; no se trata de un nombre nuevo o diferente
morphology (morfología) la forma y estructura de un organismo (p.e.: una planta)
name causing confusion (nombre que provoca confusión) un nombre que no se usa por no poder asignarse sin ambigüedad a un determinado taxón (p.e.: una especie vegetal)
native (nativo) un organismo (p.e.: una planta) que se da naturalmente en un país o región, etc.
naturalized (naturalizada) una planta alóctona (véase "introduction") o asilvestrada (véase "escape") pero que parece silvestre, y es capaz de reproducirse en su nuevo medio
nec *neque*: y no, tampoco
nom. *nomen* nombre
nom. ambig. *nomen ambiguum*: nombre ambiguo
nom. cons. prop. *nomen conservandum propositum*: nombre propuesto para la conservación con arreglo a lo dispuesto en el Código Internacional de Nomenclatura Botánica (ICBN)
nom. illeg. *nomen illegitimum*: nombre ilegítimo
nom. nud. *nomen nudum*: nombre publicado sin descripción
nomenclature (nomenclatura) rama de la ciencia que se ocupa de atribuir nombres a organismos (p.e.: plantas)
non *non*: no
only known from cultivation (sólo se conoce en cultivo) una planta que no se da en la naturaleza, únicamente en condiciones de cultivo
orthographic variant (variante ortográfica) una alternativa ortográfica del mismo nombre
p. p. *pro parte*: parcialmente, en parte
provisional name (nombre provisional) nombre asignado temporalmente hasta que se disponga de una descripción válida
sens. *sensu*: en el sentido de; la forma en que un autor interpreta o utiliza un nombre
sens. lat. *sensu lato*: en sentido amplio, un taxón (normalmente una especie) y todos sus taxa subordinados (p.e.: subspecies) y/u otros taxa a veces distintos
sic *sic*, utilizado después de una palabra que parece errónea o absurda, para dar a entender que se ha citado textualmente
spp. especies
ssp. subespecies
synonym (sinónimo) un nombre que se aplica a un taxón pero que no puede utilizarse por no ser el nombre aceptado – el sinónimo o los sinónimos forman la sinonimia
taxa plural de taxón
taxon (taxón) una determinada unidad de clasificación, p.e.: género, especie, subespecie
var. variedad

*Expresamos nuestro agradecimiento al Dr. Aaron Davis, Real Jardín Botánico de Kew, por proporcionar esta guía.
+

PART I: ALL NAMES IN CURRENT USE
All names ordered alphabetically for the genera:

Acrochaene, Bulbophyllum, Chaseella, Codonosiphon, Drymoda, Monomeria, Monosepalum, Pedilochilus, Saccoglossum, Sunipia and *Trias*

PREMIERE PARTIE: TOUS LES NOMS D'USAGE COURANT
Par ordre alphabétique de tous les noms pour les genres:

Acrochaene, Bulbophyllum, Chaseella, Codonosiphon, Drymoda, Monomeria, Monosepalum, Pedilochilus, Saccoglossum, Sunipia et *Trias*

PARTE I: TODOS LOS NOMBRES DE USO ACTUAL
Todos los nombres en orden alfabético para los géneros:

Acrochaene, Bulbophyllum, Chaseella, Codonosiphon, Drymoda, Monomeria, Monosepalum, Pedilochilus, Saccoglossum, Sunipia y *Trias*

ALL NAMES	ACCEPTED NAMES

ALPHABETICAL LISTING OF ALL NAMES FOR THE GENERA:
Acrochaene, Bulbophyllum, Chaseella, Codonosiphon, Drymoda, Monomeria, Monosepalum, Pedilochilus, Saccoglossum, Sunipia and Trias

LISTES ALPHABETIQUES DE TOUS LES NOMS POUR LES GENRES:
Acrochaene, Bulbophyllum, Chaseella, Codonosiphon, Drymoda, Monomeria, Monosepalum, Pedilochilus, Saccoglossum, Sunipia et Trias

PRESENTACION POR ORDEN ALFABETICO DE ODOS LOS NOMBRES PARA EL GENERO:
Acrochaene, Bulbophyllum, Chaseella, Codonosiphon, Drymoda, Monomeria, Monosepalum, Pedilochilus, Saccoglossum, Sunipia y Trias

ALL NAMES TOUS LES NOMS TODOS LOS NOMBRES	ACCEPTED NAME NOM ACCEPTÉS NOMBRES ACEPTADOS
Acrochaene punctata	
Acrochaene rimanni	Sunipia rimannii
Adelopetalum argyropum	**Bulbophyllum argyropus**
Adelopetalum boonjee	**Bulbophyllum boonjee**
Adelopetalum bracteatum	**Bulbophyllum bracteatum**
Adelopetalum elisae	**Bulbophyllum elizae**
Adelopetalum exiguum	**Bulbophyllum exiguum**
Adelopetalum lageniforme	**Bulbophyllum lageniforme**
Adelopetalum lilianae	**Bulbophyllum lilianae**
Adelopetalum newportii	**Bulbophyllum newportii**
Adelopetalum tuberculatum	**Bulbophyllum tuberculatum**
Adelopetalum weinthalii	**Bulbophyllum weinthalii**
Adelopetalum weinthalii subsp. *striatum*	**Bulbophyllum weinthalii** subsp. **striatum**
Adelopetalum wilkianum	**Bulbophyllum wilkianum**
Aerides radiatum	**Bulbophyllum roxburghii**
Anisopetalum careyanum	**Bulbophyllum careyanum**
Anisopetalum lasianthum	**Bulbophyllum lasianthum**
Blepharochilum macphersoni	**Bulbophyllum macphersoni**
Blepharochilum sladeanum	**Bulbophyllum macphersoni** var. **spathulatum**
Bolbophyllaria aristata	**Bulbophyllum aristatum**
Bolbophyllaria biseta	**Bulbophyllum bisetum**
Bolbophyllaria bracteolata	**Bulbophyllum bracteolatum**
Bolbophyllaria oerstedii	**Bulbophyllum oerstedii**
Bolbophyllaria pachyrhachis	**Bulbophyllum pachyrachis**
Bolbophyllaria sordida	**Bulbophyllum sordidum**
Bolbophyllaria pentasticha	**Bulbophyllum pentasticha**
Bulbophyllopsis maculosa	**Bulbophyllum umbellatum**
Bulbophyllopsis morphologorum	**Bulbophyllum umbellatum**
Bulbophyllum abbreviatum Rchb. f.	
Bulbophyllum abbreviatum Schltr.	*Bulbophyllum ormerodianum* †
Bulbophyllum abbrevilabium	
Bulbophyllum aberrans	
Bulbophyllum ablepharon	
Bulbophyllum absconditum	
Bulbophyllum absconditum subsp. **hastula**	
Bulbophyllum absconditum var. *gautierense*	**Bulbophyllum stipulaceum**
Bulbophyllum absconditum var. *neoguineese*	**Bulbophyllum stipulaceum**
Bulbophyllum acanthoglossum	
Bulbophyllum acropogon	

ALL NAMES	ACCEPTED NAMES

Bulbophyllum acuminatifolium
*Bulbophyllum acuminatum (Ridl.) Ridl.
Bulbophyllum acuminatum Schltr. .... Bulbophyllum andreeae
Bulbophyllum acutebracteatum
Bulbophyllum acutebracteatum var. rubrobrunneopapillosum
Bulbophyllum acutibrachium .... Bulbophyllum cylindrobulbum
Bulbophyllum acutiflorum
Bulbophyllum acutilingue
Bulbophyllum acutisepalum .... Bulbophyllum schimperianum
Bulbophyllum acutispicatum
Bulbophyllum acutum .... Bulbophyllum depressum
Bulbophyllum adangense
Bulbophyllum adelphidium
Bulbophyllum adenambon .... Bulbophyllum quadrangulare
Bulbophyllum adenoblepharon
Bulbophyllum adenocarpum .... Bulbophyllum lageniforme
Bulbophyllum adenopetalum .... Bulbophyllum flavescens
Bulbophyllum adenophorum .... Bulbophyllum brienianum
Bulbophyllum adiamantinum
Bulbophyllum adjungens
Bulbophyllum adolphi
Bulbophyllum adpressiscapum .... Bulbophyllum ochroleucum
Bulbophyllum aechmophorum
Bulbophyllum aemulum
Bulbophyllum aeolium
Bulbophyllum aestivale
Bulbophyllum affine
Bulbophyllum affinoides .... Bulbophyllum psittacoglossum
Bulbophyllum africanum .... Bulbophyllum nigritianum
Bulbophyllum afzelii
Bulbophyllum afzelii var. microdoron
Bulbophyllum agapethoides
Bulbophyllum agastor
Bulbophyllum agastymalayanum .... Bulbophyllum xylophyllum
Bulbophyllum aggregatum
Bulbophyllum aithorhachis
Bulbophyllum alabastraceus
Bulbophyllum alagense
Bulbophyllum alatum
Bulbophyllum albibracteum
Bulbophyllum albidostylidium
*Bulbophyllum albidum (Wight) Hook. f.
Bulbophyllum albidum De Wild. .... Bulbophyllum nigritianum
Bulbophyllum albociliatum (Liu & Su) Nakajima
Bulbophyllum albociliatum (Liu & Su) Seidenf. .... Bulbophyllum albociliatum
Bulbophyllum albociliatum var. weiminianum
Bulbophyllum albo-roseum
Bulbophyllum album .... Bulbophyllum humblottii
Bulbophyllum alcicorne
Bulbophyllum alexandrae
Bulbophyllum algidum
Bulbophyllum alinae
Bulbophyllum alkmaarense
Bulbophyllum alleizettei
Bulbophyllum allenkerrii
Bulbophyllum alliifolium
Bulbophyllum alopecurum .... Bulbophyllum triste
Bulbophyllum alsiosum
Bulbophyllum alticaule

* † ‡ For explanation see page 2 and 3, point 3
* † ‡ Voir les explications page 10 et 11, point 3
* † ‡ Para mayor explicación, véase la página 18 y 19, point 3   27

ALL NAMES	ACCEPTED NAMES

Bulbophyllum alticola
Bulbophyllum altispex.. **Bulbophyllum mutabile**
Bulbophyllum alveatum
Bulbophyllum amanicum... **Bulbophyllum josephi**
Bulbophyllum amauryae .... **Bulbophyllum intertextum**
Bulbophyllum amazonicum
Bulbophyllum ambatoavense
Bulbophyllum amblyacron
Bulbophyllum amblyanthum
Bulbophyllum amblyoglossum
Bulbophyllum ambohitrense.. **Bulbophyllum nutans** var.
                                                 variifolium
Bulbophyllum ambongense .... **Bulbophyllum rubrum**
Bulbophyllum ambreae .... **Bulbophyllum septatum**
Bulbophyllum ambrense
Bulbophyllum ambrosia
Bulbophyllum ambrosia subsp. **nepalensis**
Bulbophyllum amesianum.. **Bulbophyllum cumingii**
Bulbophyllum amoenum
Bulbophyllum amphorimorphum
Bulbophyllum amplebracteatum
Bulbophyllum amplifolium
Bulbophyllum amplistigmaticum
Bulbophyllum amplum .... **Epigeneium amplum**
Bulbophyllum amygdalinum... **Bulbophyllum ambrosia**
Bulbophyllum anaclastum
Bulbophyllum anakbaruppui
Bulbophyllum analamazoatrae
Bulbophyllum anceps
***Bulbophyllum andersonii** (Hook. f.) J. J. Sm.
Bulbophyllum andersonii* Kurz .... **Bulbophyllum flabellum-veneris
Bulbophyllum andohahelense
Bulbophyllum andongense .... **Bulbophyllum cocoinum**
Bulbophyllum andrangense.. **Bulbophyllum francoisii** var.
                                                 andrangense
Bulbophyllum andreeae
Bulbophyllum andringitranum.. **Bulbophyllum nutans**
Bulbophyllum angiense.. **Bulbophyllum cylindrobulbum**
Bulbophyllum anguipes
Bulbophyllum angulatum
Bulbophyllum anguliferum
Bulbophyllum angustatifolium .... **Bulbophyllum deltoideum**
Bulbophyllum anguste-ellipticum................................... **Bulbophyllum nigrescens**
Bulbophyllum angusteovatum
Bulbophyllum angustifolium
Bulbophyllum angustifolium var. *nanum*........................ **Bulbophyllum smithianum**
Bulbophyllum angustifolium var. *pavum* .... **Bulbophyllum smithianum**
Bulbophyllum anisopterum
Bulbophyllum anjozorobeense
Bulbophyllum ankaizinense
Bulbophyllum ankaratranum
Bulbophyllum ankylochele
Bulbophyllum ankylorhinon
Bulbophyllum annamense †
Bulbophyllum annamicum.. **Bulbophyllum umbellatum**
Bulbophyllum annandalei
Bulbophyllum antennatum
Bulbophyllum antenniferum
Bulbophyllum antenniferum... **Bulbophyllum geraense**

ALL NAMES	ACCEPTED NAMES
Bulbophyllum antioquiense	
Bulbophyllum antongilense	
Bulbophyllum apertum	
Bulbophyllum apetalum	Genyorchis apetala
Bulbophyllum aphanopetalum	
Bulbophyllum apheles	
Bulbophyllum apiculatum	
Bulbophyllum apiferum	
Bulbophyllum apodum	
Bulbophyllum apodum var. lanceolatum	
Bulbophyllum apoense	
Bulbophyllum appendiculatum	
Bulbophyllum appressicaule	
Bulbophyllum appressum	
Bulbophyllum approximatum	
Bulbophyllum arachnidium	
Bulbophyllum arachnites	Bulbophyllum korthalsii
Bulbophyllum arachnoideum	
Bulbophyllum araniferum	Bulbophyllum stormii
Bulbophyllum arcaniflorum	Bulbophyllum arachnoideum
Bulbophyllum arcuatum	Bulbophyllum chloranthum
Bulbophyllum arcutilabium	
Bulbophyllum ardjunense	
Bulbophyllum arfakense	
Bulbophyllum arfakianum	
Bulbophyllum argyropus	
Bulbophyllum arianeae	
Bulbophyllum ariel	
Bulbophyllum aristatum	
Bulbophyllum aristilabre	
Bulbophyllum aristopetalum	
Bulbophyllum armeniacum	
Bulbophyllum arminii †	
Bulbophyllum arnoldianum	Bulbophyllum falcatum var. velutinum
Bulbophyllum arrectum	
Bulbophyllum arsoanum	
Bulbophyllum artostigma	
Bulbophyllum ascochiloides	
Bulbophyllum asperilingue	
Bulbophyllum aspersum	
Bulbophyllum asperulum	
Bulbophyllum astelidum	
Bulbophyllum atratum	
Bulbophyllum atrolabium	
Bulbophyllum atropurpureum	
Bulbophyllum atrorubens	
Bulbophyllum atrosanguineum	
Bulbophyllum atroviolaceum	Bulbophyllum betchei
Bulbophyllum attenuatum	Bulbophyllum pugioniforme
Bulbophyllum aubrevillei	
Bulbophyllum aundense	
Bulbophyllum aurantiacum F. Muell.	Bulbophyllum schillerianum
Bulbophyllum aurantiacum Hook. f.	Bulbophyllum josephi
Bulbophyllum aurantiacum var. *wuttsii*	Bulbophyllum schillerianum
Bulbophyllum auratum	
Bulbophyllum aureoapex	
Bulbophyllum aureobrunneum	
Bulbophyllum aureolabellum	

* † ‡ For explanation see page 2 and 3, point 3
* † ‡ Voir les explications page 10 et 11, point 3
* † ‡ Para mayor explicación, véase la página 18 y 19, point 3

29

ALL NAMES	ACCEPTED NAMES
Bulbophyllum aureum	
Bulbophyllum auricomum	
Bulbophyllum auriculatum	
Bulbophyllum auriflorum	
Bulbophyllum auroreum	
Bulbophyllum auroreum var. *grandiflorum*.....................	**Bulbophyllum auroreum**
Bulbophyllum autumnale	**Bulbophyllum macraei** var. **autumnale**
Bulbophyllum averyanovii	
Bulbophyllum avicella..	**Bulbophyllum membranaceum**
Bulbophyllum bacilliferum	
Bulbophyllum baculiferum	
Bulbophyllum badium..	**Bulbophyllum membranifolium**
Bulbophyllum baileyi	
Bulbophyllum bakhuizenii	
Bulbophyllum bakossorum Schltr.................................	**Bulbophyllum falcatum** var. **bufo**
Bulbophyllum baladeanum	
Bulbophyllum balaeniceps †	**Bulbophyllum elegans**
Bulbophyllum balapiuense	
Bulbophyllum balfourianum..	**Bulbophyllum macrobulbum**
Bulbophyllum ballii	
Bulbophyllum bambiliense...	**Bulbophyllum scaberulum**
Bulbophyllum bambusifolium...	**Bulbophyllum piestobulbon**
Bulbophyllum bandischii	
Bulbophyllum barbatum Barb. Rodr.	
Bulbophyllum barbatum P. Royen	**Bulbophyllum aundense**
Bulbophyllum barbellatum...	**Bulbophyllum toranum**
Bulbophyllum barbigerum	
Bulbophyllum barbilabium...	**Bulbophyllum bulliferum**
Bulbophyllum bariense	
Bulbophyllum baronii	
Bulbophyllum barrinum	**Bulbophyllum flavescens**
Bulbophyllum basisetum	
Bulbophyllum bataanense	
Bulbophyllum bathieanum	
Bulbophyllum batukauense	
Bulbophyllum baucoense	**Bulbophyllum weberi**
Bulbophyllum bavonis	
Bulbophyllum beccarii	
Bulbophyllum befaonense	**Bulbophyllum sarcorhachis** var. **beforonense**
Bulbophyllum bequaerti...	**Bulbophyllum cochleatum** var. **bequaertii**
Bulbophyllum bequaerti var. *brachyanthum*	**Bulbophyllum cochleatum** var. **brachyanthum**
Bulbophyllum berenicis	
Bulbophyllum betchei	
Bulbophyllum biantennatum	
Bulbophyllum bibundiense...	**Bulbophyllum sandersonii**
Bulbophyllum bicaudatum	
*****Bulbophyllum bicolor** Lindl.	
Bulbophyllum bicolor Jum. & H. Perrier.......................	**Bulbophyllum bicoloratum**
Bulbophyllum bicolor (Lindl.) Hook. f..........................	**Sunipia bicolor**
Bulbophyllum bicoloratum	
Bulbophyllum bicornutum..	**Bulbophyllum posticum**
Bulbophyllum bidentatum	
Bulbophyllum bidenticulatum	
Bulbophyllum bidenticulatum var. **joyceae**	

ALL NAMES   **ACCEPTED NAMES**

Bulbophyllum bidi
Bulbophyllum bifarium
Bulbophyllum biflorum
Bulbophyllum bigibbosum
Bulbophyllum bigibbum
Bulbophyllum bilobipetalum.. **Bulbophyllum schefferi**
Bulbophyllum binnendijkii
Bulbophyllum birmense
Bulbophyllum birugatum
Bulbophyllum bisepalum
Bulbophyllum biseriale
Bulbophyllum bisetoides
Bulbophyllum bisetum
Bulbophyllum bismarckense
Bulbophyllum bittnerianum
Bulbophyllum bivalve
Bulbophyllum blepharadenium..................................... **Bulbophyllum minutipetalum**
Bulbophyllum blepharicardium
Bulbophyllum blephariglossum
Bulbophyllum blepharistes
Bulbophyllum blepharochilum
Bulbophyllum blepharopetalum
Bulbophyllum blepharosepalum................................... **Bulbophyllum limbatum**
Bulbophyllum bliteum
Bulbophyllum blumei
Bulbophyllum blumei var. *longicaudatum*...................... **Bulbophyllum longicaudatum**
Bulbophyllum blumei var. *pumilum*.............................. **Bulbophyllum nasica**
Bulbophyllum boiteaui
Bulbophyllum bokorense.. **Plocoglottis bokorensis**
Bulbophyllum bolaninum.. **Bulbophyllum savaiense** subsp.
                                                      gorumense
Bulbophyllum bolivianum
Bulbophyllum bolovenense... **Bulbophyllum clandestinum**
Bulbophyllum bolsteri
Bulbophyllum bomiensis
Bulbophyllum boninense (Schltr.) J. J. Sm.
Bulbophyllum boninense Makino................................... **Bulbophyllum macraei**
Bulbophyllum bontocense
Bulbophyllum boonjee
Bulbophyllum bootanense
Bulbophyllum bootanoides.. **Bulbophyllum frostii**
Bulbophyllum borneense.. **Bulbophyllum auratum**
Bulbophyllum bosseri... **Bulbophyllum reflexiflorum**
Bulbophyllum botryophorum
Bulbophyllum boudetiana
Bulbophyllum boulbetii
Bulbophyllum bowkettae
Bulbophyllum bowringianum.. **Bulbophyllum khasyanum**
Bulbophyllum braccatum.. **Bulbophyllum odoratum**
Bulbophyllum brachychilum
Bulbophyllum brachypetalum
Bulbophyllum brachyphyton
Bulbophyllum brachypodium var. *geei*........................... **Bulbophyllum emarginatum**
Bulbophyllum brachypodium var. *parviflorum*............... **Bulbophyllum yoksunense**
Bulbophyllum brachypodum .... **Bulbophyllum yoksunense**
Bulbophyllum brachystachyum
Bulbophyllum bracteatum
Bulbophyllum bracteolatum
Bulbophyllum bracteosum... **Bulbophyllum steyermarkii**

* † ‡ For explanation see page 2 and 3, point 3
* † ‡ Voir les explications page 10 et 11, point 3
* † ‡ Para mayor explicación, véase la página 18 y 19, point 3

ALL NAMES	ACCEPTED NAMES

Bulbophyllum bractescens
Bulbophyllum brassii
Bulbophyllum brastagiense
Bulbophyllum brauni .... Bulbophyllum falcipetalum
‡*Bulbophyllum breimerianum*‡
Bulbophyllum breve
Bulbophyllum brevibrachiatum
Bulbophyllum brevicolumna
Bulbophyllum brevidenticulatum .... Bulbophyllum cocoinum
Bulbophyllum breviflorum
Bulbophyllum brevilabium
Bulbophyllum brevipes
Bulbophyllum brevipetalum
Bulbophyllum brevipetalum var. *majus* .... Bulbophyllum brevipetalum
Bulbophyllum brevipetalum var. *speculiferum* .... Bulbophyllum brevipetalum
Bulbophyllum breviscapum J. J. Sm. .... Bulbophyllum desmotrichoides
Bulbophyllum breviscapum (Rolfe) J. J. Sm. .... Bulbophyllum lasiochilum
Bulbophyllum breviscapum (Rolfe) Ridl. .... Bulbophyllum lasiochilum
Bulbophyllum brevispicatum
Bulbophyllum brevistylidium
Bulbophyllum brienianum (Rolfe) Ames
Bulbophyllum brienianum (Rolfe) J. J. Sm. .... Bulbophyllum brienianum
Bulbophyllum brixhei.. Bulbophyllum falcatum var
velutinum
Bulbophyllum brookeanum .... Bulbophyllum vermiculare
Bulbophyllum brookesii .... Bulbophyllum odoratum
Bulbophyllum brunnescens .... Bulbophyllum brienianum
Bulbophyllum bryoides
Bulbophyllum buchenavianum...................................... Bulbophyllum calyptratum
Bulbophyllum bufo... Bulbophyllum falcatum var.
bufo
Bulbophyllum bulhartii †
Bulbophyllum bulliferum
Bulbophyllum buntingii... Bulbophyllum oxychilum
Bulbophyllum burfordiense
Bulbophyllum burfordiense .... Bulbophyllum grandiflorum
Bulbophyllum burkilli
Bulbophyllum burttii
Bulbophyllum cadetioides
Bulbophyllum caecilii
Bulbophyllum caecum
Bulbophyllum caeruleolineatum..................................... Bulbophyllum oxycalyx var.
rubescens
Bulbophyllum caesariatum.. Bulbophyllum lindleyanum
Bulbophyllum caespitosum
Bulbophyllum calabaricum.. Bulbophyllum pumilum
Bulbophyllum calamarioides... Bulbophyllum erectum
Bulbophyllum calamarium.. Bulbophyllum saltatorium var.
calamarium
Bulbophyllum calamarium var. *albociliatum* .... Bulbophyllum saltatorium var.
albociliatum
Bulbophyllum calceilabium
Bulbophyllum calceolus
Bulbophyllum caldericola
‡ *Bulbophyllum calimanianum* ‡
Bulbophyllum callichroma
Bulbophyllum callipes
Bulbophyllum callosum
Bulbophyllum calodictyon... Bulbophyllum griffithii

ALL NAMES   **ACCEPTED NAMES**

Bulbophyllum caloglossum
Bulbophyllum calophyllum
Bulbophyllum calothyrsus.. **Bulbophyllum callichroma**
Bulbophyllum calviventer
Bulbophyllum calvum
Bulbophyllum calyptratum
Bulbophyllum calyptratum var. **graminifolium**
Bulbophyllum calyptratum var. **lucifugum**
Bulbophyllum calyptropus
Bulbophyllum cameronense
Bulbophyllum campanulatum.. **Bulbophyllum auratum**
Bulbophyllum campanulatum var. *inconspicum*.............. **Bulbophyllum auratum**
Bulbophyllum campos-portoi
Bulbophyllum camptochilum
Bulbophyllum candidum
Bulbophyllum canlaonense
Bulbophyllum cantagallense
Bulbophyllum capilligerum
*****Bulbophyllum capillipes** C. S. P. Parish & Rchb. f.
Bulbophyllum capillipes (Guillaumin) N. Halle............. **Bulbophyllum aphanopetalum**
Bulbophyllum capitatum
Bulbophyllum capituliflorum
Bulbophyllum capuronii
Bulbophyllum caputgnomonis
Bulbophyllum cardiobulbum
Bulbophyllum cardiophyllum
Bulbophyllum careyanum
Bulbophyllum careyanum var. *crassipes*......................... **Bulbophyllum crassipes**
Bulbophyllum careyanum var. *ochraceum*...................... **Bulbophyllum careyanum**
Bulbophyllum carinatum (Teijsm. & Binn.) Naves
Bulbophyllum carinatum Ames...................................... **Bulbophyllum mearnsii**
Bulbophyllum carinatum Cogn. .... **Bulbophyllum reticulatum**
Bulbophyllum carinatum G. Will. .... **Bulbophyllum unifoliatum**
                                                  subsp. **infracarinatum**
Bulbophyllum cariniflorum
Bulbophyllum carinilabium
Bulbophyllum carnosilabium
Bulbophyllum carnosisepalum
Bulbophyllum carrianum
Bulbophyllum carunculatum
Bulbophyllum carunculilabrum...................................... **Bulbophyllum catenarium**
Bulbophyllum caryophyllum .... **Bulbophyllum baileyi**
Bulbophyllum cassideum... **Bulbophyllum orbiculare** subsp.
                                                  cassideum
Bulbophyllum cataractarum
Bulbophyllum catenarium
Bulbophyllum catenulatum
Bulbophyllum cateorum
Bulbophyllum catillus
Bulbophyllum caudatisepalum
Bulbophyllum caudatum Lindl....................................... **Bulbophyllum sterile**
Bulbophyllum caudatum L. O. Williams......................... **Bulbophyllum williamsii**
Bulbophyllum caudipetalum
Bulbophyllum cauliflorum
Bulbophyllum cauliflorum var. **sikkimense**
Bulbophyllum cavibulbum
Bulbophyllum cavipes
Bulbophyllum cavistigma... **Bulbophyllum trifilum**
Bulbophyllum centrosemiflorum

* † ‡ For explanation see page 2 and 3, point 3
* † ‡ Voir les explications page 10 et 11, point 3
* † ‡ Para mayor explicación, véase la página 18 y 19, point 3   33

ALL NAMES   ACCEPTED NAMES

Bulbophyllum cephalophorum
Bulbophyllum cerambyx
Bulbophyllum ceratostylis
Bulbophyllum ceratostyloides Ridl.
Bulbophyllum ceratostyloides (Schltr.) Schltr................ **Bulbophyllum mutabile**
Bulbophyllum cercanthum †
Bulbophyllum cercoglossum .... **Bulbophyllum renkinianum**
Bulbophyllum cerebellum
Bulbophyllum cerinum
Bulbophyllum ceriodorum
Bulbophyllum cernuum
Bulbophyllum cernuum var. **vittata**
Bulbophyllum chaetopus .... **Bulbophyllum oreodoxa**
Bulbophyllum chaetostroma
Bulbophyllum chanii
Bulbophyllum chaunobulbon
Bulbophyllum chaunobulbon var. **ctenopetalum**
Bulbophyllum cheiri
Bulbophyllum cheiropetalum
Bulbophyllum chekaense .... **Bulbophyllum puguahaanense**
Bulbophyllum chevalieri .... **Bulbophyllum scaberulum**
Bulbophyllum chimaera
Bulbophyllum chinense
Bulbophyllum chloranthum
Bulbophyllum chlorascens
Bulbophyllum chloroglossum
Bulbophyllum chloropterum
Bulbophyllum chlororhopalon
Bulbophyllum chlorostachys .... **Bulbophyllum propinquum**
Bulbophyllum chondriophorum
Bulbophyllum christophersenii .... **Bulbophyllum samoanum**
Bulbophyllum chrysendetum
Bulbophyllum chryseum (Kraenzl.) Ames
Bulbophyllum chryseum (Kraenzl.) J. J. Sm................... **Bulbophyllum chryseum**
Bulbophyllum chrysobulbum .... **Bulbophyllum nutans**
Bulbophyllum chrysocephalum
Bulbophyllum chrysochilum
Bulbophyllum chrysoglossum
Bulbophyllum chrysotes
Bulbophyllum ciliatilabrum
Bulbophyllum ciliatoides .... **Bulbophyllum membranaceum**
Bulbophyllum ciliatum (Blume) Lindl.
Bulbophyllum ciliatum Schltr... **Bulbophyllum maximum** var.
                                                                              oxypterum
Bulbophyllum ciliipetalum
Bulbophyllum cilioglossum .... **Bulbophyllum radicans**
Bulbophyllum ciliolatum
Bulbophyllum ciluliae
Bulbophyllum cimicinum
Bulbophyllum cincinnatum .... **Bulbophyllum hirtulum**
Bulbophyllum cirrhatum
Bulbophyllum cirrhoglossum
Bulbophyllum cirrhopetaloides .... **Bulbophyllum bisetum**
Bulbophyllum cirrhosum
Bulbophyllum citrellum Ridl.
Bulbophyllum citrellum J. J. Sm.................................... **Bulbophyllum citricolor**
Bulbophyllum citricolor
Bulbophyllum citrinilabre
Bulbophyllum citrinum... **Bulbophyllum purpurascens**

ALL NAMES	ACCEPTED NAMES

Bulbophyllum clandestinum

Bulbophyllum claptonense .... **Bulbophyllum lobbii**

Bulbophyllum clarkeanum .... **Bulbophyllum stenobulbon**

Bulbophyllum clarkei (Rolfe) Schltr. .... **Bulbophyllum scaberulum**

Bulbophyllum clarkei Rchb. f. .... **Bulbophyllum reptans**

Bulbophyllum clausseni

Bulbophyllum clavatum

Bulbophyllum clavigerum (Fitzg.) Dockr....................... **Bulbophyllum longiflorum**

Bulbophyllum clavigerum (Fitzg.) F. Muell .... **Bulbophyllum longiflorum**

Bulbophyllum clavigerum H. Perrier............................. **Bulbophyllum lemurense**

Bulbophyllum cleistogamum

Bulbophyllum clemensiae

Bulbophyllum clipeibulbum

Bulbophyllum coccinatum

Bulbophyllum cochinchinense.................................... **Bulbophyllum macranthum**

*****Bulbophyllum cochleatum** Lindl.

Bulbophyllum cochleatum Schltr................................... **Bulbophyllum macphersoni** var. **spathulatum**

Bulbophyllum cochleatum var. **bequaertii**

Bulbophyllum cochleatum var. **brachyanthum**

Bulbophyllum cochleatum var. *gravidum*...................... **Bulbophyllum gravidum**

Bulbophyllum cochleatum var. **tenuicaule**

Bulbophyllum cochlia

Bulbophyllum cochlioides

Bulbophyllum cocoinum

Bulbophyllum codonanthum... **Codonosiphon codonanthum**

Bulbophyllum coelochilum

Bulbophyllum coeruleum .... **Bulbophyllum bicoloratum**

Bulbophyllum cogniauxianum

Bulbophyllum coiloglossum .... **Pedilochilus coiloglossum**

Bulbophyllum collettii

Bulbophyllum colliferum

Bulbophyllum collinum

Bulbophyllum colomaculosum

Bulbophyllum coloratum

Bulbophyllum colubrimodum

Bulbophyllum colubrinum

Bulbophyllum comatum

Bulbophyllum comatum var. **inflatum**

Bulbophyllum comberi

Bulbophyllum comberipictum

Bulbophyllum cominsii

Bulbophyllum commersonii

Bulbophyllum commissibulbum

Bulbophyllum commocardium.. **Bulbophyllum discolor**

Bulbophyllum comorianum

*****Bulbophyllum comosum** Collett & Hemsl.

Bulbophyllum comosum H. Perrier................................ **Bulbophyllum reflexiflorum** subsp. **pogonochilum**

Bulbophyllum compactum.. **Bulbophyllum coriophorum**

Bulbophyllum complanatum

Bulbophyllum compressilabellatum

Bulbophyllum compressum Teijsm. & Binn.

Bulbophyllum compressum Frapp. ex Cordem............... **Bulbophyllum frappieri**

Bulbophyllum comptonii

Bulbophyllum concatenatum

Bulbophyllum concavibasalis

* † ‡ For explanation see page 2 and 3, point 3
* † ‡ Voir les explications page 10 et 11, point 3
* † ‡ Para mayor explicación, véase la página 18 y 19, point 3   **35**

ALL NAMES	ACCEPTED NAMES

Bulbophyllum concavum .... **Bulbophyllum cornutum (Blume) Rchb. f.**

‡*Bulbophyllum concavum* .... *Hamularia cornuta* ‡

Bulbophyllum conchidioides

Bulbophyllum conchiferum .... **Bulbophyllum khasyanum**

Bulbophyllum conchophyllum

Bulbophyllum concinnum

Bulbophyllum concolor

Bulbophyllum confertum .... **Bulbophyllum scabratum**

Bulbophyllum confusum †

Bulbophyllum congestiflorum

Bulbophyllum congestum .... **Bulbophyllum odoratissimum**

Bulbophyllum congolanum .... **Bulbophyllum scaberulum**

Bulbophyllum congolense .... **Bulbophyllum imbricatum**

Bulbophyllum conicum .... **Bulbophyllum clavatum**

Bulbophyllum coniferum

Bulbophyllum connatum

Bulbophyllum conspectum .... **Bulbophyllum sopoetanense**

Bulbophyllum conspersum

Bulbophyllum constrictilabre .... **Bulbophyllum cylindrobulbum**

Bulbophyllum contortisepalum

Bulbophyllum cootesii

Bulbophyllum copelandii Ames

Bulbophyllum copelandii F. M. Bailey .... **Diplocaulobium copelandii**

Bulbophyllum corallinum

Bulbophyllum cordemoyi

Bulbophyllum cordilabium .... **Bulbophyllum cylindrobulbum**

Bulbophyllum coriaceum

Bulbophyllum coriophorum

Bulbophyllum coriscense

Bulbophyllum corneri .... **Bulbophyllum vesiculosum**

Bulbophyllum cornu-cervi

Bulbophyllum cornutum Ridl. .... **Bulbophyllum forbesii**

*****Bulbophyllum cornutum (Blume) Rchb. f.**

‡*Bulbophyllum cornutum* (Blume) Rchb. f. .... *Hamularia cornuta* ‡

Bulbophyllum cornutum var. *ecornutum* .... **Bulbophyllum ecornutum**

‡*Bulbophyllum cornutum* var. *ecornutum* .... *Hamularia ecornuta* ‡

Bulbophyllum corolliferum

Bulbophyllum corolliferum var. *atropurpureum* .... **Bulbophyllum corolliferum**

Bulbophyllum correae

Bulbophyllum corticicola .... **Bulbophyllum schefferi**

Bulbophyllum corticicola var. *minor* .... **Bulbophyllum schefferi**

Bulbophyllum corythium .... **Bulbophyllum argyropus**

Bulbophyllum costatum

Bulbophyllum coursianum .... **Bulbophyllum rutenbergianum**

Bulbophyllum coweniorum

Bulbophyllum craibianum .... **Bulbophyllum shweliense**

Bulbophyllum crassicaudatum .... **Bulbophyllum odoratum**

Bulbophyllum crassifolioides .... **Bulbophyllum osyricera**

Bulbophyllum crassifolium Thwaites ex Trimen

Bulbophyllum crassifolium (Blume) J. J. Sm. .... **Bulbophyllum osyricera**

Bulbophyllum crassinervium

Bulbophyllum crassipes

Bulbophyllum crassipetalum

Bulbophyllum crassissimum

Bulbophyllum crassiusculifolium

Bulbophyllum crassulifolium .... **Bulbophyllum shepherdi**

Bulbophyllum crenilabium

Bulbophyllum crenulatum .... **Bulbophyllum coriophorum**

ALL NAMES   **ACCEPTED NAMES**

Bulbophyllum crepidiferum
Bulbophyllum cribbianum
Bulbophyllum crispatisepalum
Bulbophyllum crista-galli... **Bulbophyllum membranifolium**
Bulbophyllum croceum
Bulbophyllum crocodilus
Bulbophyllum cruciatum
Bulbophyllum cruciferum
Bulbophyllum cruentum
Bulbophyllum cruttwellii
Bulbophyllum cryptanthoides
Bulbophyllum cryptanthum Schltr.
Bulbophyllum cryptanthum Cogn................................... **Bulbophyllum clandestinum**
Bulbophyllum cryptophoranthoides Garay..................... **Bulbophyllum cryptophoranthus**
Bulbophyllum cryptophoranthoides Kraenzl.................. **Bulbophyllum membranifolium**
Bulbophyllum cryptophoranthus
Bulbophyllum cryptostachyum
Bulbophyllum cubicum
Bulbophyllum cucullatum
Bulbophyllum culex
Bulbophyllum cumingii
Bulbophyllum cuneatum
Bulbophyllum cuneifolium .... **Bulbophyllum caudatisepalum**
Bulbophyllum cuniculiforme
Bulbophyllum cupreum Lindl.
Bulbophyllum cupreum Hook. f. .... **Bulbophyllum careyanum**
Bulbophyllum cupuligerum... **Stolzia cupuligera**
Bulbophyllum curranii
*Bulbophyllum curtisii (Hook. f.) J. J. Sm..................... **Bulbophyllum corolliferum**
Bulbophyllum curtisii (Hook. f.) Ridl............................. **Bulbophyllum purpurascens**
Bulbophyllum curtisii var. *purpureum* .... **Bulbophyllum corolliferum**
Bulbophyllum curvibulbum
Bulbophyllum curvicaule
Bulbophyllum curvifolium
Bulbophyllum curvimentatum
Bulbophyllum cuspidilingue... **Bulbophyllum blumei**
Bulbophyllum cuspidipetalum J. J. Sm.
Bulbophyllum cuspidipetalum Schltr.............................. **Bulbophyllum sepikense**
Bulbophyllum cyanotriche
Bulbophyllum cyclanthum
Bulbophyllum cycloglossum
Bulbophyllum cyclopense
Bulbophyllum cyclophoroides
Bulbophyllum cyclophyllum
Bulbophyllum cyclosepalon
Bulbophyllum cylindraceum
Bulbophyllum cylindraceum var. *khasyanum*.................. **Bulbophyllum khasyanum**
Bulbophyllum cylindricum
Bulbophyllum cylindrobulbum
Bulbophyllum cylindrocarpum Frapp ex Cordem.
Bulbophyllum cylindrocarpum Schltr............................. **Bulbophyllum densibulbum**
Bulbophyllum cylindrocarpum var. **andringitrense**
Bulbophyllum cylindrocarpum var. **aurantiacum**
Bulbophyllum cylindrocarpum var. **olivacea**
Bulbophyllum cyrtopetalum .... **Bulbophyllum maximum**
Bulbophyllum dagamense
Bulbophyllum dahlemense .... **Bulbophyllum falcatum**
Bulbophyllum dalatense

* † ‡ **For explanation see page 2 and 3, point 3**
* † ‡ **Voir les explications page 10 et 11, point 3**
* † ‡ **Para mayor explicación, véase la página 18 y 19, point 3**   **37**

ALL NAMES   **ACCEPTED NAMES**

Bulbophyllum dalatensis .... Bulbophyllum vietnamense
Bulbophyllum daloaense .... **Bulbophyllum resupinatum** var.
                                                                        filiforme
Bulbophyllum danii
Bulbophyllum dasypetalum
Bulbophyllum dasyphyllum
Bulbophyllum dawongense
Bulbophyllum dayanum
Bulbophyllum dearei (Hort.) Rchb.f.
Bulbophyllum dearei Veitch .... **Bulbophyllum dearei**
Bulbophyllum debile
Bulbophyllum debrincatiae
Bulbophyllum debruynii
Bulbophyllum decarhopalon
Bulbophyllum decaryanum
Bulbophyllum decatriche
Bulbophyllum deceptum
Bulbophyllum decipiens .... **Bulbophyllum colubrinum**
Bulbophyllum decumbens
Bulbophyllum decurrentilobum
Bulbophyllum decurviscapum
Bulbophyllum decurvulum
Bulbophyllum deistelianum .... **Bulbophyllum falcatum** var.
                                                                        bufo
Bulbophyllum dekockii
Bulbophyllum delicatulum
Bulbophyllum delitescens
Bulbophyllum deltoideum
Bulbophyllum deminutum
Bulbophyllum demissum .... **Bulbophyllum smithianum**
Bulbophyllum dempoense
Bulbophyllum dendrobioides
Bulbophyllum dendrochiloides
Bulbophyllum dennisii
Bulbophyllum densibulbum
Bulbophyllum densiflorum Ridl .... **Bulbophyllum singaporeanum**
Bulbophyllum densiflorum Rolfe .... **Bulbophyllum cariniflorum**
Bulbophyllum densifolium
Bulbophyllum densissimum .... **Bulbophyllum farinulentum**
                                                                        subsp. **densissimum**
Bulbophyllum densum
Bulbophyllum denticulatum
Bulbophyllum dentiferum
Bulbophyllum dependens
Bulbophyllum depressum
Bulbophyllum derchianum .... **Bulbophyllum tokioi**
Bulbophyllum desmanthum
Bulbophyllum desmotrichoides
Bulbophyllum devangiriense .... **Bulbophyllum monanthum**
Bulbophyllum devium
Bulbophyllum devogelii
Bulbophyllum dewildei
Bulbophyllum dhaninivatii
Bulbophyllum dianthum
Bulbophyllum dibothron
Bulbophyllum diceras .... **Bulbophyllum posticum**
Bulbophyllum dichaeoides
Bulbophyllum dichilus
Bulbophyllum dichotomum

ALL NAMES	ACCEPTED NAMES
Bulbophyllum dichromum	**Sunipia dichroma**
Bulbophyllum dickasonii	
Bulbophyllum dictyoneuron	
Bulbophyllum didymotropis	
Bulbophyllum digitatum	**Drymoda digitata**
Bulbophyllum digoelense	
Bulbophyllum digoelense var. *septemtrionale*	**Bulbophyllum septemtrionale**
Bulbophyllum diplantherum	
Bulbophyllum diploncos	
Bulbophyllum dischidiifolium	
Bulbophyllum dischorense	
Bulbophyllum disciflorum	**Trias disciflora**
Bulbophyllum discilabium	
Bulbophyllum discolor	
Bulbophyllum discolor var. **cubitale**	
Bulbophyllum disjunctibulbum	**Bulbophyllum cylindrobulbum**
Bulbophyllum disjunctum	
Bulbophyllum dispersum	**Bulbophyllum kaniense**
Bulbophyllum dispersum var. *roseans*	**Bulbophyllum kaniense**
Bulbophyllum dissitiflorum	
Bulbophyllum dissolutum	
Bulbophyllum distans J. J. Sm.	**Bulbophyllum saltatorium** var. **albociliatum**
Bulbophyllum distans Lindl.	**Bulbophyllum saltatorium var. albociliatum**
Bulbophyllum distichobulbum	
Bulbophyllum distichum	
Bulbophyllum divaricatum	
Bulbophyllum dixoni	**Bulbophyllum morphologorum**
Bulbophyllum djamuense	
Bulbophyllum djumaense	**Bulbophyllum maximum**
Bulbophyllum djumaense var. *grandifolium*	**Bulbophyllum maximum**
Bulbophyllum dolabriforme	
Bulbophyllum dolichoblepharon	
Bulbophyllum dolichoglottis	
Bulbophyllum dorotheae	**Bulbophyllum pumilum**
Bulbophyllum doryphoroide	
Bulbophyllum dracunculus	
Bulbophyllum drallei	**Bulbophyllum pumilum**
Bulbophyllum dransfieldii	
Bulbophyllum drepanosepalum	
Bulbophyllum dryadum	
Bulbophyllum dryas	
Bulbophyllum drymoglossum	
Bulbophyllum dschischungarense	
Bulbophyllum dubium	
Bulbophyllum dulitense	**Bulbophyllum caudatisepalum**
Bulbophyllum dunstervillei	
Bulbophyllum dusenii	
Bulbophyllum dyeranum	**Bulbophyllum rolfei**
Bulbophyllum dyphoniae	**Bulbophyllum dayanum**
Bulbophyllum ealaense	**Bulbophyllum scaberulum**
Bulbophyllum eberhardtii	**Bulbophyllum picturatum**
Bulbophyllum ebracteolatum	
Bulbophyllum ebulbe	
Bulbophyllum ebulbum	**Bulbophyllum apodum**
Bulbophyllum eburneum	**Bulbophyllum scaberulum**
Bulbophyllum echinochilum	
Bulbophyllum echinolabium	

* † ‡ For explanation see page 2 and 3, point 3
* † ‡ Voir les explications page 10 et 11, point 3
* † ‡ Para mayor explicación, véase la página 18 y 19, point 3

ALL NAMES	ACCEPTED NAMES

Bulbophyllum echinulus
Bulbophyllum eciliatum
‡ *Bulbophyllum ecornutoides* ‡
Bulbophyllum ecornutum
Bulbophyllum ecornutum var. *daliense* .... **Bulbophyllum ecornutum**
‡*Bulbophyllum ecornutum* var. *daliense* .... *Hamularia cornuta* ‡
Bulbophyllum ecornutum var. *teloense* .... **Bulbophyllum ecornutum**
‡*Bulbophyllum ecornutum* var. *teloense* .... *Hamularia cornuta* ‡
Bulbophyllum ecuadorense .... **Bulbophyllum sordidum**
Bulbophyllum ecuadoriensis .... **Bulbophyllum sordidum**
Bulbophyllum edentatum
Bulbophyllum effusum .... **Bulbophyllum fractiflexum**
Bulbophyllum elachanthe
Bulbophyllum elachon .... **Bulbophyllum pumilum**
Bulbophyllum elaidium .... **Stolzia elaidum**
Bulbophyllum elaphoglossum
Bulbophyllum elasmatopus
Bulbophyllum elassoglossum
Bulbophyllum elassonotum
Bulbophyllum elatius .... **Bulbophyllum odoratum**
Bulbophyllum elatum
Bulbophyllum elbertii
Bulbophyllum electrinum
Bulbophyllum elegans Gardner ex Thwaites
Bulbophyllum elegans (Teijsm. & Binn.) J. J. Sm........... **Bulbophyllum pulchrum**
Bulbophyllum elegantius
Bulbophyllum elegantulum
Bulbophyllum elephantinum
Bulbophyllum elevatopunctatum
Bulbophyllum elizae
Bulbophyllum elliae
Bulbophyllum elliottii
Bulbophyllum ellipticifolium
Bulbophyllum ellipticum Schltr.
Bulbophyllum ellipticum De Wild. .... **Bulbophyllum oxychilum**
Bulbophyllum elmeri
Bulbophyllum elodeiflorum
Bulbophyllum elongatum (Blume) Hassk.
Bulbophyllum elongatum (De Wild.) De Wild. .... **Bulbophyllum ivorense**
Bulbophyllum emarginatum
Bulbophyllum emiliorum
Bulbophyllum encephalodes
Bulbophyllum endotrachys
Bulbophyllum ensiculiferum
Bulbophyllum entomonopsis
Bulbophyllum epapillosum
Bulbophyllum ephippium .... **Bulbophyllum blumei**
Bulbophyllum epibulbon
Bulbophyllum epicrianthes
Bulbophyllum epicrianthes var. **sumatranum**
Bulbophyllum epiphytum
Bulbophyllum equivestigium .... **Bulbophyllum cylindrobulbum**
Bulbophyllum erectum Thouars
Bulbophyllum erectum Ridl. .... **Bulbophyllum gracilicaule**
Bulbophyllum ericssoni .... **Bulbophyllum binnendijkii**
Bulbophyllum erinaceum
Bulbophyllum erioides
Bulbophyllum erosipetalum

ALL NAMES	ACCEPTED NAMES
Bulbophyllum errabundum..	**Bulbophyllum orbiculare** subsp. cassideum
Bulbophyllum erratum	
Bulbophyllum erythrochilum...	**Bulbophyllum simile**
Bulbophyllum erythroglossum	
Bulbophyllum erythrostachyum	
Bulbophyllum erythrostictum	
Bulbophyllum escritorii	
Bulbophyllum eublepharum..	**Bulbophyllum ligulatum**
Bulbophyllum euplepharum	
Bulbophyllum evansii	
Bulbophyllum evasum	
Bulbophyllum evrardii	
Bulbophyllum exaltatum	
Bulbophyllum exasperatum	
Bulbophyllum exiguiflorum	
Bulbophyllum exiguum	
Bulbophyllum exiguum var. *dallachyi*............................	**Bulbophyllum wilkianum**
Bulbophyllum exile	
Bulbophyllum exilipes	
Bulbophyllum exiliscapum..	**Bulbophyllum flavescens**
Bulbophyllum eximium..	**Bulbophyllum ceratostylis**
Bulbophyllum expallidum	
Bulbophyllum exquisitum	
Bulbophyllum extensum	**Bulbophyllum kaniense**
Bulbophyllum facetum	
Bulbophyllum falcatocaudatum	
Bulbophyllum falcatum	
Bulbophyllum falcatum var. **bufo** (Lindl.) Govaerts	
Bulbophyllum falcatum var. *bufo* (Lindl.) J.J. Verm.**.......**	**Bulbophyllum falcatum** var. **bufo** (Lindl.) Govaerts
Bulbophyllum falcatum var. **velutinum**	
Bulbophyllum falcibracteum	
Bulbophyllum falciferum	
Bulbophyllum falcifolium	
Bulbophyllum falcipetalum	
Bulbophyllum falculicorne	
Bulbophyllum fallax	
Bulbophyllum farinulentum	
Bulbophyllum farinulentum subsp. **densissimum**	
Bulbophyllum farreri	
Bulbophyllum fasciatum	
Bulbophyllum fasciculatum	
Bulbophyllum fasciculiferum	
Bulbophyllum fascinator..	**Bulbophyllum putidum**
Bulbophyllum fatuum..	**Bulbophyllum trifilum**
Bulbophyllum faunula	
Bulbophyllum fayi	
Bulbophyllum fenestratum	**Bulbophyllum taeniophyllum**
Bulbophyllum fenixii	
Bulbophyllum ferkoanum	
Bulbophyllum ferruginescens...	**Bulbophyllum cylindrobulbum**
Bulbophyllum fibratum (Gagnep.) Seidenf.	
Bulbophyllum fibratum (Gagnep.) Bân & D. H. Duong..	**Bulbophyllum fibratum**
Bulbophyllum fibrinum	
Bulbophyllum fibrosum..	**Bulbophyllum pileatum**
Bulbophyllum filamentosum	
Bulbophyllum filicaule J.J. Sm.	
Bulbophyllum filicoides	

* † ‡ **For explanation see page 2 and 3, point 3**
* † ‡ **Voir les explications page 10 et 11, point 3**
* † ‡ **Para mayor explicación, véase la página 18 y 19, point 3**

ALL NAMES   **ACCEPTED NAMES**

Bulbophyllum filifolium
Bulbophyllum filiforme .... **Bulbophyllum resupinatum** var.
                                                             filiforme
Bulbophyllum filisepalum... **Bulbophyllum trifilum** subsp.
                                                             filisepalum
Bulbophyllum filovagans
Bulbophyllum fimbriatum (Lindl.) Rchb. f.
Bulbophyllum fimbriatum H. Perrier............................. **Bulbophyllum peyrotii**
‡ *Bulbophyllum fimbriperianthium* ‡
Bulbophyllum finetianum .... **Bulbophyllum betchei**
Bulbophyllum finetii
Bulbophyllum finisterrae
Bulbophyllum fischeri
Bulbophyllum fissibrachium
Bulbophyllum fissipetalum Schltr.
Bulbophyllum fissipetalum Kraenzl............................... **Genoplesium calopterum**
Bulbophyllum flabellum-veneris
Bulbophyllum flagellare
Bulbophyllum flammuliferum
Bulbophyllum flavescens
Bulbophyllum flavescens var. *temelenense*..................... **Bulbophyllum flavescens**
Bulbophyllum flavescens var. *triflorum*.......................... **Bulbophyllum flavescens**
Bulbophyllum flavicolor
Bulbophyllum flavidiflorum
*Bulbophyllum flavidum Lindl...................................... **Bulbophyllum pumilum**
Bulbophyllum flavidum S. Z. Lucksom .... **Bulbophyllum pantlingii**
Bulbophyllum flavidum var. *elongatum*......................... **Bulbophyllum ivorense**
Bulbophyllum flavidum var. *purpureum*........................ **Bulbophyllum pumilum**
Bulbophyllum flaviflorum
Bulbophyllum flavisepalum Hayata.............................. **Bulbophyllum retusiusculum**
Bulbophyllum flavisepalum (Hayata) Masamune........... **Bulbophyllum retusiusculum**
Bulbophyllum flavofimbriatum
Bulbophyllum flavorubellum
Bulbophyllum flavum
Bulbophyllum flectens.. **Bulbophyllum unifoliatum**
                                                             subsp. **flectens**
Bulbophyllum fletcherianum Rolfe
Bulbophyllum fletcherianum Hort................................. **Bulbophyllum fletcherianum**
Bulbophyllum fletcherianum Pearson........................... **Bulbophyllum fletcherianum**
Bulbophyllum flexiliscapum... **Bulbophyllum saltatorium** var.
                                                             albociliatum
Bulbophyllum flexuosum
Bulbophyllum flickingerianum...................................... **Bulbophyllum peyrotii**
Bulbophyllum floribundum
Bulbophyllum florulentum
Bulbophyllum foenisecii... **Bulbophyllum auricomum**
Bulbophyllum foetidilabrum
Bulbophyllum foetidolens
Bulbophyllum foetidum
Bulbophyllum foetidum var. **grandiflorum**
Bulbophyllum folliculiferum
Bulbophyllum fonsflorum
Bulbophyllum foraminiferum
Bulbophyllum forbesii
Bulbophyllum fordii
Bulbophyllum formosanum (Rolfe) Nakajima
Bulbophyllum formosanum (Rolfe) S. S. Ying............... **Bulbophyllum formosanum**
Bulbophyllum formosanum (Rolfe) Seidenf.................. **Bulbophyllum formosanum**
Bulbophyllum formosum

ALL NAMES	ACCEPTED NAMES
Bulbophyllum forresti	
Bulbophyllum forsythianum	
Bulbophyllum foveatum..	**Bulbophyllum savaiense** subsp. **subcubicum**
Bulbophyllum fractiflexoides..	**Bulbophyllum fractiflexum**
Bulbophyllum fractiflexum J. J. Sm.	
Bulbophyllum fractiflexum Kraenzl................................	**Bulbophyllum falcatum** var. **velutinum**
Bulbophyllum fractiflexum Pabst....................................	**Bulbophyllum pabstii**
Bulbophyllum fractiflexum subsp. **salomonense**	
Bulbophyllum francoisii	
Bulbophyllum francoisii var. **andrangense**	
Bulbophyllum frappieri Schltr.	
Bulbophyllum frappieri A. D. Hawkes...........................	**Bulbophyllum frappieri**
Bulbophyllum fraudulentum	
Bulbophyllum fritillariiflorum	
Bulbophyllum frostii	
Bulbophyllum frustrans	
Bulbophyllum fruticicola	
Bulbophyllum fuerstenbergianum	**Bulbophyllum scaberulum** var. **fuerstenbergianum**
Bulbophyllum fukuyamae	
Bulbophyllum fulgens	
Bulbophyllum fulvibulbum	
Bulbophyllum funingense	
Bulbophyllum furcatum	
Bulbophyllum furciferum	**Bulbophyllum ochroleucum**
Bulbophyllum furcillatum	
Bulbophyllum fuscatum	
Bulbophyllum fuscescens	**Epigeneium fuscescens**
Bulbophyllum fusciflorum	
Bulbophyllum fuscoides	**Bulbophyllum acutebracteatum** var. **rubrobrunneopapillosum**
Bulbophyllum fusco-purpureum	
Bulbophyllum fuscum	
Bulbophyllum fuscum var. **melinostachyum**	
Bulbophyllum futile	
Bulbophyllum gabonis..	**Bulbophyllum pumilum**
Bulbophyllum gabunense	**Bulbophyllum colubrinum**
Bulbophyllum gadgarrense	
Bulbophyllum gajoense	
Bulbophyllum galactanthum	
Bulbophyllum galbinum	**Bulbophyllum uniflorum**
Bulbophyllum galeatum	**Polystachya galeata**
Bulbophyllum galliaheneum	
Bulbophyllum gamblei	**Bulbophyllum fischeri**
Bulbophyllum gamosepalum	**Bulbophyllum flabellum-veneris**
Bulbophyllum garupinum..	**Bulbophyllum infundibuliforme**
Bulbophyllum gautierense	
Bulbophyllum geminatum...	**Bulbophyllum biflorum**
Bulbophyllum geminum..	**Bulbophyllum pachytelos**
Bulbophyllum gemma-reginae	
Bulbophyllum geniculiferum	
Bulbophyllum gentilii	**Bulbophyllum schinzianum**
Bulbophyllum genybrachyum..	**Bulbophyllum fractiflexum**
Bulbophyllum geraense	
Bulbophyllum gerlandianum	
Bulbophyllum gibbilingue	**Bulbophyllum cernuum**
Bulbophyllum gibbolabium	

* † ‡ **For explanation see page 2 and 3, point 3**
* † ‡ **Voir les explications page 10 et 11, point 3**
* † ‡ **Para mayor explicación, véase la página 18 y 19, point 3**   **43**

ALL NAMES	ACCEPTED NAMES

‡*Bulbophyllum gibbolabium* .... *Hamularia gibbolabia* ‡
Bulbophyllum gibbonianum .... **Bulbophyllum membranaceum**
Bulbophyllum gibbosum
Bulbophyllum gibbsiae
Bulbophyllum gibsonii .... **Bulbophyllum khasyanum**
Bulbophyllum gigas .... **Bulbophyllum elongatum**
Bulbophyllum gilgianum
Bulbophyllum gilleti .... **Bulbophyllum imbricatum**
Bulbophyllum gilvum
Bulbophyllum gimagaanense
Bulbophyllum giriwoensc
Bulbophyllum gjellerupii
Bulbophyllum glabrilabre .... **Bulbophyllum orbiculare**
Bulbophyllum glabrum
Bulbophyllum gladiatum
Bulbophyllum glanduliferum
Bulbophyllum glandulosum
Bulbophyllum glaucifolium
Bulbophyllum glaucum
Bulbophyllum globiceps
Bulbophyllum globiceps var. **boloboense**
Bulbophyllum globuliforme
Bulbophyllum globulosum
Bulbophyllum globulus
Bulbophyllum glutinosum
Bulbophyllum gnomoniferum
Bulbophyllum gobiense
Bulbophyllum godseffianum .... **Bulbophyllum dearei**
Bulbophyllum goebelianum .... **Bulbophyllum dearei**
Bulbophyllum goliathense
Bulbophyllum gomesii
Bulbophyllum gomphreniflorum
Bulbophyllum gongshanense
Bulbophyllum gorumense .... **Bulbophyllum savaiense** subsp.
    gorumense
Bulbophyllum govidjoae .... **Bulbophyllum cylindrobulbum**
Bulbophyllum gracile Thouars
Bulbophyllum gracile (Blume) Lindl. .... **Bulbophyllum schefferi**
Bulbophyllum gracile C. S. P. Parish & Rchb. f............. **Bulbophyllum reichenbachii**
Bulbophyllum gracilicaule
Bulbophyllum gracilipes
Bulbophyllum graciliscapum Schltr.
Bulbophyllum graciliscapum Ames & Rolfe.................. **Bulbophyllum apoense**
Bulbophyllum graciliscapum H. Perrier .... *Bulbophyllum perseverans* †
Bulbophyllum graciliscapum Summerh. .... **Bulbophyllum saltatorium** var.
    albociliatum
***Bulbophyllum gracillimum** (Rolfe) Rolfe
Bulbophyllum gracillimum Hayata................................ **Bulbophyllum aureolabellum**
Bulbophyllum gramineum
Bulbophyllum graminifolium.. **Bulbophyllum calyptratum** var.
    graminifolium
Bulbophyllum grammopoma
Bulbophyllum grandiflorum Schltr.
Bulbophyllum grandiflorum Griff. .... **Bulbophyllum reptans**
Bulbophyllum grandifolium
Bulbophyllum grandilabre
Bulbophyllum grandimesense
Bulbophyllum granulosum
Bulbophyllum graveolens

44

ALL NAMES	ACCEPTED NAMES

Bulbophyllum gravidum
Bulbophyllum griffithianum .... **Bulbophyllum flabellum-veneris**
Bulbophyllum griffithii
Bulbophyllum groeneveldtii
Bulbophyllum grotianum
Bulbophyllum grudense
Bulbophyllum guamense
‡ *Bulbophyllum gunnarii* ‡
Bulbophyllum gusdorfii
Bulbophyllum gusdorfii var. *johorense* .... **Bulbophyllum gusdorfii**
Bulbophyllum gustavii.. **Bulbophyllum josephi**
Bulbophyllum guttatum
Bulbophyllum guttifilum
‡ *Bulbophyllum guttulatoides* ‡
Bulbophyllum guttulatum
Bulbophyllum gyaloglossum
Bulbophyllum gymnopus
Bulbophyllum gyrochilum
Bulbophyllum habbemense
Bulbophyllum habropus... **Bulbophyllum orbiculare**
Bulbophyllum habrotinum
Bulbophyllum hahlianum
Bulbophyllum hainanense
Bulbophyllum halconense
Bulbophyllum hamadryas
Bulbophyllum hamadryas var. **orientale**
Bulbophyllum hamatifolium... **Bulbophyllum prianganense**
Bulbophyllum hamatipes
Bulbophyllum hamelini W. Watson
Bulbophyllum hamelinii Hort. ex Rolfe.......................... **Bulbophyllum hamelini**
Bulbophyllum haniffii
Bulbophyllum hans-meyeri
Bulbophyllum hapalanthos
Bulbophyllum harposepalum
Bulbophyllum hashimotoi.. **Bulbophyllum orthosepalum**
Bulbophyllum hassalli
Bulbophyllum hastatum.. **Bulbophyllum depressum**
Bulbophyllum hastiferum
Bulbophyllum hatusimanum
Bulbophyllum hedyothyrsus .... **Bulbophyllum chloranthum**
Bulbophyllum heldiorum
Bulbophyllum helenae
Bulbophyllum heliophilum
Bulbophyllum helix
Bulbophyllum hellwigianum
Bulbophyllum hemiprionotum
Bulbophyllum hemirhachis... **Bulbophyllum falcatum**
Bulbophyllum henanense
Bulbophyllum henrici
Bulbophyllum henrici var. **rectangulare**
Bulbophyllum henryi... **Bulbophyllum andersonii**
Bulbophyllum henshallii .... **Bulbophyllum lobbii**
Bulbophyllum herbula
Bulbophyllum herminiostachys .... **Bulbophyllum pumilum**
Bulbophyllum heteroblepharon
Bulbophyllum heterorhopalon
Bulbophyllum heterosepalum
Bulbophyllum hewetii.. **Bulbophyllum uniflorum**
Bulbophyllum hexarhopalos

* † ‡ For explanation see page 2 and 3, point 3
* † ‡ Voir les explications page 10 et 11, point 3
* † ‡ Para mayor explicación, véase la página 18 y 19, point 3   **45**

ALL NAMES   ACCEPTED NAMES

Bulbophyllum hexurum
Bulbophyllum hians
Bulbophyllum hians var. **alticola**
Bulbophyllum hiepii
Bulbophyllum hildebrandtii
Bulbophyllum hiljeae
Bulbophyllum hirsutissimum Kraenzl. (1912)
Bulbophyllum hirsutissimum Kraenzl. (1914) .... **Bulbophyllum comatum**
Bulbophyllum hirsutiusculum
Bulbophyllum hirsutum
Bulbophyllum hirtulum
Bulbophyllum hirtum
Bulbophyllum hirudiniferum
Bulbophyllum hirundinis
Bulbophyllum hirundinis var. *electrinum* .... **Bulbophyllum electrinum**
Bulbophyllum hispidum .... **Bulbophyllum dayanum**
Bulbophyllum hodgsoni
Bulbophyllum hollandianum
Bulbophyllum holochilum
Bulbophyllum holochilum var. **aurantiacum**
Bulbophyllum holochilum var. **pubescens**
Bulbophyllum holttumii .... **Bulbophyllum apiferum**
Bulbophyllum hookeri
Bulbophyllum hookerianum .... **Bulbophyllum oreonastes**
Bulbophyllum horizontale
Bulbophyllum hornense .... **Bulbophyllum stenobulbon**
Bulbophyllum horridulum
Bulbophyllum hortense .... **Bulbophyllum odoratum**
Bulbophyllum hortensoides .... **Bulbophyllum odoratum**
Bulbophyllum hovarum
Bulbophyllum howcroftii
Bulbophyllum hoyifolium
Bulbophyllum humbertii
Bulbophyllum humblotianum .... **Bulbophyllum leoni**
Bulbophyllum humblottii
Bulbophyllum humile
Bulbophyllum humiligibbum
Bulbophyllum hyacinthiodorum .... **Bulbophyllum odoratissimum**
Bulbophyllum hyalinum
Bulbophyllum hydrophilum J. J. Sm.
Bulbophyllum hydrophilum Schltr. .... **Bulbophyllum kenejiense**
Bulbophyllum hymenanthum
Bulbophyllum hymenobracteum
Bulbophyllum hymenobracteum var. **giriwoense**
Bulbophyllum hymenochilum
Bulbophyllum hystricinum
Bulbophyllum ialibuense
Bulbophyllum iboense
Bulbophyllum ichthyostomum .... **Bulbophyllum pygmaeum**
Bulbophyllum icteranthum
Bulbophyllum idenburgense
Bulbophyllum igneocentrum .... **Bulbophyllum gibbosum**
Bulbophyllum igneocentrum var. *lativaginatum* .... **Bulbophyllum gibbosum**
Bulbophyllum igneum
Bulbophyllum ignevenosum
Bulbophyllum ignobile
Bulbophyllum ikongoense
Bulbophyllum illecebrum
Bulbophyllum illudens

ALL NAMES   **ACCEPTED NAMES**

Bulbophyllum imbricans
***Bulbophyllum imbricatum** Lindl.
Bulbophyllum imbricatum Griff. .... **Bulbophyllum cylindraceum**
Bulbophyllum imerinense
Bulbophyllum imitans .... **Bulbophyllum cylindrobulbum**
Bulbophyllum imitator
Bulbophyllum immobile .... **Bulbophyllum cruciatum**
Bulbophyllum imogeniae... **Bulbophyllum pumilum**
Bulbophyllum impar
Bulbophyllum implexum... **Bulbophyllum minutum**
Bulbophyllum imschootianum.. **Bulbophyllum colubrinum**
Bulbophyllum inabai.. **Bulbophyllum japonicum**
Bulbophyllum inaequale (Blume) Lindl.
Bulbophyllum inaequale Rchb. f..................................... **Bulbophyllum colubrinum**
Bulbophyllum inaequale var. **angustifolium**
Bulbophyllum inaequisepalum
Bulbophyllum inauditum Schltr. (1913)
Bulbophyllum inauditum Schltr. (1925) .... **Bulbophyllum reflexiflorum**
Bulbophyllum incarum
Bulbophyllum inciferum
Bulbophyllum incisilabrum
Bulbophyllum inclinatum
Bulbophyllum incommodum
Bulbophyllum inconspicuum
Bulbophyllum incumbens
Bulbophyllum incurvum
Bulbophyllum indragirense... **Bulbophyllum tortuosum**
Bulbophyllum iners
Bulbophyllum inflatum.. **Bulbophyllum comatum** var.
                                                                                        inflatum
Bulbophyllum infracarinatum.. **Bulbophyllum unifoliatum**
                                                                                        subsp. **infracarinatum**
Bulbophyllum infundibuliflorum .... **Bulbophyllum oreonastes**
Bulbophyllum infundibuliforme
Bulbophyllum injoloense
Bulbophyllum injoloense subsp. **pseudoxypterum**
Bulbophyllum injoloense var. *pseudoxypterum* .... **Bulbophyllum injoloense** subsp.
                                                                                        pseudoxypterum
Bulbophyllum inopinatum .... **Bulbophyllum penicillium**
Bulbophyllum inops
Bulbophyllum inornatum
Bulbophyllum inquirendum
Bulbophyllum insectiferum
Bulbophyllum insigne.. **Bulbophyllum membranifolium**
Bulbophyllum insolitum
Bulbophyllum insulare... **Bulbophyllum pulchrum**
Bulbophyllum insulsoides... **Bulbophyllum insulsum**
Bulbophyllum insulsum
Bulbophyllum intermedium F. M. Bailey
Bulbophyllum intermedium De Wild.............................. **Bulbophyllum calyptratum** var.
                                                                                        graminifolium
Bulbophyllum intersitum
Bulbophyllum intertextum
Bulbophyllum intertextum var. *parvilabium*................... **Bulbophyllum intertextum**
Bulbophyllum intervallatum... **Bulbophyllum macrochilum**
Bulbophyllum intricatum
Bulbophyllum inunctum
Bulbophyllum inversum
Bulbophyllum invisum

* † ‡ **For explanation see page 2 and 3, point 3**
* † ‡ **Voir les explications page 10 et 11, point 3**
* † ‡ **Para mayor explicación, véase la página 18 y 19, point 3**   **47**

ALL NAMES ACCEPTED NAMES

Bulbophyllum involutum
Bulbophyllum ionophyllum
Bulbophyllum ipanemense
Bulbophyllum irigaleae.. **Bulbophyllum schinzianum** var.
irigaleae
Bulbophyllum ischnopus
Bulbophyllum ischnopus var. *major*.............................. **Bulbophyllum major**
Bulbophyllum ischnopus var. **rhodoneuron**
Bulbophyllum iterans
Bulbophyllum ituriense... **Bulbophyllum lupulinum**
Bulbophyllum ivorense
Bulbophyllum jaapii
‡ *Bulbophyllum jackyi* ‡
Bulbophyllum jacobi.. **Bulbophyllum cordemoyi**
Bulbophyllum jacobsonii.. **Bulbophyllum plumatum**
Bulbophyllum jacquetii.. **Sunipia dichroma**
Bulbophyllum jadunae
Bulbophyllum jaguariahyvae
Bulbophyllum jamaicense
Bulbophyllum janus
Bulbophyllum japonicum (Makino) J. J. Sm.................. **Bulbophyllum japonicum**
Bulbophyllum japonicum (Makino) Makino
Bulbophyllum jarense.. **Bulbophyllum penduliscapum**
Bulbophyllum javanicum (Blume) J. J. Sm. .... **Bulbophyllum epicrianthes**
*Bulbophyllum javanicum Miq..................................... **Bulbophyllum pahudi**
Bulbophyllum javanicum var. *sumatranum*.................... **Bulbophyllum epicrianthes** var.
sumatranum
Bulbophyllum jensenii
Bulbophyllum jesperseni.. **Bulbophyllum scaberulum**
Bulbophyllum jiewhoei
Bulbophyllum johannis
Bulbophyllum johannis-winkleri .... **Bulbophyllum nematocaulon**
Bulbophyllum johannulii
Bulbophyllum johannum... **Bulbophyllum hildebrandtii**
Bulbophyllum johnsonii
Bulbophyllum jolandae
*Bulbophyllum josephi (Kuntze) Summerh.
Bulbophyllum josephi M.Kumar & Sequiera.................. **Bulbophyllum orezii**
Bulbophyllum josephi var. **mahonii**
Bulbophyllum jugicola... **Bulbophyllum dekockii**
Bulbophyllum jumellanum
Bulbophyllum jungwirthianum....................................... **Bulbophyllum cochleatum**
Bulbophyllum kainochiloides
Bulbophyllum kaitiense
Bulbophyllum kamerunense .... **Bulbophyllum imbricatum**
Bulbophyllum kanburiense
Bulbophyllum kaniense
Bulbophyllum katherinae .... **Bulbophyllum reticulatum**
Bulbophyllum kauloense
Bulbophyllum kautskyi
Bulbophyllum keekee
Bulbophyllum kegelii
Bulbophyllum kelelense
Bulbophyllum kempfii
Bulbophyllum kempterianum.. **Bulbophyllum cylindrobulbum**
Bulbophyllum kemulense
Bulbophyllum kenae
Bulbophyllum kenejianum
Bulbophyllum kenejiense

48

ALL NAMES   ACCEPTED NAMES

Bulbophyllum keralensis
Bulbophyllum kermesinum
Bulbophyllum kerri .... **Bulbophyllum hirtum**
Bulbophyllum kestron
Bulbophyllum kettridgei
Bulbophyllum kewense .... **Bulbophyllum falcatum** var.
                                                                        velutinum
Bulbophyllum kewense var. *purpureum* .... **Bulbophyllum falcatum** var.
                                                                        velutinum
Bulbophyllum khaoyaiense
Bulbophyllum khasyanum
Bulbophyllum kieneri
Bulbophyllum kinabaluense .... **Bulbophyllum coriaceum**
Bulbophyllum kindtianum .... **Bulbophyllum saltatorium** var.
                                                                        albociliatum
Bulbophyllum kingii .... **Acrochaene punctata**
Bulbophyllum kirkwoodae
Bulbophyllum kirroanthum
Bulbophyllum kivuense
Bulbophyllum kjellbergii
Bulbophyllum klabatense
Bulbophyllum klossii Ridl.
Bulbophyllum klossii Ridl. .... **Bulbophyllum peyerianum**
Bulbophyllum kontumense
Bulbophyllum koordersii .... **Bulbophyllum pulchrum**
Bulbophyllum korimense
Bulbophyllum korinchense
Bulbophyllum korinchense var. **grandflorum**
Bulbophyllum korinchense var. **parviflorum**
Bulbophyllum korthalsii
Bulbophyllum koyanense .... **Bulbophyllum caudatisepalum**
Bulbophyllum kraenzlinianum .... **Bulbophyllum hirsutissimum**
‡ *Bulbophyllum kuanwuensis* ‡
Bulbophyllum kupense
Bulbophyllum kusaiense
Bulbophyllum kusukusense .... **Bulbophyllum affine**
Bulbophyllum kwangtungense
Bulbophyllum labatii
Bulbophyllum laciniatum
Bulbophyllum laciniatum var. *janeirense* .... **Bulbophyllum laciniatum**
Bulbophyllum lacinulosum
Bulbophyllum laetum
Bulbophyllum lageniforme
Bulbophyllum laggiarae .... **Bulbophyllum humblottii**
Bulbophyllum lakatoense
Bulbophyllum lambii
Bulbophyllum lamelluliferum
Bulbophyllum lamii
Bulbophyllum lamingtonense
Bulbophyllum lamprobulbon .... **Bulbophyllum fractiflexum**
Bulbophyllum lamprochlamys .... **Bulbophyllum xanthochlamys**
Bulbophyllum lanceolatum .... **Bulbophyllum flavescens**
Bulbophyllum lancifolium
Bulbophyllum lancilabium
Bulbophyllum lancipetalum
Bulbophyllum lancisepalum
Bulbophyllum langbianense .... **Bulbophyllum retusiusculum**
Bulbophyllum languidum
Bulbophyllum lanuginosum

* † ‡ **For explanation see page 2 and 3, point 3**
* † ‡ **Voir les explications page 10 et 11, point 3**
* † ‡ **Para mayor explicación, véase la página 18 y 19, point 3**   **49**

ALL NAMES	ACCEPTED NAMES
Bulbophyllum lanuriense	**Bulbophyllum falcatum** var. **velutinum**
Bulbophyllum laoticum	
Bulbophyllum lasianthum	
Bulbophyllum lasiochilum	
Bulbophyllum lasioglossum	
Bulbophyllum lasiopetalum	
Bulbophyllum latibrachiatum	
Bulbophyllum latibrachiatum var. **epilosum**	
Bulbophyllum latipes	
Bulbophyllum latipetalum	
Bulbophyllum latisepalum	
Bulbophyllum laurentianum..	**Bulbophyllum imbricatum**
Bulbophyllum laxiflorum	
Bulbophyllum laxiflorum var. **celebicum**	
Bulbophyllum laxiflorum var. *taluense*...........................	**Bulbophyllum talauense**
Bulbophyllum laxum	
Bulbophyllum layardii..	**Bulbophyllum longiflorum**
Bulbophyllum leandrianum	
Bulbophyllum lecouflei	
Bulbophyllum ledermanni	**Bulbophyllum imbricatum**
Bulbophyllum ledungense	
Bulbophyllum lehmannianum	
Bulbophyllum leibergii	
Bulbophyllum lemnifolium	
Bulbophyllum lemniscatoides	
Bulbophyllum lemniscatoides var. **exappendiculatum**	
Bulbophyllum lemniscatum	
Bulbophyllum lemuraeoides	
Bulbophyllum lemurense	
Bulbophyllum leniae	
Bulbophyllum leoni	
Bulbophyllum leontoglossum	
Bulbophyllum leopardinum	
Bulbophyllum leopardinum var. **tuberculatum**	
Bulbophyllum lepantense	
Bulbophyllum lepanthiflorum	
Bulbophyllum lepanthiflorum var. *rivulare*	**Bulbophyllum lepanthiflorum**
Bulbophyllum lepidum	**Bulbophyllum flabellum-veneris**
Bulbophyllum lepidum var. *insigne*	**Bulbophyllum flabellum-veneris**
Bulbophyllum leproglossum	
Bulbophyllum leptanthum	
Bulbophyllum leptobulbon	
Bulbophyllum leptocaulon	
Bulbophyllum leptochlamys	
Bulbophyllum leptoleucum	
Bulbophyllum leptophyllum	
Bulbophyllum leptopus	
Bulbophyllum leptorrhachis...	**Bulbophyllum falcatum**
Bulbophyllum leptosepalum	
Bulbophyllum leptostachyum	
Bulbophyllum leratiae..	**Bulbophyllum baladeanum**
Bulbophyllum leratii	**Bulbophyllum gracillimum**
Bulbophyllum leucopogon...	**Bulbophyllum pumilum**
Bulbophyllum leucorhachis..	**Bulbophyllum imbricatum**
Bulbophyllum leucorhodum	
Bulbophyllum leucothyrsus	
Bulbophyllum levanae	
Bulbophyllum levanae var. **giganteum**	

ALL NAMES

ACCEPTED NAMES

Bulbophyllum levatii
Bulbophyllum levatii var. **mischanthum**
Bulbophyllum leve
Bulbophyllum levidense
Bulbophyllum levinei
Bulbophyllum levyae
Bulbophyllum lewisense
Bulbophyllum leysenianum
Bulbophyllum leytense
Bulbophyllum lichenastrum .... **Dendrobium lichenastrum**
Bulbophyllum lichenoides
Bulbophyllum lichenophylax
Bulbophyllum lichenophylax var. *microdoron* .... **Bulbophyllum afzelii** var. **microdoron**
Bulbophyllum ligulatum
Bulbophyllum ligulifolium
Bulbophyllum liliacinum
Bulbophyllum liliacinum var. *sorocianum*....................... **Bulbophyllum liliacinum**
Bulbophyllum lilianae
Bulbophyllum limbatum
Bulbophyllum linchianum .... **Bulbophyllum melanoglossum**
Bulbophyllum linderi .... **Bulbophyllum imbricatum**
Bulbophyllum lindleyanum
Bulbophyllum lindleyi .... **Bulbophyllum calyptratum**
Bulbophyllum lineare
Bulbophyllum lineariflorum
Bulbophyllum linearifolium
Bulbophyllum linearilabium
Bulbophyllum lineariligulatum
Bulbophyllum linearipetalum.. **Bulbophyllum fractiflexum**
Bulbophyllum lineatum
Bulbophyllum lineolatum
Bulbophyllum linggense
Bulbophyllum linguiforme... **Bulbophyllum humblottii**
Bulbophyllum lingulatum
Bulbophyllum lingulatum f. *microphyton*........................ **Bulbophyllum lingulatum**
Bulbophyllum liparidioides
Bulbophyllum lipense
Bulbophyllum lissoglossum
Bulbophyllum listeri... **Bulbophyllum tortuosum**
Bulbophyllum lizae
Bulbophyllum lobbii
Bulbophyllum lobbii var. *breviflorum* .... **Bulbophyllum sumatranum**
Bulbophyllum lobbii var. *colosseum*.............................. **Bulbophyllum lobbii**
Bulbophyllum lobbii var. *henshallii* .... **Bulbophyllum lobbii**
Bulbophyllum lobbii var. *nettesiae*................................ **Bulbophyllum lobbii**
Bulbophyllum lobbii var. *siamense* .... **Bulbophyllum siamense**
Bulbophyllum lobulatum... **Bulbophyllum erectum**
Bulbophyllum lockii
Bulbophyllum loherianum
Bulbophyllum lohokii
Bulbophyllum lokonense
Bulbophyllum lonchophyllum Schltr. (1913)
Bulbophyllum lonchophyllum Schltr. (1919).................. **Bulbophyllum leptophyllum**
Bulbophyllum longebracteatum
Bulbophyllum longerepens
Bulbophyllum longhutense
Bulbophyllum longibrachiatum
Bulbophyllum longibracteatum

* † ‡ For explanation see page 2 and 3, point 3
* † ‡ Voir les explications page 10 et 11, point 3
* † ‡ Para mayor explicación, véase la página 18 y 19, point 3   51

ALL NAMES	ACCEPTED NAMES
Bulbophyllum longibulbum	**Bulbophyllum falcatum** var. **bufo**
Bulbophyllum longicaudatum	
Bulbophyllum longidens	
*****Bulbophyllum longiflorum** Thouars	
Bulbophyllum longiflorum Ridl.	**Bulbophyllum inunctum**
Bulbophyllum longilabre	
Bulbophyllum longimucronatum	
Bulbophyllum longipedicellatum	
Bulbophyllum longipedicellatum var. **gjellerupii**	
Bulbophyllum longipes	**Monomeria longipes**
Bulbophyllum longipetalum	
Bulbophyllum longipetiolatum	
Bulbophyllum longipiliferum	**Bulbophyllum octarrhenipetalum**
Bulbophyllum longirostre	
Bulbophyllum longiscapum	
Bulbophyllum longisepalum	
Bulbophyllum longiserpens	**Bulbophyllum cylindrobulbum**
Bulbophyllum longispicatum Cogn.	
Bulbophyllum longispicatum Kraenzl. & Schltr.	**Bulbophyllum resupinatum** var. **filiforme**
Bulbophyllum longissimum (Ridl.) Ridl.	
Bulbophyllum longissimum (Ridl.) J. J. Sm.	**Bulbophyllum longissimum**
Bulbophyllum longistelidium	**Bulbophyllum stormii**
Bulbophyllum longivagans	
Bulbophyllum longivaginans	
Bulbophyllum lophoglottis	
Bulbophyllum lophoton	
Bulbophyllum lordoglossum	
Bulbophyllum lorentzianum	
Bulbophyllum loroglossum	
Bulbophyllum louisiadum	
Bulbophyllum loxodiphyllum	**Bulbophyllum oxycalyx** var. **rubescens**
Bulbophyllum loxophyllum	
Bulbophyllum luanii	
Bulbophyllum lubiense	**Bulbophyllum falcatum** var. **bufo**
Bulbophyllum lucidum	
Bulbophyllum lucifugum	**Bulbophyllum calyptratum** var. **lucifugum**
Bulbophyllum luciphilum	
Bulbophyllum luckraftii	
Bulbophyllum luederwaldtii	
Bulbophyllum lumbriciforme	
Bulbophyllum lundianum	
Bulbophyllum lupulinum	
Bulbophyllum luteobracteatum	
Bulbophyllum luteolabium	**Bulbophyllum humblottii**
Bulbophyllum luteopurpureum	
Bulbophyllum lutescens	**Bulbophyllum falcipetalum**
Bulbophyllum luzonense	**Bulbophyllum laxiflorum**
Bulbophyllum lygeron	
Bulbophyllum lyperocephalum	
Bulbophyllum lyperostachyum	
Bulbophyllum lyriforme	
Bulbophyllum maboroense	
Bulbophyllum macgregorii Ames	**Bulbophyllum exquisitum**

ALL NAMES	ACCEPTED NAMES
Bulbophyllum macgregorii Schltr.	**Bulbophyllum chloranthum**
Bulbophyllum machupicchuense	
Bulbophyllum macilentum	
Bulbophyllum mackeeanum	**Bulbophyllum triste**
Bulbophyllum macneiceae	
Bulbophyllum macphersoni	
Bulbophyllum macphersoni var. **spathulatum**	
Bulbophyllum macraei	
Bulbophyllum macraei var. **autumnale**	
Bulbophyllum macranthoides	
Bulbophyllum macranthum	
Bulbophyllum macranthum var. *albescens*	**Bulbophyllum hahlianum**
Bulbophyllum macrobulbum	
Bulbophyllum macrocarpum	
Bulbophyllum macroceras	
Bulbophyllum macrochilum	
Bulbophyllum macrocoleum	
Bulbophyllum macrolepis	**Bulbophyllum longiscapum**
Bulbophyllum macrophyllum	**Bulbophyllum penduliscapum**
Bulbophyllum macrorhopalon	
Bulbophyllum macrostachyum	**Bulbophyllum resupinatum** var. **filiforme**
Bulbophyllum macrourum	
Bulbophyllum maculatum Boxall ex Naves	
Bulbophyllum maculatum Jum. & H. Perrier	**Bulbophyllum hildebrandtii**
Bulbophyllum maculosum Ames	
Bulbophyllum maculosum (Lindl.) Rchb. f.	**Bulbophyllum umbellatum**
Bulbophyllum madagascariense	**Bulbophyllum hildebrandtii**
Bulbophyllum magnibracteatum	
Bulbophyllum magnivaginatum	**Bulbophyllum gibbosum**
Bulbophyllum mahakamense	
Bulbophyllum mahoni	**Bulbophyllum josephi** var. **mahonii**
Bulbophyllum maijenense	
Bulbophyllum major	
Bulbophyllum makakense	**Bulbophyllum colubrinum**
Bulbophyllum makinoanum	**Bulbophyllum macraei**
Bulbophyllum makoyanum	
Bulbophyllum makoyanum var. *brienianum*	**Bulbophyllum brienianum**
Bulbophyllum malachadenia	
Bulbophyllum malawiense	**Bulbophyllum elliottii**
Bulbophyllum malayanum	**Bulbophyllum blepharistes**
Bulbophyllum maleolens	
Bulbophyllum malleolabrum	
Bulbophyllum mamberamense	
Bulbophyllum mananjarense	
Bulbophyllum manarae	
Bulbophyllum mandibulare	
Bulbophyllum mandrakanum	**Bulbophyllum coriophorum**
Bulbophyllum mangenotii	
Bulbophyllum mangoroanum	**Bulbophyllum ophiuchus**
Bulbophyllum manifestans	**Bulbophyllum callichroma**
Bulbophyllum manipetalum	**Bulbophyllum cheiropetalum**
‡ *Bulbophyllum manipurense* ‡	
Bulbophyllum mannii Hook. f.	**Bulbophyllum cochleatum**
Bulbophyllum mannii Rchb. f.	**Bulbophyllum reichenbachianum**
Bulbophyllum manobulbum	
Bulbophyllum maquilinguense	

* † ‡ For explanation see page 2 and 3, point 3
* † ‡ Voir les explications page 10 et 11, point 3
* † ‡ Para mayor explicación, véase la página 18 y 19, point 3

53

ALL NAMES   ACCEPTED NAMES

Bulbophyllum marcidum .... **Bulbophyllum schefferi**
Bulbophyllum marginatum
Bulbophyllum marivelense
Bulbophyllum marojejiense
Bulbophyllum maromanganum
Bulbophyllum marovoense
Bulbophyllum marudiense
Bulbophyllum masaganapense
Bulbophyllum masarangicum
Bulbophyllum masarangicum var. *nanodes* .... **Bulbophyllum masarangicum**
Bulbophyllum masdevalliaceum .... **Bulbophyllum blumei**
Bulbophyllum maskeliyense
Bulbophyllum masoalanum
Bulbophyllum masonii
Bulbophyllum mastersianum
Bulbophyllum matitanense
Bulbophyllum matitanense subsp. **rostratum**
Bulbophyllum mattesii †
Bulbophyllum maudeae
Bulbophyllum mauritianum .... **Hederorkis scandens**
Bulbophyllum maxillare
Bulbophyllum maxillarioides
***Bulbophyllum maximum** (Lindl.) Rchb. f.
Bulbophyllum maximum (Ridl.) Ridl .... **Bulbophyllum virescens**
Bulbophyllum maximum var. **oxypterum**
Bulbophyllum mayae .... **Bulbophyllum peyrotii**
Bulbophyllum mayombeense
Bulbophyllum mayrii
Bulbophyllum mearnsii
Bulbophyllum mediocre
Bulbophyllum medusae
Bulbophyllum medusella .... **Bulbophyllum croceum**
Bulbophyllum megalanthum .... **Bulbophyllum cheiri**
Bulbophyllum megalonyx
Bulbophyllum melanoglossum Hayata
Bulbophyllum melanoglossum Kraenzl. .... **Bulbophyllum mona-lisae** †
Bulbophyllum melanoglossum var. *rubropunctatum* .... **Bulbophyllum melanoglossum**
Bulbophyllum melanopogon .... **Bulbophyllum hildebrandtii**
Bulbophyllum melanorrhachis .... **Bulbophyllum falcatum** var.
                                                                                              velutinum
Bulbophyllum melanoxanthum
Bulbophyllum melilotus
Bulbophyllum melinanthum
Bulbophyllum melinoglossum
Bulbophyllum melinostachyum .... **Bulbophyllum fuscum** var.
                                                                                              melinostachyum
Bulbophyllum meliphagirostrum
Bulbophyllum melleri .... **Bulbophyllum sandersonii**
Bulbophyllum melleum
Bulbophyllum melliferum .... **Bulbophyllum rugosum**
Bulbophyllum melloi
Bulbophyllum membranaceum
Bulbophyllum membranifolium
Bulbophyllum menghaiense
Bulbophyllum menglunense
Bulbophyllum mentiferum
Bulbophyllum mentosum
Bulbophyllum merguense .... **Bulbophyllum lineatum**
Bulbophyllum meridense

ALL NAMES	ACCEPTED NAMES

Bulbophyllum meristorhachis
Bulbophyllum merrittii
Bulbophyllum mesodon
Bulbophyllum metonymon
Bulbophyllum micholitzianum
*__Bulbophyllum micholitzii__ Rolfe
Bulbophyllum micholitzii (Rolfe) Ho .... **Bulbophyllum retusiusculum**
Bulbophyllum micholitzii (Rolfe) J. J. Sm. .... **Bulbophyllum retusiusculum**
Bulbophyllum micholitzii (Rolfe) Seidenf. & Smitinand .... **Bulbophyllum retusiusculum**
Bulbophyllum micranthum Barb. Rodr.
Bulbophyllum micranthum Hook. f. .... **Bulbophyllum triste**
Bulbophyllum microblepharon
Bulbophyllum microbulbon Schltr.
Bulbophyllum microbulbon (Ridl.) Ridl. .... **Bulbophyllum ruficaudatum**
Bulbophyllum microcala
Bulbophyllum microcharis Schltr. (1905) .... **Bulbophyllum cylindrobulbum**
Bulbophyllum microcharis Schltr. (1923) .... **Bulbophyllum microcala**
Bulbophyllum microdendron
Bulbophyllum microdoron .... **Bulbophyllum afzelii** var. **microdoron**
Bulbophyllum microglossum Ridl.
Bulbophyllum microglossum H. Perrier .... **Bulbophyllum moldenkeanum**
Bulbophyllum microlabium
Bulbophyllum micronesiacum
Bulbophyllum micropetaliforme
Bulbophyllum micropetalum Rchb. f.
Bulbophyllum micropetalum Barb. Rodr. .... **Bulbophyllum cribbianum**
Bulbophyllum micropetalum Lindl. .... **Genyorchis micropetala**
Bulbophyllum microphyton .... **Bulbophyllum lingulatum**
Bulbophyllum microrhombos
Bulbophyllum microsphaerum
Bulbophyllum microstele .... **Bulbophyllum tenuifolium**
Bulbophyllum microtatanthum .... **Bulbophyllum savaiense** subsp. **subcubicum**
Bulbophyllum microtepalum
Bulbophyllum microtes
Bulbophyllum microthamnus
Bulbophyllum mildbraedii .... **Bulbophyllum saltatorium** var. **albociliatum**
Bulbophyllum milesii .... **Bulbophyllum pipio**
Bulbophyllum millenii .... **Bulbophyllum falcatum** var. **velutinum**
Bulbophyllum mimiense
Bulbophyllum minahassae
Bulbophyllum minax
Bulbophyllum mindanaense
Bulbophyllum mindorense
Bulbophyllum miniatum .... **Bulbophyllum saltatorium** var. **albociliatum**
Bulbophyllum minimibulbum .... **Bulbophyllum rhizomatosum**
Bulbophyllum minus .... **Bulbophyllum sandersonii** subsp. **stenopetalum**
Bulbophyllum minutibulbum
Bulbophyllum minutiflorum .... **Bulbophyllum gibbsiae**
Bulbophyllum minutilabrum
Bulbophyllum minutipetalum
Bulbophyllum minutissimum
Bulbophyllum minutulum
Bulbophyllum minutum Thouars

* † ‡ For explanation see page 2 and 3, point 3
* † ‡ Voir les explications page 10 et 11, point 3
* † ‡ Para mayor explicación, véase la página 18 y 19, point 3

55

ALL NAMES	ACCEPTED NAMES
Bulbophyllum minutum (Rolfe) Engler............................	**Bulbophyllum falcatum** var. **velutinum**
Bulbophyllum minutum var. *purpureum*.........................	**Bulbophyllum falcatum** var. **velutinum**
Bulbophyllum mirabile	
Bulbophyllum mirandaianum	
Bulbophyllum mirandum..	**Bulbophyllum macrochilum**
Bulbophyllum mirificum	
Bulbophyllum mirum	
Bulbophyllum mischobulbon	
Bulbophyllum mishmeense..	**Sunipia cirrhata**
Bulbophyllum mobilifilum	
Bulbophyllum modestum..	**Bulbophyllum sulcatum**
Bulbophyllum modicum..	**Bulbophyllum josephi** var. **mahonii**
Bulbophyllum moirianum...	**Bulbophyllum maximum**
Bulbophyllum moldenkeanum	
Bulbophyllum moliwense	**Bulbophyllum pumilum**
Bulbophyllum molossus	
Bulbophyllum mona-lisae †	
Bulbophyllum monanthos	
Bulbophyllum monanthum	
*Bulbophyllum moniliforme C. S. P. Parish & Rchb. f.	
Bulbophyllum moniliforme F. Muell.	**Bulbophyllum minutissimum**
Bulbophyllum moniliforme R. King	**Bulbophyllum minutissimum**
Bulbophyllum monosema	
Bulbophyllum monosepalum..	**Bulbophyllum napelli**
Bulbophyllum monstrabile	
Bulbophyllum montanum	
Bulbophyllum montense	
Bulbophyllum monticolum	**Bulbophyllum gravidum**
Bulbophyllum montigenum...	**Bulbophyllum flavescens**
Bulbophyllum mooreanum	**Bulbophyllum sandersonii**
Bulbophyllum moramanganum......................................	**Bulbophyllum maromanganum**
Bulbophyllum moratii	
Bulbophyllum morenoi	
Bulbophyllum moroides	
Bulbophyllum morotaiense	
Bulbophyllum morphologorum	
Bulbophyllum moulmeinense..	**Trias oblonga**
Bulbophyllum mucronatum	
Bulbophyllum mucronifolium	
Bulbophyllum mulderae	
Bulbophyllum multiflexum	
Bulbophyllum multiflorum Ridl.	
Bulbophyllum multiflorum (Breda) Kraenzl.	**Bulbophyllum bakhuizenii**
Bulbophyllum multiligulatum	
Bulbophyllum multivaginatum	
Bulbophyllum muluense..	**Bulbophyllum anguliferum**
Bulbophyllum mundulum	**Bulbophyllum taeniophyllum**
Bulbophyllum muricatum...	**Monosepalum muricatum**
Bulbophyllum muriceum	
Bulbophyllum murkelense	
Bulbophyllum muscarirubrum	
Bulbophyllum muscicola Schltr. (1913)	
Bulbophyllum muscicola Schltr. (1913)	**Bulbophyllum subpatulum**
Bulbophyllum muscicolum...	**Bulbophyllum retusiusculum**
Bulbophyllum musciferum..	**Bulbophyllum coniferum**
Bulbophyllum muscohaerens	

ALL NAMES	ACCEPTED NAMES

Bulbophyllum mutabile
Bulbophyllum mutabile var. *ceratostyloides*................... **Bulbophyllum mutabile**
Bulbophyllum mutabile var. **obesum**
Bulbophyllum mutatum
Bulbophyllum myolaense
Bulbophyllum myon
Bulbophyllum myrianthum... **Bulbophyllum clandestinum**
Bulbophyllum myrmecochilum
Bulbophyllum myrtillus
Bulbophyllum mysorense
Bulbophyllum mystax
Bulbophyllum mystrochilum
Bulbophyllum mystrophyllum
Bulbophyllum nabawanense
Bulbophyllum nagelii
Bulbophyllum namoronae
Bulbophyllum nannodes
Bulbophyllum nanobulbon... **Bulbophyllum ruficaudatum**
Bulbophyllum nanopetalum
Bulbophyllum nanum... **Bulbophyllum pumilum**
Bulbophyllum napelli
Bulbophyllum napelloides
Bulbophyllum nasica
Bulbophyllum nasilabium
Bulbophyllum nasseri
Bulbophyllum nasutum.. **Trias nasuta**
Bulbophyllum navicula
Bulbophyllum navigioliferum.. **Bulbophyllum antennatum**
Bulbophyllum nebularum
Bulbophyllum neglectum
Bulbophyllum negrosianum
Bulbophyllum neilgherrense
Bulbophyllum nematocaulon
Bulbophyllum nematopodum
Bulbophyllum nematorhizis
Bulbophyllum nemorale
Bulbophyllum nemorosum (Barb. Rodr.) Cogn.
Bulbophyllum nemorosum Schltr. .... **Bulbophyllum singuliflorum**
Bulbophyllum neo-caledonicum
Bulbophyllum neo-pommeranicum
Bulbophyllum neoebudicus †
Bulbophyllum neoguinense
Bulbophyllum nephropetalum
Bulbophyllum nervulosum
Bulbophyllum nesiotes
Bulbophyllum newportii
Bulbophyllum ngoclinhensis
Bulbophyllum ngoyense
Bulbophyllum nieuwenhuisii
Bulbophyllum nigericum
Bulbophyllum nigrescens Rolfe
Bulbophyllum nigrescens Schltr.................................... **Bulbophyllum rubromaculatum**
Bulbophyllum nigriflorum
Bulbophyllum nigrilabium Schltr.
Bulbophyllum nigrilabium H. Perrier............................ **Bulbophyllum maudeae**
Bulbophyllum nigripetalum
Bulbophyllum nigritianum
Bulbophyllum nigromaculatum...................................... **Bulbophyllum tenuifolium**
Bulbophyllum nigropurpureum

* † ‡ **For explanation see page 2 and 3, point 3**
* † ‡ **Voir les explications page 10 et 11, point 3**
* † ‡ **Para mayor explicación, véase la página 18 y 19, point 3**   **57**

ALL NAMES   **ACCEPTED NAMES**

Bulbophyllum nigroscapum .... **Bulbophyllum ebulbe**
Bulbophyllum nipondhii
Bulbophyllum nitens
Bulbophyllum nitens var. **intermedium**
Bulbophyllum nitens var. **majus**
Bulbophyllum nitens var. **minus**
Bulbophyllum nitens var. **pulverulentum**
Bulbophyllum nitidum
Bulbophyllum niveo-sulphureum.................................... **Bulbophyllum colliferum**
Bulbophyllum niveum... **Bulbophyllum odoratum**
Bulbophyllum nodosum
Bulbophyllum noeanum... **Bulbophyllum farinulentum**
Bulbophyllum notabilipetalum
Bulbophyllum novaciae
Bulbophyllum novae-hiberniae
Bulbophyllum nubigenum
Bulbophyllum nubinatum
Bulbophyllum nudiscapum.. **Bulbophyllum saltatorium** var.
                                                                   albociliatum
Bulbophyllum nummularia
Bulbophyllum nummularioides
Bulbophyllum nuruanum.. **Bulbophyllum membranaceum**
Bulbophyllum nutans (Thouars) Thouars
Bulbophyllum nutans (Lindl.) Rchb. f. .... **Bulbophyllum othonis**
Bulbophyllum nutans var. **variifolium**
Bulbophyllum nyassanum... **Bulbophyllum maximum**
Bulbophyllum nymphopolitanum
Bulbophyllum obanense... **Bulbophyllum fuscum** var.
                                                                   melinostachyum
Bulbophyllum oblanceolatum King & Pantl.
Bulbophyllum oblanceolatum Schltr. .... **Bulbophyllum mattesii** †
Bulbophyllum obliquum
Bulbophyllum oblongum .... **Trias oblonga**
Bulbophyllum obovatifolium
Bulbophyllum obrienianum.. **Bulbophyllum brienianum**
Bulbophyllum obscuriflorum
Bulbophyllum obscurum .... **Bulbophyllum coniferum**
Bulbophyllum obtusatum
Bulbophyllum obtusiangulum
Bulbophyllum obtusilabium
Bulbophyllum obtusipetalum
Bulbophyllum obtusum (Blume) Lindl.
Bulbophyllum obtusum Jum. & H. Perrier..................... **Bulbophyllum obtusatum**
Bulbophyllum obtusum var. *robustum*........................... **Bulbophyllum flavidiflorum**
Bulbophyllum obyrnei
Bulbophyllum occidentale... **Dinema polybulbon**
Bulbophyllum occlusum
Bulbophyllum occultum
Bulbophyllum ochraceum (Barb. Rodr.) Cogn.
Bulbophyllum ochraceum (Ridl.) Ridl. .... **Bulbophyllum**
                                                                   serratotruncatum
Bulbophyllum ochranthum.. **Bulbophyllum sulcatum**
Bulbophyllum ochrochlamys Schltr. (1924)
‡*Bulbophyllum ochrochlamys* Schltr. (1924)................ ‡ *Bulbophyllum vakonae* † ‡
Bulbophyllum ochrochlamys Schltr. (1913)................... **Bulbophyllum absconditum**
Bulbophyllum ochroleucum
Bulbophyllum ochthochilum
Bulbophyllum ochthodes
Bulbophyllum octarrhenipetalum

ALL NAMES	ACCEPTED NAMES
Bulbophyllum octorhopalon	
Bulbophyllum oculatum	**Bulbophyllum cochlia**
Bulbophyllum odoardi	
Bulbophyllum odontoglossum	
Bulbophyllum odontopetalum	
Bulbophyllum odoratissimum	
Bulbophyllum odoratissimum var. **racemosum**	
Bulbophyllum odoratissimum var. *rubrolabellum*	**Bulbophyllum rubrolabellum**
Bulbophyllum odoratum	
Bulbophyllum odoratum var. **grandiflorum**	
Bulbophyllum odoratum var. *niveum*	**Bulbophyllum odoratum**
Bulbophyllum odoratum var. **obtusisepalum**	
Bulbophyllum odoratum var. **polyarachne**	
Bulbophyllum oeneum	**Bulbophyllum grudense**
Bulbophyllum oerstedii	
Bulbophyllum ogoouense	**Bulbophyllum fuscum**
Bulbophyllum oliganthum	
Bulbophyllum oligoblepharon	
Bulbophyllum oligochaete	
Bulbophyllum oligoglossum	
Bulbophyllum olivinum	
Bulbophyllum olivinum subsp. **linguiferum**	
Bulbophyllum olorinum	
Bulbophyllum ombrophilum	**Bulbophyllum reptans**
Bulbophyllum omerandrum	
Bulbophyllum oncidiochilum	**Dendrobium bifalce**
Bulbophyllum onivense	
Bulbophyllum oobulbum	
Bulbophyllum ophiuchus	
Bulbophyllum ophiuchus var. *ankaizinensis*	**Bulbophyllum ankaizinense**
Bulbophyllum ophiuchus var. **baronianum**	
Bulbophyllum ophiuchus var. *pallens*	**Bulbophyllum pallens**
Bulbophyllum orbiculare	
Bulbophyllum orbiculare subsp. **cassideum**	
Bulbophyllum oreas	**Bulbophyllum nematocaulon**
Bulbophyllum orectopetalum	
Bulbophyllum oreocharis	
Bulbophyllum oreodorum	
Bulbophyllum oreodoxa	
Bulbophyllum oreogenes	**Bulbophyllum retusiusculum** var. **oreogenes**
Bulbophyllum oreogenum	
Bulbophyllum oreonastes	
Bulbophyllum orezii	
Bulbophyllum orientale	
Bulbophyllum origami	
Bulbophyllum ormerodianum †	
Bulbophyllum ornatissimum	
Bulbophyllum ornatum	
Bulbophyllum ornithoglossum	**Bulbophyllum zebrinum**
Bulbophyllum ornithorhynchum	
Bulbophyllum orohense	
Bulbophyllum orsidice	
Bulbophyllum ortalis	
Bulbophyllum orthoglossum	
Bulbophyllum orthosepalum	
Bulbophyllum osyricera	
Bulbophyllum osyriceroides	
Bulbophyllum othonis	

* † ‡ For explanation see page 2 and 3, point 3
* † ‡ Voir les explications page 10 et 11, point 3
* † ‡ Para mayor explicación, véase la página 18 y 19, point 3

ALL NAMES   ACCEPTED NAMES

Bulbophyllum otochilum
Bulbophyllum otoglossum
Bulbophyllum ovale
Bulbophyllum ovalifolium
Bulbophyllum ovalifolium.. **Bulbophyllum clandestinum**
Bulbophyllum ovalitepalum
Bulbophyllum ovatilabellum
Bulbophyllum ovatolanceatum
Bulbophyllum ovatum
Bulbophyllum oxyanthum
Bulbophyllum oxycalyx
Bulbophyllum oxycalyx var. **rubescens**
Bulbophyllum oxychilum
Bulbophyllum oxyodon.. **Bulbophyllum falcatum**
Bulbophyllum oxypterum .... **Bulbophyllum maximum** var.
                                                              oxypterum
Bulbophyllum oxypterum var. *mosambicense*.................. **Bulbophyllum maximum**
Bulbophyllum oxysepaloides
Bulbophyllum pabstii
Bulbophyllum pachyacris
Bulbophyllum pachyanthum
Bulbophyllum pachybulbum.. **Bulbophyllum graveolens**
Bulbophyllum pachyglossum
Bulbophyllum pachyneuron
Bulbophyllum pachyphyllum.. **Bulbophyllum wrayi**
Bulbophyllum pachypus
Bulbophyllum pachyrachis
Bulbophyllum pachytelos
Bulbophyllum pahudi
Bulbophyllum paleaceum.. **Sunipia cirrhata**
Bulbophyllum paleiferum
Bulbophyllum palilabre
Bulbophyllum pallens
Bulbophyllum pallescens.. **Bulbophyllum bifarium**
Bulbophyllum pallidiflavum.. **Bulbophyllum cylindrobulbum**
Bulbophyllum pallidiflorum
Bulbophyllum pallidum
Bulbophyllum pampangense
Bulbophyllum pan
Bulbophyllum pandanetorum
Bulbophyllum pandurella
Bulbophyllum pangerangi... **Bulbophyllum gibbosum**
Bulbophyllum paniculatum .... **Ridleyella paniculata**
Bulbophyllum panigraphianum...................................... **Bulbophyllum sarcophyllum**
Bulbophyllum paniscus
Bulbophyllum pantlingii
Bulbophyllum pantoblepharon
Bulbophyllum pantoblepharon var. **vestitum**
Bulbophyllum papangense
Bulbophyllum papilio
Bulbophyllum papillatum
Bulbophyllum papillipetalum
Bulbophyllum papillosefilum
Bulbophyllum papillosum Finet .... **Bulbophyllum pumilum**
Bulbophyllum papillosum J. J. Sm. .... **Bulbophyllum papillatum**
Bulbophyllum papillosum (Rolfe) Seidenf. & Smitinand .... **Bulbophyllum thaiorum**
Bulbophyllum papuliferum
Bulbophyllum papuliglossum
Bulbophyllum papulilabium.. **Bulbophyllum colliferum**

ALL NAMES   ACCEPTED NAMES

Bulbophyllum papulipetalum
Bulbophyllum papulosum
Bulbophyllum parabates
Bulbophyllum paranaense
Bulbophyllum paranaense var. pauloense
Bulbophyllum pardalinum
Bulbophyllum pardalotum
Bulbophyllum parryae.................................... **Bulbophyllum suavissimum**
Bulbophyllum parviflorum
Bulbophyllum parvilabium Schltr. (1919)...................... **Bulbophyllum microlabium**
Bulbophyllum parvilabium Schltr. (1911)...................... **Bulbophyllum obtusum**
Bulbophyllum parvimentatum.. **Bulbophyllum pumilum**
Bulbophyllum parvulum (Hook. f.) J. J. Sm. .... **Bulbophyllum rolfei**
Bulbophyllum parvulum Lindl. .... **Bulbophyllum ovalifolium**
Bulbophyllum parvum
Bulbophyllum patella
Bulbophyllum patens
Bulbophyllum pauciflorum... **Bulbophyllum mutabile**
Bulbophyllum paucisetum
Bulbophyllum paullum.. **Bulbophyllum sulcatum**
Bulbophyllum paululum
Bulbophyllum pavimentatum... **Bulbophyllum pumilum**
Bulbophyllum pechei... **Bulbophyllum cupreum**
Bulbophyllum pectenveneris
Bulbophyllum pectinatum
Bulbophyllum pectinatum var. *transarisanense*.............. **Bulbophyllum pectinatum**
Bulbophyllum pedicellatum.. **Bulbophyllum laxiflorum**
Bulbophyllum pelicanopsis
Bulbophyllum pelma .... **Bulbophyllum stipulaceum**
Bulbophyllum pelma var. *gautierense* .... **Bulbophyllum stipulaceum**
Bulbophyllum peltopus
Bulbophyllum pemae
Bulbophyllum penduliscapum
Bulbophyllum pendulum
Bulbophyllum penicillium
Bulbophyllum peniculus... **Bulbophyllum rutenbergianum**
Bulbophyllum peninsulare
Bulbophyllum pensile... **Bulbophyllum macrourum**
Bulbophyllum pentaneurum
Bulbophyllum pentasticha
Bulbophyllum peperomiifolium
Bulbophyllum peperomioides... **Stolzia peperomioides**
Bulbophyllum perakense... **Bulbophyllum purpurascens**
Bulbophyllum peramoenum
Bulbophyllum percorniculatum
Bulbophyllum perductum
Bulbophyllum perductum var. sebesiense
Bulbophyllum perexiguum
Bulbophyllum perforans
Bulbophyllum pergracile... **Bulbophyllum caudatisepalum**
Bulbophyllum perii
Bulbophyllum perlongum.. **Bulbophyllum cylindrobulbum**
Bulbophyllum perparvulum
Bulbophyllum perpendiculare
Bulbophyllum perpusillum Wendl. & Kraenzl.
Bulbophyllum perpusillum Ridl...................................... **Bulbophyllum perparvulum**
Bulbophyllum perreflexum
Bulbophyllum perrieri
Bulbophyllum pertenue .... **Bulbophyllum intertextum**

* † ‡ For explanation see page 2 and 3, point 3
* † ‡ Voir les explications page 10 et 11, point 3
* † ‡ Para mayor explicación, véase la página 18 y 19, point 3

ALL NAMES   ACCEPTED NAMES

Bulbophyllum perseverans †
Bulbophyllum pervillei
Bulbophyllum petiolare
Bulbophyllum petiolatum
‡ *Bulbophyllum petrae* ‡
Bulbophyllum peyerianum
Bulbophyllum peyrotii
Bulbophyllum phaeanthum
Bulbophyllum phaeoglossum
Bulbophyllum phaeoneuron
Bulbophyllum phaeopogon... **Bulbophyllum schinzianum** var.
                                                  phaeopogon
Bulbophyllum phaeorhabdos
Bulbophyllum phalaenopsis
Bulbophyllum phayamense
Bulbophyllum philippinense
Bulbophyllum phillipsianum
Bulbophyllum pholidotoides... **Bulbophyllum cochleatum**
Bulbophyllum phormion
Bulbophyllum phreatiopse
Bulbophyllum phymatum
Bulbophyllum physocoryphum
Bulbophyllum pictum Schltr... **Bulbophyllum crenilabium**
*Bulbophyllum pictum C. S. P. Parish & Rchb. f. .... **Trias picta**
Bulbophyllum picturatum
Bulbophyllum pidacanthum
Bulbophyllum piestobulbon
Bulbophyllum piestoglossum
Bulbophyllum pileatum
Bulbophyllum pileolatum .... **Bulbophyllum elegantulum**
Bulbophyllum piliferum
Bulbophyllum pilosum
Bulbophyllum piluliferum
Bulbophyllum pingtungense .... **Bulbophyllum wightii**
Bulbophyllum pinicolum
Bulbophyllum pipio
Bulbophyllum pisibulbum
Bulbophyllum piundensis .... **Bulbophyllum ochroleucum**
Bulbophyllum placochilum
Bulbophyllum plagiatum
Bulbophyllum plagiopetalum
Bulbophyllum planiaxe .... **Bulbophyllum oreonastes**
Bulbophyllum planibulbe
Bulbophyllum planibulbe var. **sumatranum**
Bulbophyllum planifolium.. **Bulbophyllum desmotrichoides**
Bulbophyllum planilabre.. **Bulbophyllum peltopus**
Bulbophyllum planitiae
Bulbophyllum platirachis... **Bulbophyllum acutebracteatum**
Bulbophyllum platypodum
Bulbophyllum platyrhachis (Rolfe) Schltr...................... **Bulbophyllum maximum**
Bulbophyllum platyrrhachis Ridl.................................. **Bulbophyllum janus**
Bulbophyllum pleiopterum
Bulbophyllum pleurothallianthum
Bulbophyllum pleurothalloides Ames
Bulbophyllum pleurothalloides Schltr........................... **Bulbophyllum conchidioides**
Bulbophyllum pleurothallopsis
Bulbophyllum plicatum
Bulbophyllum plumatum
Bulbophyllum plumosum

ALL NAMES	ACCEPTED NAMES
Bulbophyllum plumula	
Bulbophyllum pobeguenii	Bulbophyllum scaberulum
Bulbophyllum pocillum	
Bulbophyllum poekilon	
Bulbophyllum pogonochilum	Bulbophyllum reflexiflorum subsp. **pogonochilum**
Bulbophyllum poilanei	
Bulbophyllum pokapindjangense	Bulbophyllum mutabile
Bulbophyllum polliculosum	
Bulbophyllum polyarachne	Bulbophyllum odoratum
Bulbophyllum polyblepharon	
Bulbophyllum polycyclum	
Bulbophyllum polygaliflorum	
Bulbophyllum polyphyllum	
Bulbophyllum polypodioides	Bulbophyllum ebulbe
Bulbophyllum polyrhizum	
*Bulbophyllum polystictum Ridl.	Bulbophyllum lobbii
Bulbophyllum polystictum Schltr.	Bulbophyllum erythrostictum
Bulbophyllum ponapense	Bulbophyllum betchei
Bulbophyllum popayanense	
Bulbophyllum porphyroglossum	Bulbophyllum pumilum
Bulbophyllum porphyrostachys	
Bulbophyllum porphyrotriche	
Bulbophyllum posticum	
Bulbophyllum potamophila	
Bulbophyllum praealtum	Bulbophyllum longiscapum
Bulbophyllum praestans	
Bulbophyllum praetervisum	
Bulbophyllum prenticei	Dendrobium lichenastrum
Bulbophyllum prianganense	
Bulbophyllum prismaticum	
Bulbophyllum pristis	
Bulbophyllum proboscideum	
Bulbophyllum procerum	
Bulbophyllum proculcastris	
Bulbophyllum proencai	
Bulbophyllum profusum	
Bulbophyllum propinquum	
Bulbophyllum prorepens	
Bulbophyllum protectum	
Bulbophyllum proteranthum	Dendrobium proteranthum
Bulbophyllum protractum	
Bulbophyllum proudlockii	
Bulbophyllum proximum	Bulbophyllum pachytelos
Bulbophyllum pseudofilicaule	
Bulbophyllum pseudonutans	Bulbophyllum brachystachyum
Bulbophyllum pseudopelma	
Bulbophyllum pseudopicturatum †	
Bulbophyllum pseudoserrulatum	
Bulbophyllum pseudotrias	
Bulbophyllum psilorhopalon	
Bulbophyllum psittacoglossum	
Bulbophyllum psittacoides	Bulbophyllum gracillimum
Bulbophyllum psychoon	
Bulbophyllum pteriphilum	Dendrochilum pallidiflavens
Bulbophyllum pteroglossum	Bulbophyllum monanthum
Bulbophyllum ptiloglossum	
Bulbophyllum ptilotes	
Bulbophyllum ptychantyx	

* † ‡ For explanation see page 2 and 3, point 3
* † ‡ Voir les explications page 10 et 11, point 3
* † ‡ Para mayor explicación, véase la página 18 y 19, point 3   63

ALL NAMES	ACCEPTED NAMES
Bulbophyllum puberulum	**Bulbophyllum flavescens**
Bulbophyllum pubiflorum	
Bulbophyllum pugilanthum	
Bulbophyllum pugioniforme	
Bulbophyllum puguahaanense	
Bulbophyllum pulchellum	
Bulbophyllum pulchellum var. **brachysepalum**	
Bulbophyllum pulchellum var. *purpureum*	**Bulbophyllum corolliferum**
Bulbophyllum pulchrum (N. E. Br.) J. J. Sm.	
Bulbophyllum pulchrum Schltr.	**Bulbophyllum bulhartii** †
Bulbophyllum pulchrum var. *cliftonii*	**Bulbophyllum pulchrum**
Bulbophyllum pulvinatum	
*****Bulbophyllum pumilio** C. S. P. Parish & Rchb. f.	
*Bulbophyllum pumilio Ridl.	**Bulbophyllum bidi**
Bulbophyllum pumilum	
Bulbophyllum punamense	
Bulbophyllum punctatissimum	**Bulbophyllum taeniophyllum**
Bulbophyllum punctatum Barb. Rodr.	
Bulbophyllum punctatum Fitzg.	**Bulbophyllum baileyi**
Bulbophyllum punctatum Ridl.	**Bulbophyllum dearei**
Bulbophyllum pungens	
Bulbophyllum pungens var. **pachyphyllum**	
Bulbophyllum puntjakense	
*****Bulbophyllum purpurascens** Teijsm. & Binn.	
Bulbophyllum purpurascens F. M. Bailey	**Bulbophyllum macphersoni**
Bulbophyllum purpureifolium	
Bulbophyllum purpurellum	
Bulbophyllum purpureorhachis	
Bulbophyllum purpureum Thwaites	
Bulbophyllum purpureum Naves	**Bulbophyllum macranthum**
Bulbophyllum pusillum Thouars	
Bulbophyllum pusillum (Rolfe) De Wild.	**Bulbophyllum sandersonii**
Bulbophyllum pustulatum	
Bulbophyllum putidum	
Bulbophyllum putii	
Bulbophyllum pygmaeum	
Bulbophyllum pyridion	
Bulbophyllum pyroglossum	
Bulbophyllum quadrangulare	
Bulbophyllum quadrangulare var. **latisepalum**	
Bulbophyllum quadrangulum	
Bulbophyllum quadrans	**Bulbophyllum octarrhenipetalum**
Bulbophyllum quadratum	**Bulbophyllum savaiense** subsp. **subcubicum**
Bulbophyllum quadrialatum	
Bulbophyllum quadricarinum	
Bulbophyllum quadricaudatum	
Bulbophyllum quadrichaete	
Bulbophyllum quadricolor	
Bulbophyllum quadrifalciculatum	
Bulbophyllum quadrifarium	
Bulbophyllum quadrisetum	
Bulbophyllum quadrisubulatum	
Bulbophyllum quasimodo	
Bulbophyllum quinquecornutum	**Bulbophyllum lichenophylax**
Bulbophyllum quinquelobum	
Bulbophyllum quinquelobum var. **lancilabium**	
Bulbophyllum quintasii	**Bulbophyllum intertextum**

ALL NAMES	ACCEPTED NAMES
Bulbophyllum racemosum Rolfe	
Bulbophyllum racemosum Hayata.....................................	**Bulbophyllum insulsum**
Bulbophyllum radiatum..	**Bulbophyllum laxiflorum**
Bulbophyllum radicans	
Bulbophyllum rajanum	
Bulbophyllum ramosii...	**Bulbophyllum flavescens**
Bulbophyllum ramosum	**Bulbophyllum ochroleucum**
Bulbophyllum ramulicola	
Bulbophyllum ranomafanae	
Bulbophyllum rariflorum	
Bulbophyllum rarum	
Bulbophyllum rauhii	
Bulbophyllum rauhii var. **andranobeense**	
Bulbophyllum raui	
Bulbophyllum reclusum	
Bulbophyllum rectilabre	
Bulbophyllum recurviflorum	
Bulbophyllum recurvilabre	
Bulbophyllum recurvimarginatum	**Bulbophyllum trifilum**
Bulbophyllum recurvum..	**Bulbophyllum pumilum**
Bulbophyllum reductum	
Bulbophyllum reevei	
Bulbophyllum reflexiflorum	
Bulbophyllum reflexiflorum subsp. **pogonochilum**	
Bulbophyllum reflexum	
Bulbophyllum refractilingue	
Bulbophyllum refractoides...	**Bulbophyllum wallichi**
Bulbophyllum refractum	
Bulbophyllum regnelli	
Bulbophyllum reichenbachianum	
Bulbophyllum reichenbachii	
Bulbophyllum reifii †	
Bulbophyllum reilloi	
Bulbophyllum reinwardtii	**Bulbophyllum uniflorum**
Bulbophyllum remiferum	
Bulbophyllum remotifolium (Fukuyama) Nakajima	**Bulbophyllum hirundinis**
Bulbophyllum remotifolium (Fukuyama) S. S. Ying	**Bulbophyllum hirundinis**
Bulbophyllum remotum..	**Bulbophyllum cylindrobulbum**
Bulbophyllum renipetalum	
Bulbophyllum renkinianum	
Bulbophyllum repens	
Bulbophyllum reptans	
Bulbophyllum reptans var. *acuta*....................................	**Bulbophyllum reptans**
Bulbophyllum reptans var. *subracemosa*.........................	**Bulbophyllum reptans**
Bulbophyllum restrepia	
Bulbophyllum resupinatum	
Bulbophyllum resupinatum var. **filiforme**	
Bulbophyllum reticosum	**Bulbophyllum dearei**
Bulbophyllum reticulatum	
Bulbophyllum retusiusculum	
Bulbophyllum retusiusculum var. **oreogenes**	
Bulbophyllum retusiusculum var. **tigridum**	
Bulbophyllum revolutum..	**Bulbophyllum lilianae**
Bulbophyllum rheedei	
Bulbophyllum rhizomatosum Ames & C. Schweinf.	
Bulbophyllum rhizomatosum Schltr. (1924)...................	**Bulbophyllum obtusilabium**
Bulbophyllum rhizomatosum Schltr. (1923)..................	**Bulbophyllum reifii** †
Bulbophyllum rhizophorae...	**Bulbophyllum falcatum** var. **velutinum**

* † ‡ **For explanation see page 2 and 3, point 3**
* † ‡ **Voir les explications page 10 et 11, point 3**
* † ‡ **Para mayor explicación, véase la página 18 y 19, point 3**   **65**

ALL NAMES	ACCEPTED NAMES
Bulbophyllum rhizophoreti..	**Bulbophyllum purpurascens**
Bulbophyllum rhodoglossum	
Bulbophyllum rhodoleucum	
Bulbophyllum rhodoneuron	
Bulbophyllum rhodopetalum...	**Bulbophyllum sandersonii** subsp. **stenopetalum**
Bulbophyllum rhodosepalum	
Bulbophyllum rhodostachys	
Bulbophyllum rhodostictum	
Bulbophyllum rhombifolium..	**Bulbophyllum hymenochilum**
Bulbophyllum rhomboglossum	
Bulbophyllum rhopaloblepharon	
Bulbophyllum rhopalochilum...	**Bulbophyllum oreonastes**
Bulbophyllum rhopalophorum	
Bulbophyllum rhynchoglossum Schltr. (1910)	
Bulbophyllum rhynchoglossum Schltr. (1912)................	**Bulbophyllum wechsbergii** †
Bulbophyllum ricaldonei	
Bulbophyllum rictorium...	**Bulbophyllum lucidum**
Bulbophyllum ridleyanum...	**Bulbophyllum virescens**
Bulbophyllum ridleyi...	**Bulbophyllum multiflorum**
Bulbophyllum rienanense	
Bulbophyllum rigidifilum	
Bulbophyllum rigidipes	
Bulbophyllum rigidum	
Bulbophyllum ringens...	**Bulbophyllum lindleyanum**
Bulbophyllum riparium	
Bulbophyllum rivulare	
Bulbophyllum riyanum	
Bulbophyllum robustum..	**Bulbophyllum coriophorum**
Bulbophyllum rojasii	
Bulbophyllum rolfeanum	
Bulbophyllum rolfei	
Bulbophyllum romburghii	
Bulbophyllum roraimense	
Bulbophyllum rosemarianum	
Bulbophyllum roseopunctatum	
Bulbophyllum roseum..	**Trias rosea**
Bulbophyllum rostratum...	**Bulbophyllum ochroleucum**
Bulbophyllum rostriceps	
Bulbophyllum rostriferum...	**Bulbophyllum oxycalyx** var. **rubescens**
Bulbophyllum rothschildianum	
Bulbophyllum rotundatum...	**Epigeneium rotundatum**
Bulbophyllum roxburghii	
Bulbophyllum rubescens...	**Bulbophyllum oxycalyx** var. **rubescens**
Bulbophyllum rubescens var. *meizobulbon*	**Bulbophyllum oxycalyx**
Bulbophyllum rubiferum	
Bulbophyllum rubiginosum	
Bulbophyllum rubipetalum	
Bulbophyllum rubrobrunneopapillosum	**Bulbophyllum acutebracteatum** var. **rubrobrunneopapillosum**
Bulbophyllum rubroguttatum	
Bulbophyllum rubrolabellum	
Bulbophyllum rubrolabium	
Bulbophyllum rubrolineatum	
Bulbophyllum rubromaculatum	
Bulbophyllum rubropunctatum	**Bulbophyllum melanoglossum**

ALL NAMES   **ACCEPTED NAMES**

Bulbophyllum rubroviolaceum.. **Bulbophyllum resupinatum** var.
                                                                 filiforme
Bulbophyllum rubrum
Bulbophyllum ruficaudatum
Bulbophyllum rufilabrum
Bulbophyllum rufinum
Bulbophyllum ruginosum
Bulbophyllum rugosibulbum
Bulbophyllum rugosisepalum
Bulbophyllum rugosum
Bulbophyllum rugulosum
Bulbophyllum rupestre
Bulbophyllum rupicola
Bulbophyllum rupicolum.. **Bulbophyllum taeniophyllum**
Bulbophyllum rupincola.. **Bulbophyllum saltatorium** var.
                                                                 calamarium
Bulbophyllum rutenbergianum
Bulbophyllum saccatum.. **Bulbophyllum apodum**
Bulbophyllum saccolabioides
Bulbophyllum salaccense
Bulbophyllum salebrosum
Bulbophyllum saltatorium
Bulbophyllum saltatorium var. **albociliatum**
Bulbophyllum saltatorium var. **calamarium**
Bulbophyllum sambiranense
Bulbophyllum sambiranense var. **ankiranense**
Bulbophyllum sambiranense var. **latibracteatum**
Bulbophyllum samoanum
Bulbophyllum sanderianum
Bulbophyllum sandersonii
Bulbophyllum sandersonii subsp. **stenopetalum**
Bulbophyllum sandrangatense
Bulbophyllum sangae
Bulbophyllum sanguineomaculatum .... **Bulbophyllum membranifolium**
Bulbophyllum sanguineopunctatum
Bulbophyllum sanguineum
Bulbophyllum sanitii
Bulbophyllum santoense
Bulbophyllum santosii
Bulbophyllum sapphirinum
Bulbophyllum sarasinorum
Bulbophyllum sarcanthiforme
Bulbophyllum sarcanthoides.. **Biermannia sarcanthoides**
Bulbophyllum sarcodanthum
Bulbophyllum sarcophylloides
Bulbophyllum sarcophyllum
Bulbophyllum sarcorhachis
Bulbophyllum sarcorhachis var. **beforonense**
Bulbophyllum sarcorhachis var. **flavemarginatum**
Bulbophyllum sarcoscapum
Bulbophyllum saronae
Bulbophyllum saruwatarii... **Bulbophyllum umbellatum**
Bulbophyllum sauguetiense
Bulbophyllum saurocephalum
Bulbophyllum savaiense
Bulbophyllum savaiense subsp. **gorumense**
Bulbophyllum savaiense subsp. **subcubicum**
Bulbophyllum sawiense
Bulbophyllum scaberulum

* † ‡ For explanation see page 2 and 3, point 3
* † ‡ Voir les explications page 10 et 11, point 3
* † ‡ Para mayor explicación, véase la página 18 y 19, point 3   **67**

ALL NAMES	ACCEPTED NAMES
Bulbophyllum scaberulum var. **album**	
Bulbophyllum scaberulum var. **crotalicaudatum**	
Bulbophyllum scaberulum var. **fuerstenbergianum**	
Bulbophyllum scabratum	
Bulbophyllum scabrum	
Bulbophyllum scandens Kraenzl.	**Bulbophyllum membranifolium**
Bulbophyllum scandens Rolfe..	**Hederorkis seychellensis**
Bulbophyllum scaphiforme	
Bulbophyllum scaphosepalum	
Bulbophyllum scariosum	
Bulbophyllum sceliphron	
Bulbophyllum sceptrum..	**Bulbophyllum elongatum**
Bulbophyllum schefferi	
Bulbophyllum schillerianum	
Bulbophyllum schimperianum	
Bulbophyllum schinzianum	
Bulbophyllum schinzianum var. **irigaleae**	
Bulbophyllum schinzianum var. **phaeopogon**	
Bulbophyllum schistopetalum	
Bulbophyllum schizopetalum	
Bulbophyllum schlechteri De Wild.	**Bulbophyllum josephi**
Bulbophyllum schlechteri H. Perrier	**Bulbophyllum metonymon**
Bulbophyllum schlechteri Kraenzl.	**Bulbophyllum metonymon**
Bulbophyllum schmidii	
Bulbophyllum schmidtianum	
Bulbophyllum schwarzii	
Bulbophyllum sciadanthum...	**Epiblastus sciadanthus**
Bulbophyllum sciaphile	
Bulbophyllum scintilla	
Bulbophyllum scitulum	
Bulbophyllum scopa	
Bulbophyllum scopula	
Bulbophyllum scotifolium	
Bulbophyllum scotinochiton	
Bulbophyllum scrobiculilabre	
Bulbophyllum sculptum..	**Bulbophyllum cylindrobulbum**
Bulbophyllum scutiferum	
Bulbophyllum scyphochilus	
Bulbophyllum scyphochilus var. **phaeanthum**	
Bulbophyllum secundum	
Bulbophyllum seidenfadenii	
Bulbophyllum selangorense	**Bulbophyllum gibbosum**
Bulbophyllum semiasperum	
Bulbophyllum semibifidum...	**Bulbophyllum acuminatum**
Bulbophyllum semipellucidum	**Bulbophyllum mutabile**
Bulbophyllum semiteres	
Bulbophyllum semiteretifolium	
Bulbophyllum semperflorens..	**Bulbophyllum flavescens**
Bulbophyllum sempiternum	
Bulbophyllum sennii..	**Bulbophyllum josephi**
Bulbophyllum sensile	
Bulbophyllum sepikense	
Bulbophyllum septatum	
Bulbophyllum septemtrionale	
Bulbophyllum sereti	**Bulbophyllum falcatum** var. **bufo**
Bulbophyllum serpens Lindl..	**Bulbophyllum nutans**
Bulbophyllum serpens Schltr...	**Bulbophyllum minutibulbum**
Bulbophyllum serra	

ALL NAMES	ACCEPTED NAMES
Bulbophyllum serratotruncatum	
Bulbophyllum serratum..	**Bulbophyllum septatum**
Bulbophyllum serripetalum	
Bulbophyllum serrulatifolium	
Bulbophyllum serrulatum	
Bulbophyllum sessile (J. Koenig) J. J. Sm......................	**Bulbophyllum clandestinum**
Bulbophyllum sessile Hochr..	**Bulbophyllum clandestinum**
Bulbophyllum sessiliflorum..	**Bulbophyllum molossus**
Bulbophyllum setaceum	
Bulbophyllum setiferum	**Bulbophyllum reichenbachianum**
Bulbophyllum setigerum	
Bulbophyllum setipes..	**Bulbophyllum graciliscapum**
Bulbophyllum setuliferum	
Bulbophyllum seychellarum..	**Bulbophyllum intertextum**
Bulbophyllum shanicum	
Bulbophyllum shephardi var. *intermedium*.....................	**Bulbophyllum intermedium**
Bulbophyllum shepherdi	
Bulbophyllum shweliense	
Bulbophyllum siamense	
Bulbophyllum sibuyanense	
Bulbophyllum sicyobulbon	
Bulbophyllum siederi	
Bulbophyllum sigaldiae	
Bulbophyllum sigilliforme...	**Bulbophyllum complanatum**
Bulbophyllum sigmoideum	
Bulbophyllum signatum	
Bulbophyllum sikapingense	
Bulbophyllum sikkimense	
Bulbophyllum silentvalliensis	
Bulbophyllum sillemianum	
Bulbophyllum similare	
Bulbophyllum simile	
Bulbophyllum similissimum	
Bulbophyllum simillinum	**Bulbophyllum taeniophyllum**
Bulbophyllum simmondsii	
Bulbophyllum simondii	
Bulbophyllum simoni	**Bulbophyllum falcatum** var. **velutinum**
Bulbophyllum simplex	
Bulbophyllum simplicilabellum	
Bulbophyllum simulacrum Schltr.	
Bulbophyllum simulacrum Ames	**Bulbophyllum flavescens**
Bulbophyllum sinapis	
Bulbophyllum singaporeanum	
Bulbophyllum singulare	
Bulbophyllum singuliflorum	
Bulbophyllum skeatianum	
Bulbophyllum sladeanum...	**Bulbophyllum macphersoni** var. **spathulatum**
Bulbophyllum smithianum	
Bulbophyllum smitinandii	
Bulbophyllum sociale	
Bulbophyllum solheidi	**Bulbophyllum falcatum** var. **velutinum**
Bulbophyllum solteroi	
Bulbophyllum solutisepalum..	**Bulbophyllum chloranthum**
Bulbophyllum somai..	**Bulbophyllum drymoglossum**
Bulbophyllum sopoetanense	

* † ‡ **For explanation see page 2 and 3, point 3**
* † ‡ **Voir les explications page 10 et 11, point 3**
* † ‡ **Para mayor explicación, véase la página 18 y 19, point 3**   **69**

ALL NAMES	ACCEPTED NAMES

Bulbophyllum sordidum
Bulbophyllum sororculum
Bulbophyllum spadiciflorum
Bulbophyllum spaerobulbum
Bulbophyllum sparsifolium .... **Bulbophyllum clandestinum**
Bulbophyllum spathaceum
Bulbophyllum spathilingue
Bulbophyllum spathipetalum
Bulbophyllum spathulatum
Bulbophyllum spathulifolium .... **Bulbophyllum rutenbergianum**
Bulbophyllum speciosum
Bulbophyllum spectabile .... **Bulbophyllum pectinatum**
Bulbophyllum sphaeracron
Bulbophyllum sphaericum
Bulbophyllum sphaerobulbum
Bulbophyllum spiesii
Bulbophyllum spinulipes .... **Bulbophyllum obtusipetalum**
Bulbophyllum spissum
Bulbophyllum spongiola
Bulbophyllum squamiferum .... **Bulbophyllum chloranthum**
Bulbophyllum squamipetalum .... **Bulbophyllum minutipetalum**
Bulbophyllum stabile
Bulbophyllum steffensii
Bulbophyllum stelis
Bulbophyllum stelis var. **humile**
Bulbophyllum stella .... **Bulbophyllum macrochilum**
Bulbophyllum stellatum
Bulbophyllum stellula
Bulbophyllum stenobulbon
Bulbophyllum stenochilum
Bulbophyllum stenopetalum .... **Bulbophyllum sandersonii**
subsp. **stenopetalum**
Bulbophyllum stenophyllum Schltr.
Bulbophyllum stenophyllum Ridl. .... **Bulbophyllum acuminatum**
Bulbophyllum stenorhachis .... **Bulbophyllum imbricatum**
Bulbophyllum stenorhopalos
Bulbophyllum stenurum
Bulbophyllum sterile
Bulbophyllum steyermarkii
Bulbophyllum stictanthum
Bulbophyllum stictosepalum
Bulbophyllum stipitatibulbum
Bulbophyllum stipulaceum
Bulbophyllum stolleanum
Bulbophyllum stolzii
Bulbophyllum stormii
Bulbophyllum stormii var. *pengadangaense* .... **Bulbophyllum stormii**
Bulbophyllum stramineum .... **Bulbophyllum cumingii**
Bulbophyllum stramineum var. *purpuratum* .... **Bulbophyllum flabellum-veneris**
Bulbophyllum strangularium .... **Bulbophyllum pulchrum**
Bulbophyllum streptosepalum .... **Bulbophyllum falcibracteum**
Bulbophyllum streptotriche
Bulbophyllum striatellum
Bulbophyllum striatitepalum .... **Bulbophyllum striatum**
Bulbophyllum striatum (Griff.) Rchb. f.
Bulbophyllum striatum (Liu & Su) Nakajima .... **Bulbophyllum melanoglossum**
Bulbophyllum strigosum †
Bulbophyllum strobiliferum .... **Bulbophyllum imbricatum**
Bulbophyllum sturmhoefelii

ALL NAMES	ACCEPTED NAMES
Bulbophyllum suave	**Bulbophyllum hirtum**
Bulbophyllum suavissimum	
Bulbophyllum subaequale	
Bulbophyllum subalpinum	**Bulbophyllum cylindrobulbum**
Bulbophyllum subapetalum	
Bulbophyllum subapproximatum	
Bulbophyllum subbullatum	
Bulbophyllum subclausum	
Bulbophyllum subclavatum	
Bulbophyllum subcoriaceum	**Bulbophyllum maximum**
Bulbophyllum subcrenulatum	
Bulbophyllum subcubicum	**Bulbophyllum savaiense** subsp. subcubicum
Bulbophyllum subcubicum var. *coccineum*	**Bulbophyllum savaiense** subsp. subcubicum
Bulbophyllum subebulbum	
Bulbophyllum subligaculiferum	
Bulbophyllum submarmoratum	
Bulbophyllum subparviflorum	**Bulbophyllum secundum**
Bulbophyllum subpatulum	
Bulbophyllum subsecundum	
Bulbophyllum subsessile	
Bulbophyllum subtenellum	
Bulbophyllum subtrilobatum	
Bulbophyllum subuliferum	
Bulbophyllum subulifolium	
Bulbophyllum subumbellatum	
Bulbophyllum subverticillatum	
Bulbophyllum succedaneum	
Bulbophyllum sukhakulii	
Bulbophyllum sulawesii	
Bulbophyllum sulcatum	
Bulbophyllum sulfureum	
Bulbophyllum sumatranum	
Bulbophyllum summerhayesii	**Bulbophyllum scaberulum**
Bulbophyllum superfluum	
Bulbophyllum superpositum	
Bulbophyllum supervacaenum	
Bulbophyllum surigaense	
Bulbophyllum sutepense	
Bulbophyllum syllectum	**Bulbophyllum laxiflorum**
Bulbophyllum systenochilum	
Bulbophyllum tacitum	**Bulbophyllum trifolium**
Bulbophyllum taeniophyllum	
Bulbophyllum taeter	
Bulbophyllum tahanense	
Bulbophyllum tahitense	
Bulbophyllum taichungianum	**Bulbophyllum albociliatum**
Bulbophyllum taiwanense (Fukuyama) Nakajima	
Bulbophyllum taiwanense (Fukuyama) S. S. Ying	**Bulbophyllum taiwanense**
Bulbophyllum taiwanense (Fukuyama) Seidenf.	**Bulbophyllum taiwanense**
Bulbophyllum talauense	
Bulbophyllum talbothi	**Bulbophyllum cochleatum**
Bulbophyllum tampoketsense	
Bulbophyllum tanegasimense	**Bulbophyllum macraei**
Bulbophyllum tanystiche	
Bulbophyllum tapirus	**Bulbophyllum stormii**
Bulbophyllum tarantula	
Bulbophyllum tardeflorens	

* † ‡ For explanation see page 2 and 3, point 3
* † ‡ Voir les explications page 10 et 11, point 3
* † ‡ Para mayor explicación, véase la página 18 y 19, point 3   **71**

ALL NAMES	ACCEPTED NAMES
Bulbophyllum taylori	**Cadetia taylori**
Bulbophyllum tectipes	
Bulbophyllum tectipetalum	
Bulbophyllum tectipetalum var. **longisepalum**	
Bulbophyllum tectipetalum var. **maximum**	
Bulbophyllum tekuense	
Bulbophyllum tenasserimense J. J. Sm...........................	**Bulbophyllum purpurascens**
Bulbophyllum tenegasimense Masam.............................	**Bulbophyllum macraei**
Bulbophyllum tenellum	
Bulbophyllum tenerum	**Bulbophyllum planibulbe**
Bulbophyllum tengchongense	
Bulbophyllum tenompokense	
Bulbophyllum tentaculatum Schltr. (1913)	
Bulbophyllum tentaculatum Schltr. (1913)....................	**Bulbophyllum tentaculiferum**
Bulbophyllum tentaculiferum	
Bulbophyllum tentaculigerum	**Bulbophyllum sandersonii**
Bulbophyllum tenue	
Bulbophyllum tenuicaule...	**Bulbophyllum cochleatum** var. **tenuicaule**
Bulbophyllum tenuifolium	
Bulbophyllum tenuipes	
Bulbophyllum teres	
Bulbophyllum teresense	
Bulbophyllum teretibulbum	
Bulbophyllum teretifolium	
Bulbophyllum teretilabre	
Bulbophyllum ternatense	
Bulbophyllum tetragonum	
Bulbophyllum teysmannii	
Bulbophyllum thailandicum	**Bulbophyllum thaiorum**
Bulbophyllum thaiorum	
Bulbophyllum theiochlamys	**Bulbophyllum bicoloratum**
Bulbophyllum theioglossum	
Bulbophyllum thelantyx	
Bulbophyllum therezienii	
Bulbophyllum thersites	
Bulbophyllum theunissenii	
Bulbophyllum thiurum	
Bulbophyllum thomense	**Bulbophyllum cochleatum** var. **tenuicaule**
Bulbophyllum thompsonii	
Bulbophyllum thomsoni...	**Bulbophyllum parviflorum**
Bulbophyllum thomsonii	**Bulbophyllum fischeri**
Bulbophyllum thouarsii...	**Bulbophyllum gracile**
Bulbophyllum thrixspermiflorum	
Bulbophyllum thrixspermoides	
Bulbophyllum thwaitesii	
Bulbophyllum thymophorum	
Bulbophyllum tiagii..	**Bulbophyllum monanthum**
Bulbophyllum tibeticum	**Bulbophyllum umbellatum**
Bulbophyllum tigridium	
Bulbophyllum tinea	**Bulbophyllum ovalifolium**
Bulbophyllum tinekeae	
Bulbophyllum tingabarinum Garay, Hamer & Siegerist...	**Bulbophyllum pectenveneris**
Bulbophyllum titanea	
Bulbophyllum tixieri	
Bulbophyllum tjadasmalangense	
Bulbophyllum toilliezae	
Bulbophyllum tokioi	

ALL NAMES   **ACCEPTED NAMES**

Bulbophyllum tollenoniferum.. **Bulbophyllum macranthoides**
Bulbophyllum toppingii
Bulbophyllum toranum
Bulbophyllum toressae.. **Dendrobium toressae**
Bulbophyllum torquatum
Bulbophyllum torricellense
Bulbophyllum tortisepalum.. **Bulbophyllum umbellatum**
Bulbophyllum tortum
Bulbophyllum tortuosum
Bulbophyllum tothastes.. **Trias tothastes**
Bulbophyllum touranense.. **Bulbophyllum retusiusculum**
Bulbophyllum trachyanthum
Bulbophyllum trachybracteum
Bulbophyllum trachyglossum
Bulbophyllum trachypus
Bulbophyllum transarisanense...................................... **Bulbophyllum pectinatum**
Bulbophyllum tremulum
Bulbophyllum triadenium
Bulbophyllum triandrum
Bulbophyllum triaristella
Bulbophyllum triaristellum.. **Bulbophyllum intertextum**
Bulbophyllum tricanaliferum
Bulbophyllum tricarinatum
Bulbophyllum trichaete
Bulbophyllum trichambon
Bulbophyllum trichocephalum
Bulbophyllum trichocephalum var. **wallongense**
Bulbophyllum trichochlamys
Bulbophyllum trichoglottis.. **Bulbophyllum hirtulum**
Bulbophyllum trichoglottis var. *sumatranum*.................. **Bulbophyllum hirtulum**
Bulbophyllum trichopus... **Bulbophyllum laxum**
Bulbophyllum trichorhachis
Bulbophyllum trichromum
Bulbophyllum triclavigerum
Bulbophyllum tricolor
Bulbophyllum tricorne
Bulbophyllum tricornoides
Bulbophyllum tridentatum Kraenzl.
Bulbophyllum tridentatum Rolfe.................................... **Bulbophyllum tristelidium**
Bulbophyllum trifarium
Bulbophyllum trifilum
Bulbophyllum trifilum subsp. **filisepalum**
Bulbophyllum triflorum
Bulbophyllum trifolium
Bulbophyllum trigonidioides
Bulbophyllum trigonobulbum
Bulbophyllum trigonocarpum
Bulbophyllum trigonopus.. **Bulbophyllum abbreviatum**
Bulbophyllum trigonosepalum
Bulbophyllum trilineatum
Bulbophyllum trilobum.. **Bulbophyllum lingulatum**
Bulbophyllum trimeni
Bulbophyllum trinervium
Bulbophyllum tripaleum
Bulbophyllum tripetaloides.. **Bulbophyllum auricomum**
Bulbophyllum tripetalum
Bulbophyllum tripudians
Bulbophyllum tripudians var. *pumilum*.......................... **Bulbophyllum khaoyaiense**
Bulbophyllum trirhopalon

* † ‡ **For explanation see page 2 and 3, point 3**
* † ‡ **Voir les explications page 10 et 11, point 3**
* † ‡ **Para mayor explicación, véase la página 18 y 19, point 3**   **73**

ALL NAMES   **ACCEPTED NAMES**

Bulbophyllum trisetosum.. **Bulbophyllum clandestinum**
Bulbophyllum trisetum.. **Bulbophyllum longiflorum**
*Bulbophyllum triste Rchb. f.
Bulbophyllum triste (Rolfe) Schltr. .... **Bulbophyllum imbricatum**
Bulbophyllum tristelidium
Bulbophyllum tristriatum... **Bulbophyllum stormii**
Bulbophyllum triurum
Bulbophyllum triviale
Bulbophyllum trulliferum
Bulbophyllum truncatisepalum .... **Bulbophyllum orbiculare** subsp.
                                                                                               cassideum
Bulbophyllum truncatum
Bulbophyllum truncicola
Bulbophyllum tryssum
Bulbophyllum tseanum
Bulbophyllum tsinjoarivense... **Bulbophyllum nutans**
Bulbophyllum tuberculatum
Bulbophyllum tubilabrum
Bulbophyllum tumidum
Bulbophyllum tumoriferum
Bulbophyllum turgidum
Bulbophyllum turkii
Bulbophyllum turpis
Bulbophyllum tylophorum
Bulbophyllum uduense .... **Bulbophyllum cylindrobulbum**
Bulbophyllum ugandae .... **Bulbophyllum falcatum**
Bulbophyllum ulcerosum
Bulbophyllum umbellatum Lindl.
Bulbophyllum umbellatum J. J. Sm. .... **Bulbophyllum longiflorum**
Bulbophyllum umbellatum var. *bergemanni*.................... **Bulbophyllum guttulatum**
Bulbophyllum umbellatum var. **fuscescens**
Bulbophyllum umbonatum
Bulbophyllum umbraticola
Bulbophyllum uncinatum
Bulbophyllum unciniferum
Bulbophyllum undatilabre
Bulbophyllum undecifilum
Bulbophyllum unguiculatum
Bulbophyllum unguilabium
Bulbophyllum unicaudatum
Bulbophyllum unicaudatum var. **xanthospaerum**
*Bulbophyllum uniflorum (Blume) Hassk.
Bulbophyllum uniflorum Griff.. **Bulbophyllum monanthum**
Bulbophyllum uniflorum var. *pluriflorum* .... **Bulbophyllum uniflorum**
Bulbophyllum uniflorum var. *rubrum*............................. **Bulbophyllum uniflorum**
Bulbophyllum uniflorum var. *variabile* .... **Bulbophyllum uniflorum**
Bulbophyllum unifoliatum
Bulbophyllum unifoliatum subsp. **flectens**
Bulbophyllum unifoliatum subsp. **infracarinatum**
Bulbophyllum unigibbum.. **Bulbophyllum xanthochlamys**
Bulbophyllum unitubum
Bulbophyllum univenum
Bulbophyllum uraiense .... **Bulbophyllum macraei**
Bulbophyllum urbanianum.. **Bulbophyllum lupulinum**
Bulbophyllum urceolatum † .... **Mediocalcar paradoxum** var.
                                                                                               robustum
Bulbophyllum uroglossum
Bulbophyllum urosepalum
Bulbophyllum usambarae.. **Bulbophyllum intertextum**

ALL NAMES	ACCEPTED NAMES
Bulbophyllum ustusfortiter	
Bulbophyllum uviflorum	
Bulbophyllum vaccinioides	
Bulbophyllum vagans	
Bulbophyllum vagans var. *angustum*	**Bulbophyllum vagans**
Bulbophyllum vagans var. *linearifolium*	**Bulbophyllum vagans**
Bulbophyllum vaginatum	
Bulbophyllum vaginulosum	**Bulbophyllum apodum**
‡*Bulbophyllum vakonae*†‡	
Bulbophyllum valeryi	
Bulbophyllum validum	
Bulbophyllum vanessa †	**Bulbophyllum macrochilum**
Bulbophyllum vanoverberghii	**Bulbophyllum dasypetalum**
Bulbophyllum vanum	
Bulbophyllum vanvuurenii	
Bulbophyllum vareschii	
Bulbophyllum variabile	
Bulbophyllum variabile var. *rubrum*	**Bulbophyllum uniflorum**
Bulbophyllum variegatum	
Bulbophyllum variifolium	**Bulbophyllum nutans** var. variifolium
Bulbophyllum vaughanii	
Bulbophyllum velutinum	**Bulbophyllum falcatum** var. velutinum
Bulbophyllum ventriosum	
Bulbophyllum venustum	**Bulbophyllum coriaceum**
Bulbophyllum verecundum	**Bulbophyllum pumilum**
Bulbophyllum vermiculare	
Bulbophyllum verrucibracteum	**Bulbophyllum bulliferum**
Bulbophyllum verruciferum	
Bulbophyllum verruciferum var. **carinatisepalum**	
Bulbophyllum verrucirhachis	**Bulbophyllum orbiculare**
Bulbophyllum verruculatum	
Bulbophyllum verruculiferum	
Bulbophyllum versteegii	
Bulbophyllum vesiculosum	
Bulbophyllum vestitum	
Bulbophyllum vestitum var. **meridionale**	
Bulbophyllum vexillarium	
Bulbophyllum victoris	**Bulbophyllum resupinatum** var. filiforme
Bulbophyllum vidalii	**Bulbophyllum apodum**
Bulbophyllum vietnamense	
Bulbophyllum viguieri	
Bulbophyllum vinaceum	
Bulbophyllum vinculibulbum	**Bulbophyllum montense**
Bulbophyllum vinosum	**Bulbophyllum pachyrachis**
Bulbophyllum violaceolabellum	
Bulbophyllum violaceum (Blume) Lindl.	
Bulbophyllum violaceum (Blume) Rchb. f.	**Bulbophyllum cochlia**
Bulbophyllum virens	**Sunipia virens**
Bulbophyllum virescens	
Bulbophyllum viride	**Bulbophyllum intertextum**
Bulbophyllum viridescens	
Bulbophyllum viridiflorum (Hook. f.) Schltr.	
Bulbophyllum viridiflorum Hayata	**Bulbophyllum pectinatum**
Bulbophyllum viridiflorum (Hook. f.) J. J. Sm.	**Bulbophyllum viridiflorum**
Bulbophyllum viscidum	**Bulbophyllum flabellum-veneris**
Bulbophyllum vitellinum	

* † ‡ **For explanation see page 2 and 3, point 3**
* † ‡ **Voir les explications page 10 et 11, point 3**
* † ‡ **Para mayor explicación, véase la página 18 y 19, point 3**　　　　　75

ALL NAMES	ACCEPTED NAMES
Bulbophyllum vitiense	**Bulbophyllum cocoinum**
Bulbophyllum vittatum Teijsm. & Binn	**Bulbophyllum cernuum** var. **vittata**
Bulbophyllum vittatum Rchb. f. & Warm	**Bulbophyllum warmingianum**
Bulbophyllum volkensii	
Bulbophyllum vulcanicum	
Bulbophyllum vulcanorum	
Bulbophyllum vutimenaense	
Bulbophyllum wadsworthii	
Bulbophyllum wagneri	
Bulbophyllum wakoi	
Bulbophyllum wallichi	
Bulbophyllum wangkaense	
Bulbophyllum wanjurum	**Bulbophyllum newportii**
Bulbophyllum warianum	
Bulbophyllum warmingianum	
Bulbophyllum watsonianum	**Bulbophyllum ambrosia**
Bulbophyllum waughense	**Bulbophyllum bowkettae**
Bulbophyllum weberbauerianum	
Bulbophyllum weberbauerianum var. **angustius**	
Bulbophyllum weberi	
Bulbophyllum wechsbergii †	
Bulbophyllum weddelii	
Bulbophyllum weinthalii	
Bulbophyllum weinthalii subsp. **striatum**	
Bulbophyllum wendlandianum (Kraenzl.) Dammer	
Bulbophyllum wendlandianum (Kraenzl.) J. J. Sm	**Bulbophyllum wendlandianum**
Bulbophyllum wenzelii	
Bulbophyllum werneri	
Bulbophyllum whiteanum	**Bulbophyllum vaginatum**
Bulbophyllum whitei	
Bulbophyllum whitfordii	
Bulbophyllum wightii	
Bulbophyllum wilkianum	
Bulbophyllum williamsii	
Bulbophyllum winckelii	**Bulbophyllum hamatipes**
Bulbophyllum windsorense	
Bulbophyllum winkleri Schltr. (1914)	**Bulbophyllum josephi**
Bulbophyllum winkleri Schltr. (1906)	**Bulbophyllum pumilum**
Bulbophyllum woelfliae	
Bulbophyllum wolfei	
Bulbophyllum wollastonii	
Bulbophyllum wrayi	
Bulbophyllum wrightii	**Bulbophyllum tetragonum**
Bulbophyllum wuzhishanensis	
Bulbophyllum x chikukwa	**Hybrid**
Bulbophyllum x cipoense	**Hybrid**
Bulbophyllum xantanthum	
Bulbophyllum xanthoacron	
Bulbophyllum xanthobulbum	
Bulbophyllum xanthochlamys	
Bulbophyllum xanthoglossum	**Bulbophyllum schimperianum**
Bulbophyllum xanthophaeum	
Bulbophyllum xanthornis	
Bulbophyllum xanthotes	
Bulbophyllum xanthum	
Bulbophyllum xenosum	
Bulbophyllum xiphion	
Bulbophyllum xylocarpi	

ALL NAMES

ACCEPTED NAMES

Bulbophyllum xylophyllum
Bulbophyllum yangambiense.. Bulbophyllum pumilum
Bulbophyllum yoksunense
Bulbophyllum yoksunense var. *geei*.............................. **Bulbophyllum emarginatum**
Bulbophyllum yoksunense var. *parviflorum* .... **Bulbophyllum yoksunense**
Bulbophyllum youngsayeanum...................................... **Bulbophyllum stenobulbon**
Bulbophyllum yuanyangense
Bulbophyllum yunnanense
Bulbophyllum zambalense
Bulbophyllum zamboangense
Bulbophyllum zaratananae Schltr. (1924)
Bulbophyllum zaratananae Schltr. (1925)...................... **Bulbophyllum metonymon**
Bulbophyllum zaratananae subsp. **disjunctum**
Bulbophyllum zebrinum
Bulbophyllum zenkerianum... **Bulbophyllum oreonastes**
Bulbophyllum zobiaense .... **Bulbophyllum scaberulum**
Callista brachypetala.. **Bulbophyllum mutabile**
Canacorchis lophoglottis... **Bulbophyllum lophoglottis**
Carparomorchis baileyi... **Bulbophyllum baileyi**
Carparomorchis macracantha..................................... **Bulbophyllum macranthum**
Chaseella pseudohydra
Cirrhopetalum abbreviatum.. **Bulbophyllum abbreviatum**
Cirrhopetalum acuminatum .... **Bulbophyllum acuminatum**
Cirrhopetalum acutiflorum .... **Bulbophyllum acutiflorum**
Cirrhopetalum adenophorum... **Bulbophyllum brienianum**
Cirrhopetalum aemulum .... **Bulbophyllum forresti**
Cirrhopetalum africanum.. **Bulbophyllum longiflorum**
Cirrhopetalum albidum... **Bulbophyllum albidum**
Cirrhopetalum albociliatum... **Bulbophyllum albociliatum**
Cirrhopetalum amesianum.. **Bulbophyllum cumingii**
Cirrhopetalum amplifolium.. **Bulbophyllum amplifolium**
Cirrhopetalum andersoni.. **Bulbophyllum andersonii**
Cirrhopetalum andersonii... **Bulbophyllum flabellum-veneris**
Cirrhopetalum annamense.. **Bulbophyllum annamense** †
Cirrhopetalum annamicum .... **Bulbophyllum umbellatum**
Cirrhopetalum antenniferum Lindl. .... **Bulbophyllum antenniferum**
Cirrhopetalum appendiculatum .... **Bulbophyllum appendiculatum**
Cirrhopetalum appendiculatum var. *fascinator*............. **Bulbophyllum putidum**
Cirrhopetalum asperulum .... **Bulbophyllum asperulum**
Cirrhopetalum aurantiacum.. **Bulbophyllum electrinum**
Cirrhopetalum auratum .... **Bulbophyllum auratum**
Cirrhopetalum aureum.. **Bulbophyllum aureum**
Cirrhopetalum autumnale .... **Bulbophyllum macraei**
Cirrhopetalum baucoense .... **Bulbophyllum weberi**
Cirrhopetalum bicolor .... **Bulbophyllum bicolor**
Cirrhopetalum biflorum .... **Bulbophyllum biflorum**
Cirrhopetalum blepharistes .... **Bulbophyllum blepharistes**
Cirrhopetalum blumii.. **Bulbophyllum blumei**
Cirrhopetalum boninense Makino................................. **Bulbophyllum macraei**
Cirrhopetalum boninense Schltr. .... **Bulbophyllum boninense**
Cirrhopetalum bootanensis... **Bulbophyllum umbellatum**
Cirrhopetalum bootanoides .... **Bulbophyllum frostii**
Cirrhopetalum borneense... **Bulbophyllum auratum**
Cirrhopetalum brevibrachiatum..................................... **Bulbophyllum brevibrachiatum**
Cirrhopetalum brevipes .... **Bulbophyllum yoksunense**
Cirrhopetalum breviscapum.. **Bulbophyllum lasiochilum**
Cirrhopetalum brienianum.. **Bulbophyllum brienianum**
Cirrhopetalum brunnescens.. **Bulbophyllum brienianum**
Cirrhopetalum caespitosum .... **Bulbophyllum scabratum**

* † ‡ **For explanation see page 2 and 3, point 3**
* † ‡ **Voir les explications page 10 et 11, point 3**
* † ‡ **Para mayor explicación, véase la página 18 y 19, point 3**   **77**

ALL NAMES

ACCEPTED NAMES

Cirrhopetalum campanulatum .... **Bulbophyllum auratum**
Cirrhopetalum capillipes .... **Bulbophyllum aphanopetalum**
Cirrhopetalum capitatum .... **Bulbophyllum pahudi**
Cirrhopetalum carinatum Teijsm. & Binn. .... **Bulbophyllum carinatum**
Cirrhopetalum caudatum (Lindl.) King & Pantl. .... **Bulbophyllum sterile**
Cirrhopetalum caudatum Wight .... **Bulbophyllum vaginatum**
Cirrhopetalum cercanthum .... **Bulbophyllum cercanthum** †
Cirrhopetalum chinense .... **Bulbophyllum chinense**
Cirrhopetalum chondriophorum .... **Bulbophyllum chondriophorum**
Cirrhopetalum chryseum .... **Bulbophyllum chryseum**
Cirrhopetalum ciliatum .... **Bulbophyllum flabellum-veneris**
Cirrhopetalum citrinum .... **Bulbophyllum purpurascens**
Cirrhopetalum clavigerum .... **Bulbophyllum longiflorum**
Cirrhopetalum cogniauxianum .... **Bulbophyllum cogniauxianum**
Cirrhopetalum collettianum .... **Bulbophyllum wendlandianum**
Cirrhopetalum collettii .... **Bulbophyllum wendlandianum**
Cirrhopetalum compactum .... **Bulbophyllum purpurascens**
Cirrhopetalum compressum .... **Bulbophyllum uniflorum**
Cirrhopetalum concinnum .... **Bulbophyllum pulchellum**
Cirrhopetalum concinnum var. *purpureum* .... **Bulbophyllum corolliferum**
Cirrhopetalum confusum .... **Bulbophyllum confusum** †
Cirrhopetalum cornutum .... **Bulbophyllum helenae**
Cirrhopetalum cumingii .... **Bulbophyllum cumingii**
Cirrhopetalum curtisii .... **Bulbophyllum corolliferum**
Cirrhopetalum curtisii var. *lutescens* .... **Bulbophyllum corolliferum**
Cirrhopetalum curtisii var. *purpureum* .... **Bulbophyllum corolliferum**
Cirrhopetalum cyclosepalon .... **Bulbophyllum cyclosepalon**
Cirrhopetalum delitescens .... **Bulbophyllum delitescens**
Cirrhopetalum dentiferum .... **Bulbophyllum dentiferum**
Cirrhopetalum distans .... **Bulbophyllum blepharistes**
Cirrhopetalum dolichoblepharon .... **Bulbophyllum dolichoblepharon**
Cirrhopetalum dyerianum .... **Bulbophyllum rolfei**
Cirrhopetalum eberhardtii .... **Bulbophyllum picturatum**
Cirrhopetalum elatum .... **Bulbophyllum elatum**
Cirrhopetalum elegans .... **Bulbophyllum pulchrum**
Cirrhopetalum elegantulum .... **Bulbophyllum elegantulum**
Cirrhopetalum elisae .... **Bulbophyllum elizae**
Cirrhopetalum elliae .... **Bulbophyllum elliae**
Cirrhopetalum elongatum .... **Bulbophyllum elongatum**
Cirrhopetalum emarginatum .... **Bulbophyllum emarginatum**
Cirrhopetalum farreri .... **Bulbophyllum farreri**
Cirrhopetalum fascinator .... **Bulbophyllum putidum**
Cirrhopetalum fastuosum .... **Bulbophyllum wendlandianum**
Cirrhopetalum fenestratum .... **Bulbophyllum taeniophyllum**
Cirrhopetalum fibratum .... **Bulbophyllum fibratum**
Cirrhopetalum fimbriatum .... **Bulbophyllum fimbriatum**
Cirrhopetalum flabellum-veneris .... **Bulbophyllum flabellum-veneris**
Cirrhopetalum flagelliforme .... **Bulbophyllum pahudi**
Cirrhopetalum flaviflorum .... **Bulbophyllum flaviflorum**
Cirrhopetalum flavisepalum .... **Bulbophyllum retusiusculum**
Cirrhopetalum fletcheranum .... **Bulbophyllum fletcherianum**
Cirrhopetalum fordii .... **Bulbophyllum fordii**
Cirrhopetalum formosanum .... **Bulbophyllum formosanum**
Cirrhopetalum frostii .... **Bulbophyllum frostii**
Cirrhopetalum gagnepainii .... **Bulbophyllum flabellum-veneris**
Cirrhopetalum gamblei .... **Bulbophyllum fischeri**
Cirrhopetalum gamosepalum .... **Bulbophyllum flabellum-veneris**
Cirrhopetalum gamosepalum var. *angustum* .... **Bulbophyllum flabellum-veneris**
Cirrhopetalum gonshanense .... **Bulbophyllum gongshanense**

ALL NAMES	ACCEPTED NAMES
Cirrhopetalum gracillimum	**Bulbophyllum gracillimum**
Cirrhopetalum grandiflorum	**Bulbophyllum wightii**
Cirrhopetalum graveolens	**Bulbophyllum graveolens**
Cirrhopetalum gusdorfii	**Bulbophyllum gusdorfii**
Cirrhopetalum gusdorfii var. *johorense*	**Bulbophyllum gusdorfii**
Cirrhopetalum guttulatum	**Bulbophyllum guttulatum**
Cirrhopetalum habrotinum	**Bulbophyllum habrotinum**
Cirrhopetalum henryi	**Bulbophyllum andersonii**
Cirrhopetalum hirundinis	**Bulbophyllum hirundinis**
Cirrhopetalum hookeri	**Bulbophyllum hookeri**
Cirrhopetalum inabai	**Bulbophyllum japonicum**
Cirrhopetalum insulsum	**Bulbophyllum insulsum**
Cirrhopetalum japonicum	**Bulbophyllum japonicum**
Cirrhopetalum kenejianum	**Bulbophyllum longiflorum**
Cirrhopetalum koordersii	**Bulbophyllum pulchrum**
Cirrhopetalum lasiochilum	**Bulbophyllum lasiochilum**
Cirrhopetalum layardi	**Bulbophyllum longiflorum**
Cirrhopetalum lendyanum	**Bulbophyllum purpurascens**
Cirrhopetalum leopardinum	**Bulbophyllum binnendijkii**
Cirrhopetalum lepidum var. *angustum*	**Bulbophyllum flabellum-veneris**
Cirrhopetalum lepidum	**Bulbophyllum flabellum-veneris**
Cirrhopetalum leratii Kraenzl.	**Bulbophyllum baladeanum**
Cirrhopetalum leratii Schltr.	**Bulbophyllum gracillimum**
Cirrhopetalum linchianum	**Bulbophyllum melanoglossum**
Cirrhopetalum linearifolium	**Bulbophyllum acuminatum**
Cirrhopetalum lineatum	**Bulbophyllum lineatum**
Cirrhopetalum loherianum	**Bulbophyllum loherianum**
Cirrhopetalum longescapum	**Bulbophyllum blepharistes**
Cirrhopetalum longibrachiatum	**Bulbophyllum longibrachiatum**
Cirrhopetalum longidens	**Bulbophyllum longidens**
Cirrhopetalum longiflorum	**Bulbophyllum longiflorum**
Cirrhopetalum longissimum	**Bulbophyllum longissimum**
Cirrhopetalum macgregorii	**Bulbophyllum chloranthum**
Cirrhopetalum macraei	**Bulbophyllum macraei**
Cirrhopetalum maculosum	**Bulbophyllum umbellatum**
Cirrhopetalum maculosum var. *annamicum*	**Bulbophyllum umbellatum**
Cirrhopetalum maculosum var. *fuscescens*	**Bulbophyllum umbellatum** var. **fuscescens**
Cirrhopetalum makinoanum	**Bulbophyllum macraei**
Cirrhopetalum makoyanum var. *brienianum*	**Bulbophyllum brienianum**
Cirrhopetalum makoyanum	**Bulbophyllum makoyanum**
Cirrhopetalum mannii	**Bulbophyllum reichenbachianum**
Cirrhopetalum masonii	**Bulbophyllum masonii**
Cirrhopetalum mastersianum	**Bulbophyllum mastersianum**
Cirrhopetalum maxillare	**Bulbophyllum maxillare**
Cirrhopetalum maximum	**Bulbophyllum virescens**
Cirrhopetalum medusae	**Bulbophyllum medusae**
Cirrhopetalum melanoglossum	**Bulbophyllum melanoglossum**
Cirrhopetalum melinanthum	**Bulbophyllum electrinum**
Cirrhopetalum merguense	**Bulbophyllum lineatum**
Cirrhopetalum micholitzii	**Bulbophyllum retusiusculum**
Cirrhopetalum microbulbon	**Bulbophyllum ruficaudatum**
Cirrhopetalum miniatum	**Bulbophyllum pectenveneris**
Cirrhopetalum minutiflorum	**Bulbophyllum puguahaanense**
Cirrhopetalum mirificum	**Bulbophyllum delitescens**
Cirrhopetalum mirum	**Bulbophyllum mirum**
Cirrhopetalum morotaiense	**Bulbophyllum morotaiense**
Cirrhopetalum mundulum	**Bulbophyllum taeniophyllum**

* † ‡ **For explanation see page 2 and 3, point 3**
* † ‡ **Voir les explications page 10 et 11, point 3**
* † ‡ **Para mayor explicación, véase la página 18 y 19, point 3**   **79**

ALL NAMES	ACCEPTED NAMES
Cirrhopetalum mysorense	**Bulbophyllum mysorense**
Cirrhopetalum neilgherrense	**Bulbophyllum kaitiense**
Cirrhopetalum nodosum	**Bulbophyllum nodosum**
Cirrhopetalum nutans	**Bulbophyllum othonis**
Cirrhopetalum ochraceum	**Bulbophyllum serratotruncatum**
Cirrhopetalum omerandrum	**Bulbophyllum omerandrum**
Cirrhopetalum oreogenes	**Bulbophyllum retusiusculum** var. **oreogenes**
Cirrhopetalum ornatissimum	**Bulbophyllum ornatissimum**
Cirrhopetalum ornithorhynchius	**Bulbophyllum ornithorhynchum**
Cirrhopetalum pachybulbum	**Bulbophyllum graveolens**
Cirrhopetalum pahudi	**Bulbophyllum pahudi**
Cirrhopetalum pallidum	**Bulbophyllum purpurascens**
Cirrhopetalum panigraphianum	**Bulbophyllum sarcophyllum**
Cirrhopetalum papillosum	**Bulbophyllum thaiorum**
Cirrhopetalum parvulum	**Bulbophyllum rolfei**
Cirrhopetalum pectenveneris	**Bulbophyllum pectenveneris**
Cirrhopetalum peyerianum	**Bulbophyllum peyerianum**
Cirrhopetalum picturatum	**Bulbophyllum picturatum**
Cirrhopetalum pileolatum	**Bulbophyllum elegantulum**
Cirrhopetalum pingtungense	**Bulbophyllum wightii**
Cirrhopetalum planibulbe	**Bulbophyllum planibulbe**
Cirrhopetalum polliculosum	**Bulbophyllum polliculosum**
Cirrhopetalum proboscideum	**Bulbophyllum proboscideum**
Cirrhopetalum proliferum	**Bulbophyllum wendlandianum**
Cirrhopetalum proudlockii	**Bulbophyllum proudlockii**
Cirrhopetalum pseudopicturatum	**Bulbophyllum pseudopicturatum** †
Cirrhopetalum psittacoides	**Bulbophyllum gracillimum**
Cirrhopetalum puguahaanense	**Bulbophyllum puguahaanense**
Cirrhopetalum pulchrum	**Bulbophyllum pulchrum**
Cirrhopetalum pumilio	**Bulbophyllum pumilio**
Cirrhopetalum punctatissimum	**Bulbophyllum taeniophyllum**
Cirrhopetalum putidum	**Bulbophyllum putidum**
Cirrhopetalum racemosum	**Bulbophyllum insulsum**
Cirrhopetalum refractum	**Bulbophyllum refractum**
Cirrhopetalum refractum var. *laciniatum*	**Bulbophyllum wallichi**
Cirrhopetalum remotifolium	**Bulbophyllum hirundinis**
Cirrhopetalum restrepia	**Bulbophyllum restrepia**
Cirrhopetalum retusiusculum	**Bulbophyllum retusiusculum**
Cirrhopetalum rhombifolium	**Bulbophyllum hymenochilum**
Cirrhopetalum rivesii	**Bulbophyllum andersonii**
Cirrhopetalum robustum	**Bulbophyllum graveolens**
Cirrhopetalum roseopunctatum	**Bulbophyllum schwarzii**
Cirrhopetalum roseum	**Bulbophyllum elliae**
Cirrhopetalum rothschildianum	**Bulbophyllum rothschildianum**
Cirrhopetalum roxburghii	**Bulbophyllum roxburghii**
Cirrhopetalum rubroguttatum	**Bulbophyllum rubroguttatum**
Cirrhopetalum sarcophyllum	**Bulbophyllum sarcophyllum**
Cirrhopetalum sarcophyllum var. *minor*	**Bulbophyllum sarcophylloides**
Cirrhopetalum saruwatarii	**Bulbophyllum umbellatum**
Cirrhopetalum semibifidum	**Bulbophyllum acuminatum**
Cirrhopetalum setiferum	**Bulbophyllum reichenbachianum**
Cirrhopetalum siamense	**Bulbophyllum flabellum-veneris**
Cirrhopetalum sibuyanense	**Bulbophyllum sibuyanense**
Cirrhopetalum sigaldii	**Bulbophyllum vietnamense**

ALL NAMES

ACCEPTED NAMES

Cirrhopetalum sikkimense...	**Bulbophyllum sikkimense**
Cirrhopetalum simillimum..	**Bulbophyllum taeniophyllum**
Cirrhopetalum simullinum..	**Bulbophyllum taeniophyllum**
Cirrhopetalum skeatianum...	**Bulbophyllum skeatianum**
Cirrhopetalum spathulatum..	**Bulbophyllum spathulatum**
Cirrhopetalum spicatum	**Bulbophyllum blepharistes**
Cirrhopetalum stragularium...	**Bulbophyllum taeniophyllum**
Cirrhopetalum stramineum...	**Bulbophyllum vaginatum**
Cirrhopetalum stramineum var. *purpureum*....................	**Bulbophyllum flabellum-veneris**
Cirrhopetalum strangularium..	**Bulbophyllum pulchrum**
Cirrhopetalum striatum..	**Bulbophyllum melanoglossum**
Cirrhopetalum sutepense	**Bulbophyllum sutepense**
Cirrhopetalum taeniophyllum.......................................	**Bulbophyllum taeniophyllum**
Cirrhopetalum taiwanense...	**Bulbophyllum taiwanense**
Cirrhopetalum tanegasimense.......................................	**Bulbophyllum macraei**
Cirrhopetalum tenuicaule..	**Bulbophyllum flabellum-veneris**
Cirrhopetalum thomsonii...	**Bulbophyllum fischeri**
Cirrhopetalum thouarsii	**Bulbophyllum longiflorum**
Cirrhopetalum thouarsii var. *concolor*..........................	**Bulbophyllum longiflorum**
Cirrhopetalum thwaitesii	**Bulbophyllum thwaitesii**
Cirrhopetalum tigridum...	**Bulbophyllum tigridium**
Cirrhopetalum touranense	**Bulbophyllum retusiusculum**
Cirrhopetalum trichocephalum......................................	**Bulbophyllum trichocephalum**
Cirrhopetalum trigonopus..	**Bulbophyllum abbreviatum**
Cirrhopetalum trimeni	**Bulbophyllum trimeni**
Cirrhopetalum tripudians...	**Bulbophyllum tripudians**
Cirrhopetalum trisetum..	**Bulbophyllum longiflorum**
Cirrhopetalum tseanum..	**Bulbophyllum tseanum**
*Cirrhopetalum umbellatum(G. Forst.) Frappier ex Cordem.	**Bulbophyllum longiflorum**
*Cirrhopetalum umbellatum(G. Forst.) Hook. & Arn.....	**Bulbophyllum longiflorum**
*Cirrhopetalum umbellatum (Lindl.) Linden..................	**Bulbophyllum umbellatum**
Cirrhopetalum umbellatum Schltr.................................	**Bulbophyllum longiflorum**
Cirrhopetalum unciniferum..	**Bulbophyllum unciniferum**
Cirrhopetalum uniflorum	**Bulbophyllum baladeanum**
Cirrhopetalum uraiense	**Bulbophyllum macraei**
Cirrhopetalum vaginatum...	**Bulbophyllum vaginatum**
Cirrhopetalum viridiflorum..	**Bulbophyllum viridiflorum**
Cirrhopetalum viscidum...	**Bulbophyllum flabellum-veneris**
Cirrhopetalum walkerianum...	**Bulbophyllum macraei**
Cirrhopetalum wallichii J. Graham...............................	**Bulbophyllum fibratum**
Cirrhopetalum wallichii Lindl.	**Bulbophyllum retusiusculum**
Cirrhopetalum warianum..	**Bulbophyllum gracillimum**
Cirrhopetalum weberi..	**Bulbophyllum weberi**
Cirrhopetalum wendlandianum......................................	**Bulbophyllum wendlandianum**
Cirrhopetalum whiteanum	**Bulbophyllum vaginatum**
Cirrhopetalum wightii..	**Bulbophyllum elliae**
Cirrhopetalum zamboangense.......................................	**Bulbophyllum zamboangense**
Cirrhopetalum zygoglossum..	**Bulbophyllum pulchrum**
Cochlia violacea	**Bulbophyllum cochlia**
Codonosiphon campanulatum	
Codonosiphon codonanthum	
Codonosiphon papuanum	
Cymbidium reptans..	**Bulbophyllum nutans**
Cymbidium umbellatum	**Bulbophyllum longiflorum**
Dactylorhynchus flavescens...	**Bulbophyllum latipes**
Dendrobium aurantiacum...	**Bulbophyllum schillerianum**
Dendrobium bolbophylli	**Bulbophyllum griffithii**
Dendrobium brachypetalum..	**Bulbophyllum mutabile**
Dendrobium caleyi..	**Bulbophyllum exiguum**

* † ‡ **For explanation see page 2 and 3, point 3**
* † ‡ **Voir les explications page 10 et 11, point 3**
* † ‡ **Para mayor explicación, véase la página 18 y 19, point 3**   **81**

ALL NAMES	ACCEPTED NAMES
Dendrobium exiguum	**Bulbophyllum exiguum**
Dendrobium grandiflorum	**Bulbophyllum uniflorum**
Dendrobium leopardinum	**Bulbophyllum leopardinum**
Dendrobium minutissimum	**Bulbophyllum minutissimum**
Dendrobium psychrophilum	**Pedilochilus psychrophilum**
Dendrobium pumilum	**Bulbophyllum pumilum**
Dendrobium pygmaeum	**Bulbophyllum pygmaeum**
Dendrobium reptans	**Bulbophyllum nutans**
Dendrobium shepherdii	**Bulbophyllum shepherdi**
Dendrobium shepherdii var. *platyphyllum*	**Bulbophyllum schillerianum**
Dendrobium striatum	**Bulbophyllum striatum**
Dendrobium tripetaloides	**Bulbophyllum auricomum**
Dendrochilum occultum	**Bulbophyllum occultum**
Didactyle antennifera	**Bulbophyllum geraense**
Didactyle atropurpurea	**Bulbophyllum atropurpureum**
Didactyle bidentata	**Bulbophyllum bidentatum**
Didactyle cantagellense	**Bulbophyllum cantagallense**
Didactyle clausseni	**Bulbophyllum clausseni**
Didactyle exaltata	**Bulbophyllum exaltatum**
Didactyle galeata	**Bulbophyllum chloroglossum**
Didactyle gladiata	**Bulbophyllum meridense**
Didactyle glutinosa	**Bulbophyllum glutinosum**
Didactyle granulosa	**Bulbophyllum granulosum**
Didactyle laciniata	**Bulbophyllum laciniatum**
Didactyle laciniata var. *janeirense*	**Bulbophyllum laciniatum**
Didactyle meridensis	**Bulbophyllum meridense**
Didactyle micropetala	**Bulbophyllum cribbianum**
Didactyle nemorosa	**Bulbophyllum nemorosum**
Didactyle ochracea	**Bulbophyllum ochraceum**
Didactyle plumosa	**Bulbophyllum plumosum**
Didactyle quadricolor	**Bulbophyllum quadricolor**
Didactyle regnellii	**Bulbophyllum regnelli**
Didactyle tripetala	**Bulbophyllum tripetalum**
Didactyle weddelii	**Bulbophyllum weddelii**
Diphyes angustifolia	**Bulbophyllum angustifolium**
Diphyes capitata	**Bulbophyllum capitatum**
Diphyes cernua	**Bulbophyllum cernuum**
Diphyes ciliata	**Bulbophyllum ciliatum**
Diphyes crassifolia	**Bulbophyllum coniferum**
Diphyes crocea	**Bulbophyllum croceum**
Diphyes flavescens	**Bulbophyllum flavescens**
Diphyes gibbosa	**Bulbophyllum gibbosum**
Diphyes gracilis	**Bulbophyllum schefferi**
Diphyes hirsuta	**Bulbophyllum hirsutum**
Diphyes inaequalis	**Bulbophyllum inaequale**
Diphyes laxiflora	**Bulbophyllum laxiflorum**
Diphyes mucronata	**Bulbophyllum mucronatum**
Diphyes mutabilis	**Bulbophyllum mutabile**
Diphyes obtusa	**Bulbophyllum obtusum**
Diphyes occulta	**Bulbophyllum occultum**
Diphyes odorata	**Bulbophyllum odoratum**
Diphyes ovalifolia	**Bulbophyllum ovalifolium**
Diphyes pusilla	**Bulbophyllum ovalifolium**
Diphyes sulcata	**Bulbophyllum sulcatum**
Diphyes tenella	**Bulbophyllum tenellum**
Diphyes tenuifolia	**Bulbophyllum tenuifolium**
Diphyes tortuosa	**Bulbophyllum tortuosum**
Diphyes violacea	**Bulbophyllum violaceum**
Dipodium khasyanum	**Sunipia bicolor**

Part I: All Names / Tous les Noms / Todos los Nombres

ALL NAMES	ACCEPTED NAMES

Drymoda digitata

Drymoda gymnopus .... **Bulbophyllum gymnopus**

Drymoda latisepala .... **Bulbophyllum capillipes**

Drymoda picta

Drymoda siamensis

Ephippium capitatum .... **Bulbophyllum pahudi**

Ephippium ciliatum .... **Bulbophyllum blumei**

Ephippium cornutum .... **Bulbophyllum cornutum**

‡ *Ephippium cornutum*   *Hamularia cornuta* ‡

Ephippium elongatum .... **Bulbophyllum elongatum**

Ephippium grandiflorum .... **Bulbophyllum grandiflorum**

Ephippium lepidum .... **Bulbophyllum flabellum-veneris**

Ephippium masdevalliaceum .... **Bulbophyllum blumei**

Ephippium uniflorum .... **Bulbophyllum uniflorum**

Epicranthes abbrevilabia .... **Bulbophyllum abbrevilabium**

Epicranthes adangensis .... **Bulbophyllum adangense**

Epicranthes annamensis .... **Bulbophyllum abbrevilabium**

Epicranthes barbata .... **Monomeria barbata**

Epicranthes cheiropetalum .... **Bulbophyllum cheiropetalum**

Epicranthes chlororhopalon .... **Bulbophyllum chlororhopalon**

Epicranthes cimicina .... **Bulbophyllum cimicinum**

Epicranthes conchophylla .... **Bulbophyllum conchophyllum**

Epicranthes corneri .... **Bulbophyllum vesiculosum**

Epicranthes decarhopalon .... **Bulbophyllum decarhopalon**

Epicranthes flavofimbriata .... **Bulbophyllum flavofimbriatum**

Epicranthes haniffii .... **Bulbophyllum haniffii**

Epicranthes heterorhopalon .... **Bulbophyllum heterorhopalon**

Epicranthes hexarhopalon .... **Bulbophyllum hexarhopalos**

Epicranthes hirudinifera .... **Bulbophyllum hirudiniferum**

Epicranthes javanica .... **Bulbophyllum epicrianthes**

Epicranthes johannulii .... **Bulbophyllum johannulii**

Epicranthes macrorhopalon .... **Bulbophyllum macrorhopalon**

Epicranthes mobilifilum .... **Bulbophyllum mobilifilum**

Epicranthes octorhopalon .... **Bulbophyllum octorhopalon**

Epicranthes papillosofilum .... **Bulbophyllum papillosefilum**

Epicranthes phymatum .... **Bulbophyllum phymatum**

Epicranthes psilorhopalon .... **Bulbophyllum psilorhopalon**

Epicranthes rigidifilum .... **Bulbophyllum rigidifilum**

Epicranthes stenorhopalon .... **Bulbophyllum stenorhopalos**

Epicranthes trirhopalon .... **Bulbophyllum trirhopalon**

Epicranthes undecifila .... **Bulbophyllum undecifilum**

Epicranthes vesiculosa .... **Bulbophyllum vesiculosum**

Epidendrum flabellum-veneris .... **Bulbophyllum flabellum-veneris**

Epidendrum sessile .... **Bulbophyllum clandestinum**

Epidendrum sterile .... **Bulbophyllum sterile**

Epidendrum sterile var. *gamma* .... **Bulbophyllum rheedei**

Epidendrum umbellatum .... **Bulbophyllum longiflorum**

Epigeneium triadenium .... **Bulbophyllum triadenium**

Eria ambrosia .... **Bulbophyllum ambrosia**

Ferruminaria brastagiense .... **Bulbophyllum brastagiense**

Ferruminaria lohokii .... **Bulbophyllum lohokii**

Ferruminaria melinantha .... **Bulbophyllum melinanthum**

Fruticicola albopunctata .... **Bulbophyllum fruticicola**

Fruticicola radicans .... **Bulbophyllum radicans**

Genyorchis pumila .... **Bulbophyllum pumilum**

‡ *Hamularia cornuta* ‡

‡ *Hamularia ecornuta* ‡

‡ *Hamularia gibbolabia* ‡

‡ *Hamularia puluongensis* ‡

ALL NAMES	ACCEPTED NAMES
Hapalochilus acanthoglossus	**Bulbophyllum acanthoglossum**
Hapalochilus algidus	**Bulbophyllum algidum**
Hapalochilus alkmaarensis	**Bulbophyllum alkmaarense**
Hapalochilus alticola	**Bulbophyllum alticola**
Hapalochilus arfakensis	**Bulbophyllum arfakense**
Hapalochilus aristilabris	**Bulbophyllum aristilabre**
Hapalochilus aureoapex	**Bulbophyllum aureoapex**
Hapalochilus bandischii	**Bulbophyllum arminii** †
Hapalochilus brevis	**Bulbophyllum breve**
Hapalochilus callipes	**Bulbophyllum callipes**
Hapalochilus caudatipetalum	**Bulbophyllum caudipetalum**
Hapalochilus chrysochilus	**Bulbophyllum chrysochilum**
Hapalochilus chrysoglossus	**Bulbophyllum chrysoglossum**
Hapalochilus collinus	**Bulbophyllum collinum**
Hapalochilus coloratus	**Bulbophyllum coloratum**
Hapalochilus concavibasalis	**Bulbophyllum concavibasalis**
Hapalochilus concolor	**Bulbophyllum concolor**
Hapalochilus cruciatus	**Bulbophyllum cruciatum**
Hapalochilus cucullatus	**Bulbophyllum cucullatum**
Hapalochilus cuniculiformis	**Bulbophyllum cuniculiforme**
Hapalochilus decurvulus	**Bulbophyllum decurvulum**
Hapalochilus dolichoglottis	**Bulbophyllum dolichoglottis**
Hapalochilus fasciatus	**Bulbophyllum fasciatum**
Hapalochilus fibrinus	**Bulbophyllum fibrinum**
Hapalochilus formosus	**Bulbophyllum formosum**
Hapalochilus frustrans	**Bulbophyllum frustrans**
Hapalochilus geniculifer	**Bulbophyllum geniculiferum**
Hapalochilus gobiensis	**Bulbophyllum gobiense**
Hapalochilus holochilus	**Bulbophyllum holochilum**
Hapalochilus humilis	**Bulbophyllum humile**
Hapalochilus immobilis	**Bulbophyllum cruciatum**
Hapalochilus inclinatus	**Bulbophyllum inclinatum**
Hapalochilus jadunae	**Bulbophyllum jadunae**
Hapalochilus jensenii	**Bulbophyllum jensenii**
Hapalochilus kelelensis	**Bulbophyllum kelelense**
Hapalochilus kermesinus	**Bulbophyllum kermesinum**
Hapalochilus kusaiense	**Bulbophyllum kusaiense**
Hapalochilus leontoglossus	**Bulbophyllum leontoglossum**
Hapalochilus leucorhodus	**Bulbophyllum leucorhodum**
Hapalochilus lohokii	**Bulbophyllum lohokii**
Hapalochilus longilabris	**Bulbophyllum longilabre**
Hapalochilus melinoglossus	**Bulbophyllum melinoglossum**
Hapalochilus microrhombos	**Bulbophyllum microrhombos**
Hapalochilus monosema	**Bulbophyllum monosema**
Hapalochilus mutatus	**Bulbophyllum mutatum**
Hapalochilus mystrochilus	**Bulbophyllum mystrochilum**
Hapalochilus neoebudicus	**Bulbophyllum neoebudicus** †
Hapalochilus nitidus	**Bulbophyllum nitidum**
Hapalochilus novae-hiberniae	**Bulbophyllum novae-hiberniae**
Hapalochilus olorinus	**Bulbophyllum olorinum**
Hapalochilus oxyanthus	**Bulbophyllum oxyanthum**
Hapalochilus pemae	**Bulbophyllum pemae**
Hapalochilus quadricaudatus	**Bulbophyllum quadricaudatum**
Hapalochilus rectilabris	**Bulbophyllum rectilabre**
Hapalochilus reflexus	**Bulbophyllum wechsbergii** †
Hapalochilus scaphosepalum	**Bulbophyllum scaphosepalum**
Hapalochilus schizopetalum	**Bulbophyllum schizopetalum**
Hapalochilus scitulus	**Bulbophyllum scitulum**
Hapalochilus scyphochilus	**Bulbophyllum scyphochilus**

ALL NAMES	ACCEPTED NAMES
Hapalochilus speciosus	**Bulbophyllum speciosum**
Hapalochilus stabilis	**Bulbophyllum stabile**
Hapalochilus stellula	**Bulbophyllum stellula**
Hapalochilus stenophyton	**Bulbophyllum stenophyllum**
Hapalochilus stictanthus	**Bulbophyllum stictanthum**
Hapalochilus striatus	**Bulbophyllum bulhartii**
Hapalochilus torricellensis	**Bulbophyllum torricellense**
Hapalochilus trachyglossus	**Bulbophyllum trachyglossum**
Hapalochilus trichromus	**Bulbophyllum trichromum**
Hapalochilus trigonocarpus	**Bulbophyllum trigonocarpum**
Hapalochilus warianus	**Bulbophyllum warianum**
Hapalochilus xanthoacron	**Bulbophyllum xanthoacron**
Hapalochilus xanthophaeus	**Bulbophyllum xanthophaeum**
Henosis longipes	**Monomeria longipes**
Hippoglossum umbellatum	**Bulbophyllum umbellatum**
Hyalosema antenniferum	**Bulbophyllum antenniferum**
Hyalosema arfakianum	**Bulbophyllum arfakianum**
Hyalosema bandischii	**Bulbophyllum bandischii**
Hyalosema biantennatum	**Bulbophyllum biantennatum**
Hyalosema burfordiense	**Bulbophyllum burfordiense**
Hyalosema cominsii	**Bulbophyllum cominsii**
Hyalosema dennisii	**Bulbophyllum dennisii**
Hyalosema elbertii	**Bulbophyllum elbertii**
Hyalosema elephantinum	**Bulbophyllum elephantinum**
Hyalosema fraudulentum	**Bulbophyllum fraudulentum**
Hyalosema fritillariflorum	**Bulbophyllum fritillariiflorum**
Hyalosema grandiflorum	**Bulbophyllum grandiflorum**
Hyalosema klossii	**Bulbophyllum klossii**
Hyalosema leysenianum	**Bulbophyllum leysenianum**
Hyalosema longisepalum	**Bulbophyllum longisepalum**
Hyalosema micholitzii	**Bulbophyllum micholitzii**
Hyalosema minahassae	**Bulbophyllum minahassae**
Hyalosema ornithorhynchum	**Bulbophyllum ornithorhynchum**
Hyalosema saronae	**Bulbophyllum saronae**
Hyalosema schmidii	**Bulbophyllum schmidii**
Hyalosema singulare	**Bulbophyllum singulare**
Hyalosema trachyanthum	**Bulbophyllum trachyanthum**
Hyalosema tricanaliferum	**Bulbophyllum tricanaliferum**
Ichthyostomum pygmaeum	**Bulbophyllum pygmaeum**
Ione andersoni	**Sunipia andersonii**
Ione andersoni var. *flavescens*	**Sunipia andersonii**
Ione angustipetala	**Sunipia angustipetala**
Ione annamensis	**Sunipia annamensis**
Ione australis	**Sunipia australis**
Ione bicolor	**Sunipia bicolor**
Ione bifurcatoflorens	**Sunipia andersonii**
Ione candida	**Sunipia candida**
Ione cirrhata	**Sunipia cirrhata**
Ione cumberlegei	**Sunipia cumberlegei**
Ione dichroma	**Sunipia dichroma**
Ione flavescens	**Sunipia andersonii**
Ione fusco-purpurea	**Sunipia cirrhata**
Ione grandiflora	**Sunipia grandiflora**
Ione intermedia	**Sunipia intermedia**
Ione jainii	**Sunipia jainii**
Ione kachinensis	**Sunipia kachinensis**
Ione khasiana	**Sunipia bicolor**
Ione minor	**Sunipia minor**

* † ‡ **For explanation see page 2 and 3, point 3**
* † ‡ **Voir les explications page 10 et 11, point 3**
* † ‡ **Para mayor explicación, véase la página 18 y 19, point 3**   **85**

ALL NAMES	ACCEPTED NAMES
Ione paleacea	**Sunipia cirrhata**
Ione pallida	**Sunipia pallida**
Ione purpurata	**Sunipia andersonii**
Ione racemosa	**Bulbophyllum reptans**
Ione rimannii	**Sunipia rimannii**
Ione salweenensis	**Sunipia rimannii**
Ione sasakii	**Sunipia andersonii**
Ione scariosa	**Sunipia scariosa**
Ione scariosa var. *magnibracteatum*	**Sunipia racemosa**
Ione siamensis	**Sunipia scariosa**
Ione soidaoensis	**Sunipia soidaoensis**
Ione thailandica	**Sunipia thailandica**
Ione virens	**Sunipia virens**
Ione viridis	**Sunipia viridis**
Jejosephia pusilla	**Trias pusilla**
Karorchis evasa	**Bulbophyllum evasum**
Kaurorchis evasa	**Bulbophyllum evasum**
Lyraea prismatica	**Bulbophyllum prismaticum**
Macrolepis longiscapa	**Bulbophyllum longiscapum**
Malachadenia clavata	**Bulbophyllum malachadenia**
Mastigion appendiculatum	**Bulbophyllum putidum**
Mastigion fascinator	**Bulbophyllum putidum**
Mastigion ornatissimum	**Bulbophyllum ornatissimum**
Mastigion proboscideum	**Bulbophyllum proboscideum**
Mastigion putidum	**Bulbophyllum putidum**
Megaclinium angustum	**Bulbophyllum falcatum** var. **velutinum**
Megaclinium arnoldianum	**Bulbophyllum falcatum** var. **velutinum**
Megaclinium bambiliense	**Bulbophyllum scaberulum**
Megaclinium bequaerti	**Bulbophyllum cochleatum** var. **bequaertii**
Megaclinium brixhei	**Bulbophyllum falcatum** var. **velutinum**
Megaclinium buchenavianum	**Bulbophyllum calyptratum**
Megaclinium bufo	**Bulbophyllum falcatum** var. **bufo**
Megaclinium chevalieri	**Bulbophyllum scaberulum**
Megaclinium clarkei	**Bulbophyllum scaberulum**
Megaclinium colubrinum	**Bulbophyllum colubrinum**
Megaclinium congolense	**Bulbophyllum imbricatum**
Megaclinium deistelianum	**Bulbophyllum falcatum** var. **bufo**
Megaclinium djumaense var. *grandifolium*	**Bulbophyllum maximum**
Megaclinium djumaensis	**Bulbophyllum maximum**
Megaclinium ealaense	**Bulbophyllum scaberulum**
Megaclinium eburneum	**Bulbophyllum scaberulum**
Megaclinium endotrachys	**Bulbophyllum falcatum**
Megaclinium falcatum	**Bulbophyllum falcatum**
Megaclinium flaccidum	**Bulbophyllum calyptratum**
Megaclinium fuerstenbergianum	**Bulbophyllum scaberulum** var. **fuerstenbergianum**
Megaclinium gentilii	**Bulbophyllum falcatum** var. **bufo**
Megaclinium gilletii	**Bulbophyllum imbricatum**
Megaclinium hebetatum	**Bulbophyllum imbricatum**
Megaclinium hemirhachis	**Bulbophyllum falcatum**
Megaclinium imbricatum	**Bulbophyllum imbricatum**
Megaclinium imschootianum	**Bulbophyllum colubrinum**

ALL NAMES

ACCEPTED NAMES

Megaclinium inaequale	**Bulbophyllum colubrinum**
Megaclinium injoloense	**Bulbophyllum injoloense**
Megaclinium intermedium	**Bulbophyllum calyptratum** var. **graminifolium**
Megaclinium jesperseni	**Bulbophyllum scaberulum**
Megaclinium lanuriense	**Bulbophyllum falcatum** var. **velutinum**
Megaclinium lasianthum	**Bulbophyllum falcatum** var. **velutinum**
Megaclinium laurentianum	**Bulbophyllum imbricatum**
Megaclinium ledermannii	**Bulbophyllum imbricatum**
Megaclinium lepturum	**Bulbophyllum calyptratum**
Megaclinium leucorhachis	**Bulbophyllum imbricatum**
Megaclinium lindleyi	**Bulbophyllum calyptratum**
Megaclinium lutescens	**Bulbophyllum falcipetalum**
Megaclinium maximum	**Bulbophyllum maximum**
Megaclinium melanorrhachis	**Bulbophyllum falcatum** var. **velutinum**
Megaclinium melleri	**Bulbophyllum sandersonii**
Megaclinium millenii	**Bulbophyllum falcatum** var. **velutinum**
Megaclinium minor	**Bulbophyllum sandersonii** subsp. **stenopetalum**
Megaclinium minus	**Bulbophyllum sandersonii** subsp. **stenopetalum**
Megaclinium minutum	**Bulbophyllum falcatum** var. **velutinum**
Megaclinium minutum var. *purpureum*	**Bulbophyllum falcatum** var. **velutinum**
Megaclinium nummularia	**Bulbophyllum nummularia**
Megaclinium oxyodon	**Bulbophyllum falcatum**
Megaclinium oxypterum	**Bulbophyllum maximum** var. **oxypterum**
Megaclinium oxypterum var. *mozambicense*	**Bulbophyllum maximum**
Megaclinium platyrhachis	**Bulbophyllum maximum**
Megaclinium pobeguinii	**Bulbophyllum scaberulum**
Megaclinium purpuratum	**Bulbophyllum maximum**
Megaclinium purpureorachis	**Bulbophyllum purpureorhachis**
Megaclinium pusillum	**Bulbophyllum sandersonii**
Megaclinium renkinianum	**Bulbophyllum renkinianum**
Megaclinium sandersoni	**Bulbophyllum sandersonii**
Megaclinium scaberulum	**Bulbophyllum scaberulum**
Megaclinium sereti	**Bulbophyllum falcatum** var. **bufo**
Megaclinium solheidi	**Bulbophyllum falcatum** var. **velutinum**
Megaclinium strobiliferum	**Bulbophyllum imbricatum**
Megaclinium subcoriaceum	**Bulbophyllum maximum**
Megaclinium tentaculigerum	**Bulbophyllum sandersonii**
Megaclinium triste	**Bulbophyllum imbricatum**
Megaclinium ugandae	**Bulbophyllum falcatum**
Megaclinium velutinum	**Bulbophyllum falcatum** var. **velutinum**
Megaclinium zobiaense	**Bulbophyllum scaberulum**
Monomeria barbata	
Monomeria crabro	**Monomeria barbata**
Monomeria dichroma	**Sunipia dichroma**
Monomeria digitata	**Drymoda digitata**
Monomeria gymnopus	**Bulbophyllum gymnopus**

* † ‡ **For explanation see page 2 and 3, point 3**
* † ‡ **Voir les explications page 10 et 11, point 3**
* † ‡ **Para mayor explicación, véase la página 18 y 19, point 3**

87

ALL NAMES   **ACCEPTED NAMES**

Monomeria longipes (Rchb. f.) Aver.
Monomeria longipes (Rchb. f.) Garay, Hamer & Siegerist..... **Monomeria longipes**
Monomeria punctata .... **Acrochaene punctata**
Monomeria rimannii .... **Sunipia rimannii**
Monosepalum dischorense
Monosepalum muricatum
Monosepalum torricellense
Odontostyles minor .... **Bulbophyllum bakhuizenii**
Odontostyles multiflora .... **Bulbophyllum bakhuizenii**
Odontostyles triflora .... **Bulbophyllum triflorum**
Oncophyllum globuliforme .... **Bulbophyllum globuliforme**
Oncophyllum minutissimum .... **Bulbophyllum minutissimum**
Osyricera crassifolia .... **Bulbophyllum osyricera**
Osyricera erosipetala .... **Bulbophyllum erosipetalum**
Osyricera osyriceroides .... **Bulbophyllum osyriceroides**
Osyricera ovata .... **Bulbophyllum obyrnei**
Osyricera purpurascens .... **Bulbophyllum macphersoni**
Osyricera spadiciflora .... **Bulbophyllum spadiciflorum**
Oxysepala gadgarrensis .... **Bulbophyllum gadgarrense**
Oxysepala grandimesensis .... **Bulbophyllum grandimesense**
Oxysepala intermedia .... **Bulbophyllum intermedium**
Oxysepala lamingtonensis .... **Bulbophyllum lamingtonense**
Oxysepala lewisense .... **Bulbophyllum lewisense**
Oxysepala ovalifolia .... **Bulbophyllum clandestinum**
Oxysepala schilleriana .... **Bulbophyllum schillerianum**
Oxysepala shepherdii .... **Bulbophyllum shepherdi**
Oxysepala wadsworthii .... **Bulbophyllum wadsworthii**
Oxysepala windsorensis .... **Bulbophyllum windsorense**
Papulipetalum angustifolium .... **Bulbophyllum papulipetalum**
Papulipetalum nematopodum .... **Bulbophyllum nematopodum**
Pedilochilus alpinum
Pedilochilus alpinum var. **fasciculatum**
Pedilochilus angustifolius
Pedilochilus augustifolium
Pedilochilus bantaengensis
Pedilochilus brachiatus
Pedilochilus brachypus
Pedilochilus ciliolatum
Pedilochilus clemensiae
Pedilochilus coiloglossum
Pedilochilus cyatheicola
Pedilochilus dischorense
Pedilochilus flavum
Pedilochilus grandifolium
Pedilochilus guttulatum
Pedilochilus hermonii
Pedilochilus humile
Pedilochilus kermesinostriatum
Pedilochilus longipes
Pedilochilus macrorrhinum
Pedilochilus majus
Pedilochilus montana
Pedilochilus obovatum
Pedilochilus oreadum .... **Pedilochilus obovatum**
Pedilochilus papuanum
Pedilochilus parvulum
Pedilochilus perpusillum
Pedilochilus petiolatum
Pedilochilus petrophilum

ALL NAMES	ACCEPTED NAMES
Pedilochilus piundaundense	
Pedilochilus psychrophilum	
Pedilochilus pumilio	
Pedilochilus pusillum	
Pedilochilus sarawakatensis	
Pedilochilus stictanthum	
Pedilochilus subalpinum	
Pedilochilus sulphureum	
Pedilochilus terrestre	
Pelma absconditum	**Bulbophyllum absconditum**
Pelma neocaledonicum	**Bulbophyllum neo-caledonicum**
Peltopus aechmophorus	**Bulbophyllum aechmophorum**
Peltopus alveatus	**Bulbophyllum alveatum**
Peltopus ankylochele	**Bulbophyllum ankylochele**
Peltopus aphanopetalum	**Bulbophyllum aphanopetalum**
Peltopus artostigma	**Bulbophyllum artostigma**
Peltopus bliteus	**Bulbophyllum bliteum**
Peltopus brachypetalus	**Bulbophyllum brachypetalum**
Peltopus brassii	**Bulbophyllum brassii**
Peltopus calviventer	**Bulbophyllum calviventer**
Peltopus cubitalis	**Bulbophyllum discolor** var. **cubitale**
Peltopus cycloglossum	**Bulbophyllum cycloglossum**
Peltopus discolor	**Bulbophyllum discolor**
Peltopus greuterianus	**Bulbophyllum peltopus**
Peltopus hiljeae	**Bulbophyllum hiljeae**
Peltopus inciferus	**Bulbophyllum inciferum**
Peltopus intersitus	**Bulbophyllum intersitum**
Peltopus kenae	**Bulbophyllum kenae**
Peltopus lophotos	**Bulbophyllum lophoton**
Peltopus loroglossus	**Bulbophyllum loroglossum**
Peltopus minutipetalus	**Bulbophyllum minutipetalum**
Peltopus octarrhenipetalus	**Bulbophyllum octarrhenipetalum**
Peltopus origami	**Bulbophyllum origami**
Peltopus ortalis	**Bulbophyllum ortalis**
Peltopus patella	**Bulbophyllum patella**
Peltopus planilabris	**Bulbophyllum peltopus**
Peltopus plicatus	**Bulbophyllum plicatum**
Peltopus ptychantyx	**Bulbophyllum ptychantyx**
Peltopus reevei	**Bulbophyllum reevei**
Peltopus rhodoleucus	**Bulbophyllum rhodoleucum**
Peltopus santoensis	**Bulbophyllum santoense**
Peltopus scutifer	**Bulbophyllum scutiferum**
Peltopus subapetalus	**Bulbophyllum subapetalum**
Peltopus systenochilus	**Bulbophyllum systenochilum**
Peltopus thelantyx	**Bulbophyllum thelantyx**
Phreatia globulosa	**Bulbophyllum globulosum**
Phyllorkis acutiflora	**Bulbophyllum acutiflorum**
Phyllorkis adenopetala	**Bulbophyllum flavescens**
Phyllorkis affinis	**Bulbophyllum affine**
Phyllorkis albida	**Bulbophyllum albidum**
Phyllorkis alcicornis	**Bulbophyllum alcicorne**
Phyllorkis alopecurus	**Bulbophyllum triste**
Phyllorkis andersonii	**Bulbophyllum andersonii**
Phyllorkis angustifolia	**Bulbophyllum angustifolium**
Phyllorkis antennifera	**Bulbophyllum antenniferum**
Phyllorkis apoda	**Bulbophyllum apodum**
Phyllorkis argyropus	**Bulbophyllum argyropus**

* † ‡ **For explanation see page 2 and 3, point 3**
* † ‡ **Voir les explications page 10 et 11, point 3**
* † ‡ **Para mayor explicación, véase la página 18 y 19, point 3**   **89**

ALL NAMES

ACCEPTED NAMES

Phyllorkis aristata	**Bulbophyllum aristatum**
Phyllorkis aurantiaca	**Bulbophyllum schillerianum**
Phyllorkis aurata	**Bulbophyllum auratum**
Phyllorkis aurea	**Bulbophyllum aureum**
Phyllorkis auricoma	**Bulbophyllum auricomum**
Phyllorkis baileyi	**Bulbophyllum baileyi**
Phyllorkis barbigera	**Bulbophyllum barbigerum**
Phyllorkis baronii	**Bulbophyllum baronii**
Phyllorkis beccarii	**Bulbophyllum beccarii**
Phyllorkis bicolor	**Bulbophyllum bicolor**
Phyllorkis bifaria	**Bulbophyllum bifarium**
Phyllorkis biflora	**Bulbophyllum biflorum**
Phyllorkis biseta	**Bulbophyllum bisetum**
Phyllorkis blepharistes	**Bulbophyllum blepharistes**
Phyllorkis blumei	**Bulbophyllum blumei**
Phyllorkis bootanensis	**Bulbophyllum umbellatum**
Phyllorkis bowkettae	**Bulbophyllum bowkettae**
Phyllorkis bracteolata	**Bulbophyllum bracteolatum**
Phyllorkis brevipes	**Bulbophyllum brevipes**
Phyllorkis bufo	**Bulbophyllum falcatum** var. **bufo**
Phyllorkis bulbophylli	**Bulbophyllum griffithii**
Phyllorkis caespitosa	**Bulbophyllum caespitosum**
Phyllorkis calamaria	**Bulbophyllum saltatorium** var. **calamarium**
Phyllorkis candida	**Bulbophyllum candidum**
Phyllorkis capillipes	**Bulbophyllum capillipes**
Phyllorkis capitata	**Bulbophyllum capitatum**
Phyllorkis caudata	**Bulbophyllum sterile**
Phyllorkis cauliflora	**Bulbophyllum cauliflorum**
Phyllorkis cernua	**Bulbophyllum cernuum**
Phyllorkis cheiri	**Bulbophyllum cheiri**
Phyllorkis chinensis	**Bulbophyllum chinense**
Phyllorkis chloroptera	**Bulbophyllum chloropterum**
Phyllorkis ciliata	**Bulbophyllum ciliatum**
Phyllorkis cirrhata	**Bulbophyllum cirrhatum**
Phyllorkis clavata	**Bulbophyllum clavatum**
Phyllorkis clavigera	**Bulbophyllum longiflorum**
Phyllorkis cochleata	**Bulbophyllum cochleatum**
Phyllorkis collettii	**Bulbophyllum wendlandianum**
Phyllorkis colubrina	**Bulbophyllum colubrinum**
Phyllorkis comata	**Bulbophyllum comatum**
Phyllorkis commersonii	**Bulbophyllum commersonii**
Phyllorkis comosa	**Bulbophyllum comosum**
Phyllorkis conchifera	**Bulbophyllum khasyanum**
Phyllorkis concinna	**Bulbophyllum concinnum**
Phyllorkis conferta	**Bulbophyllum scabratum**
Phyllorkis coriscensis	**Bulbophyllum coriscense**
Phyllorkis cornuta	**Bulbophyllum helenae**
Phyllorkis crassifolia	**Bulbophyllum crassifolium**
Phyllorkis crassipes	**Bulbophyllum crassipes**
Phyllorkis crocea	**Bulbophyllum croceum**
Phyllorkis cumingii	**Bulbophyllum cumingii**
Phyllorkis cuprea	**Bulbophyllum cupreum**
Phyllorkis cylindracea	**Bulbophyllum cylindraceum**
Phyllorkis dayana	**Bulbophyllum dayanum**
Phyllorkis dearei	**Bulbophyllum dearei**
Phyllorkis densa	**Bulbophyllum densum**
Phyllorkis diphyes	**Bulbophyllum capitatum**

ALL NAMES	ACCEPTED NAMES
Phyllorkis distans	**Bulbophyllum saltatorium** var. **albociliatum**
Phyllorkis elata	**Bulbophyllum elatum**
Phyllorkis elegans	**Bulbophyllum elegans**
Phyllorkis elisae	**Bulbophyllum elizae**
Phyllorkis elliae	**Bulbophyllum elliae**
Phyllorkis elongata	**Bulbophyllum elongatum**
Phyllorkis erecta	**Bulbophyllum erectum**
Phyllorkis eublephara	**Bulbophyllum euplepharum**
Phyllorkis exaltata	**Bulbophyllum exaltatum**
Phyllorkis exigua	**Bulbophyllum exiguum**
Phyllorkis falcata	**Bulbophyllum falcatum**
Phyllorkis falcipetala	**Bulbophyllum falcipetalum**
Phyllorkis fimbriata	**Bulbophyllum fimbriatum**
Phyllorkis flavescens	**Bulbophyllum flavescens**
Phyllorkis flavida	**Bulbophyllum pumilum**
Phyllorkis fusca	**Bulbophyllum fuscum**
Phyllorkis fuscopurpurea	**Bulbophyllum fusco-purpureum**
Phyllorkis gamblei	**Bulbophyllum fischeri**
Phyllorkis gamosepala	**Bulbophyllum flabellum-veneris**
Phyllorkis geraensis	**Bulbophyllum geraense**
Phyllorkis gibbosa	**Bulbophyllum gibbosum**
Phyllorkis gladiata	**Bulbophyllum gladiatum**
Phyllorkis globulus	**Bulbophyllum globulus**
Phyllorkis gracilis	**Bulbophyllum gracile**
Phyllorkis grandiflora	**Bulbophyllum grandiflorum**
Phyllorkis gravida	**Bulbophyllum gravidum**
Phyllorkis guttulata	**Bulbophyllum guttulatum**
Phyllorkis gymnopus	**Bulbophyllum gymnopus**
Phyllorkis helenae	**Bulbophyllum helenae**
Phyllorkis herminostachys	**Bulbophyllum pumilum**
Phyllorkis hildebrandtii	**Bulbophyllum hildebrandtii**
Phyllorkis hirsuta	**Bulbophyllum hirsutum**
Phyllorkis hirta	**Bulbophyllum hirtum**
Phyllorkis hookeri	**Bulbophyllum fischeri**
Phyllorkis hymenantha	**Bulbophyllum hymenanthum**
Phyllorkis imbricata	**Bulbophyllum imbricatum**
Phyllorkis inaequalis	**Bulbophyllum inaequale**
Phyllorkis incurva	**Bulbophyllum incurvum**
Phyllorkis iners	**Bulbophyllum iners**
Phyllorkis intertexta	**Bulbophyllum intertextum**
Phyllorkis javanica	**Bulbophyllum epicrianthes**
Phyllorkis josephi	**Bulbophyllum josephi**
Phyllorkis kaitiensis	**Bulbophyllum kaitiense**
Phyllorkis kingii	**Acrochaene punctata**
Phyllorkis lasiantha	**Bulbophyllum lasianthum**
Phyllorkis lasiochila	**Bulbophyllum lasiochilum**
Phyllorkis laxiflora	**Bulbophyllum laxiflorum**
Phyllorkis lemniscata	**Bulbophyllum lemniscatum**
Phyllorkis leopardina	**Bulbophyllum leopardinum**
Phyllorkis leptantha	**Bulbophyllum leptanthum**
Phyllorkis leptosepala	**Bulbophyllum leptosepalum**
Phyllorkis limbata	**Bulbophyllum limbatum**
Phyllorkis lindleyana	**Bulbophyllum lindleyanum**
Phyllorkis lobbii	**Bulbophyllum lobbii**
Phyllorkis longiflora	**Bulbophyllum longiflorum**
Phyllorkis longiscapa	**Bulbophyllum blepharistes**
Phyllorkis lundiana	**Bulbophyllum lundianum**
Phyllorkis lupulina	**Bulbophyllum lupulinum**

* † ‡ For explanation see page 2 and 3, point 3
* † ‡ Voir les explications page 10 et 11, point 3
* † ‡ Para mayor explicación, véase la página 18 y 19, point 3   91

Part I: All Names / Tous les Noms / Todos los Nombres

ALL NAMES	ACCEPTED NAMES
Phyllorkis macraei	**Bulbophyllum macraei**
Phyllorkis macrantha	**Bulbophyllum macranthum**
Phyllorkis maculosa	**Bulbophyllum umbellatum**
Phyllorkis mannii	**Bulbophyllum cochleatum**
Phyllorkis maxillaris	**Bulbophyllum blumei**
Phyllorkis maxima	**Bulbophyllum maximum**
Phyllorkis medusae	**Bulbophyllum medusae**
Phyllorkis megalantha	**Bulbophyllum cheiri**
Phyllorkis megalonyx	**Bulbophyllum megalonyx**
Phyllorkis membranacea	**Bulbophyllum membranaceum**
Phyllorkis membranifolia	**Bulbophyllum membranifolium**
Phyllorkis merguensis	**Bulbophyllum lineatum**
Phyllorkis meridensis	**Bulbophyllum meridense**
Phyllorkis micrantha	**Bulbophyllum triste**
Phyllorkis microtepala	**Bulbophyllum microtepalum**
Phyllorkis minuta	**Bulbophyllum minutum**
Phyllorkis minutissima	**Bulbophyllum minutissimum**
Phyllorkis mischmeensis	**Sunipia cirrhata**
Phyllorkis modesta	**Bulbophyllum sulcatum**
Phyllorkis monantha	**Bulbophyllum monanthum**
Phyllorkis moniliformis	**Bulbophyllum moniliforme**
Phyllorkis monticola	**Bulbophyllum gravidum**
Phyllorkis mucronata	**Bulbophyllum mucronatum**
Phyllorkis mucronifolia	**Bulbophyllum mucronifolium**
Phyllorkis multiflora	**Bulbophyllum multiflorum**
Phyllorkis mutabilis	**Bulbophyllum mutabile**
Phyllorkis napelli	**Bulbophyllum napelli**
Phyllorkis nasuta	**Trias nasuta**
Phyllorkis neilgherensis	**Bulbophyllum neilgherrense**
Phyllorkis nematopoda	**Bulbophyllum nematopodum**
Phyllorkis nutans	**Bulbophyllum nutans**
Phyllorkis obtusa	**Bulbophyllum obtusum**
Phyllorkis occlusa	**Bulbophyllum occlusum**
Phyllorkis occulta	**Bulbophyllum occultum**
Phyllorkis odorata	**Bulbophyllum odoratum**
Phyllorkis odoratissima	**Bulbophyllum odoratissimum**
Phyllorkis oerstedtii	**Bulbophyllum oerstedii**
Phyllorkis oligoglossa	**Bulbophyllum oligoglossum**
Phyllorkis oreonastes	**Bulbophyllum oreonastes**
Phyllorkis ornatissima	**Bulbophyllum ornatissimum**
Phyllorkis othonis	**Bulbophyllum othonis**
Phyllorkis ovalifolia	**Bulbophyllum ovalifolium**
Phyllorkis oxyptera	**Bulbophyllum maximum** var. **oxypterum**
Phyllorkis pachyrhachis	**Bulbophyllum pachyrachis**
Phyllorkis paleacea	**Sunipia cirrhata**
Phyllorkis parviflora	**Bulbophyllum parviflorum**
Phyllorkis parvula	**Bulbophyllum ovalifolium**
Phyllorkis patens	**Bulbophyllum patens**
Phyllorkis pavimenta	**Bulbophyllum pumilum**
Phyllorkis pendula	**Bulbophyllum pendulum**
Phyllorkis penicillium	**Bulbophyllum penicillium**
Phyllorkis petiolaris	**Bulbophyllum petiolare**
Phyllorkis picturata	**Bulbophyllum picturatum**
Phyllorkis pileata	**Bulbophyllum pileatum**
Phyllorkis pipio	**Bulbophyllum pipio**
Phyllorkis polyrhiza	**Bulbophyllum polyrhizum**
Phyllorkis prismatica	**Bulbophyllum prismaticum**
Phyllorkis protracta	**Bulbophyllum protractum**

ALL NAMES	ACCEPTED NAMES
Phyllorkis psittacoglossa	**Bulbophyllum psittacoglossum**
Phyllorkis psychoon	**Bulbophyllum psychoon**
Phyllorkis pumilio	**Bulbophyllum pumilio**
Phyllorkis punctata	**Bulbophyllum baileyi**
Phyllorkis purpurascens	**Bulbophyllum macphersoni**
Phyllorkis purpurea	**Bulbophyllum careyanum**
Phyllorkis pusilla	**Bulbophyllum pusillum**
Phyllorkis pygmaea	**Bulbophyllum pygmaeum**
Phyllorkis quadriseta	**Bulbophyllum quadrisetum**
Phyllorkis radiata	**Bulbophyllum laxiflorum**
Phyllorkis recurva	**Bulbophyllum pumilum**
Phyllorkis refracta	**Bulbophyllum refractum**
Phyllorkis reichenbachii	**Bulbophyllum reichenbachii**
Phyllorkis reinwardtii	**Bulbophyllum uniflorum**
Phyllorkis repens	**Bulbophyllum repens**
Phyllorkis reptans	**Bulbophyllum reptans**
Phyllorkis reticulata	**Bulbophyllum reticulatum**
Phyllorkis retusiuscula	**Bulbophyllum retusiusculum**
Phyllorkis rhizophorae	**Bulbophyllum falcatum** var. velutinum
Phyllorkis ridleyana	**Bulbophyllum pulchellum**
Phyllorkis ringens	**Bulbophyllum lindleyanum**
Phyllorkis rolfei	**Bulbophyllum rolfei**
Phyllorkis rostriceps	**Bulbophyllum rostriceps**
Phyllorkis roxburghii	**Bulbophyllum roxburghii**
Phyllorkis rufilabra	**Bulbophyllum rufilabrum**
Phyllorkis rufina	**Bulbophyllum rufinum**
Phyllorkis saltatoria	**Bulbophyllum saltatorium**
Phyllorkis sceptrum	**Bulbophyllum elongatum**
Phyllorkis schefferi	**Bulbophyllum schefferi**
Phyllorkis schmidtiana	**Bulbophyllum schmidtianum**
Phyllorkis secunda	**Bulbophyllum secundum**
Phyllorkis sessilis	**Bulbophyllum clandestinum**
Phyllorkis setigera	**Bulbophyllum setigerum**
Phyllorkis seychellarum	**Bulbophyllum intertextum**
Phyllorkis shepherdii	**Bulbophyllum shepherdi**
Phyllorkis sicyobulbon	**Bulbophyllum sicyobulbon**
Phyllorkis silleniana	**Bulbophyllum sillemianum**
Phyllorkis simillima	**Bulbophyllum taeniophyllum**
Phyllorkis sordida	**Bulbophyllum sordidum**
Phyllorkis stenobulbon	**Bulbophyllum stenobulbon**
Phyllorkis striata	**Bulbophyllum striatum**
Phyllorkis striatella	**Bulbophyllum striatellum**
Phyllorkis suavissima	**Bulbophyllum suavissimum**
Phyllorkis sulcata	**Bulbophyllum sulcatum**
Phyllorkis taeniophylla	**Bulbophyllum taeniophyllum**
Phyllorkis tenella	**Bulbophyllum tenellum**
Phyllorkis tenuicaulis	**Bulbophyllum cochleatum** var. tenuicaule
Phyllorkis tenuifolia	**Bulbophyllum tenuifolium**
Phyllorkis thompsonii	**Bulbophyllum thompsonii**
Phyllorkis thomsonii	**Bulbophyllum parviflorum**
Phyllorkis thwaitesii	**Bulbophyllum thwaitesii**
Phyllorkis tortuosa	**Bulbophyllum tortuosum**
Phyllorkis tremula	**Bulbophyllum tremulum**
Phyllorkis triadenia	**Bulbophyllum triadenium**
Phyllorkis triflora	**Bulbophyllum triflorum**
Phyllorkis trimenii	**Bulbophyllum trimeni**
Phyllorkis tripetala	**Bulbophyllum tripetalum**

* † ‡ **For explanation see page 2 and 3, point 3**
* † ‡ **Voir les explications page 10 et 11, point 3**
* † ‡ **Para mayor explicación, véase la página 18 y 19, point 3**

ALL NAMES	ACCEPTED NAMES
Phyllorkis tristis	**Bulbophyllum triste**
Phyllorkis umbellata	**Bulbophyllum umbellatum**
Phyllorkis unguiculata	**Bulbophyllum unguiculatum**
Phyllorkis uniflora	**Bulbophyllum uniflorum**
Phyllorkis vaginata	**Bulbophyllum vaginatum**
Phyllorkis variegata	**Bulbophyllum variegatum**
Phyllorkis velutina	**Bulbophyllum falcatum** var. **velutinum**
Phyllorkis vermicularis	**Bulbophyllum vermiculare**
Phyllorkis violacea	**Bulbophyllum violaceum**
Phyllorkis virens	**Sunipia virens**
Phyllorkis viridiflora	**Bulbophyllum viridiflorum**
Phyllorkis vittata	**Bulbophyllum warmingianum**
Phyllorkis wallichii	**Bulbophyllum refractum**
Phyllorkis weddellii	**Bulbophyllum weddelii**
Phyllorkis wightii	**Bulbophyllum wightii**
Phyllorkis wrayi	**Bulbophyllum wrayi**
Phyllorkis xylophylla	**Bulbophyllum xylophyllum**
Pleurothallis pachyrrhachis	**Bulbophyllum pachyrachis**
Pleurothallis purpurea	**Bulbophyllum careyanum**
Rhytionanthos aemulum	**Bulbophyllum forresti**
Rhytionanthos bootanense	**Bulbophyllum bootanense**
Rhytionanthos cornutum	**Bulbophyllum helenae**
Rhytionanthos mirum	**Bulbophyllum mirum**
Rhytionanthos nodosum	**Bulbophyllum nodosum**
Rhytionanthos plumatum	**Bulbophyllum plumatum**
Rhytionanthos rheedei	**Bulbophyllum rheedei**
Rhytionanthos spathulatum	**Bulbophyllum spathulatum**
Rhytionanthos sphaericum	**Bulbophyllum sphaericum**
Rhytionanthos strigosum	**Bulbophyllum strigosum** †
Rhytionanthos unciniferum	**Bulbophyllum unciniferum**
Saccoglossum lanceolatum	
Saccoglossum maculatum	
Saccoglossum papuanum	
Saccoglossum takeuchii	
Saccoglossum verrucosum	
Sarcobodium lobbii	**Bulbophyllum lobbii**
Sarcochilus newportii	**Bulbophyllum newportii**
Sarcopodium affine	**Bulbophyllum affine**
Sarcopodium cheiri	**Bulbophyllum cheiri**
Sarcopodium dearei	**Bulbophyllum dearei**
Sarcopodium grandiflorum	**Bulbophyllum grandiflorum**
Sarcopodium griffithii	**Bulbophyllum griffithii**
Sarcopodium leopardinum	**Bulbophyllum leopardinum**
Sarcopodium lobbii	**Bulbophyllum lobbii**
Sarcopodium macranthum	**Bulbophyllum macranthum**
Sarcopodium megalanthum	**Bulbophyllum cheiri**
Sarcopodium pileatum	**Bulbophyllum pileatum**
Sarcopodium psittacoglossum	**Bulbophyllum psittacoglossum**
Sarcopodium purpureum	**Bulbophyllum macranthum**
Sarcopodium reinwardtii	**Bulbophyllum uniflorum**
Sarcopodium striatum	**Bulbophyllum striatum**
Sarcopodium triadenium	**Bulbophyllum triadenium**
Sarcopodium uniflorum	**Bulbophyllum monanthum**
Serpenticaulis bowkettiae	**Bulbophyllum bowkettae**
Serpenticaulis johnsonii	**Bulbophyllum johnsonii**
Serpenticaulis kirkwoodiae	**Bulbophyllum kirkwoodae**
Serpenticaulis whitei	**Bulbophyllum whitei**
Serpenticaulis wolfei	**Bulbophyllum wolfei**

ALL NAMES   ACCEPTED NAMES

Sestochilos uniflorum.. **Bulbophyllum lobbii**
Spilorchis weinthalii .... **Bulbophyllum weinthalii**
Spilorchis weinthalii subsp. *striatum* .... **Bulbophyllum weinthalii** subsp.
                                                                 striatum
Stelis caudata.. **Bulbophyllum odoratissimum**
Stelis hirta.. **Bulbophyllum hirtum**
Stelis odoratissima.. **Bulbophyllum odoratissimum**
Stelis racemosa .... **Sunipia racemosa**
Sunipia andersonii
Sunipia angustipetala
Sunipia annamensis
Sunipia australis
Sunipia bicolor
Sunipia bifurcatoflorens... **Sunipia andersonii**
Sunipia candida
Sunipia cirrhata
Sunipia cumberlegei
Sunipia dichroma
Sunipia flavescens.. **Sunipia andersonii**
Sunipia fuscopurpurea .... **Sunipia cirrhata**
Sunipia grandiflora
Sunipia hainanensis
Sunipia intermedia
Sunipia jainii
Sunipia kachinensis
Sunipia minor
Sunipia paleacea... **Sunipia cirrhata**
Sunipia pallida
Sunipia purpurata... **Sunipia andersonii**
Sunipia racemosa
Sunipia rimannii
Sunipia salweenensis .... **Sunipia rimannii**
Sunipia sasakii.. **Sunipia andersonii**
Sunipia scariosa
Sunipia soidaoensis
Sunipia thailandica
Sunipia virens
Sunipia viridis
Synarmosepalum heldiorum... **Bulbophyllum heldiorum**
Synarmosepalum kittredgei... **Bulbophyllum kettridgei**
Tapeinoglossum centrosemiflorum................................ **Bulbophyllum**
                                                                 centrosemiflorum
Tapeinoglossum nannodes .... **Bulbophyllum nannodes**
Taurostalix herminiostachys... **Bulbophyllum pumilum**
Thelychiton argyropus .... **Bulbophyllum argyropus**
Trias antheae
Trias bonaccordensis
Trias cambodiana
Trias crassifolia... **Bulbophyllum crassifolium**
Trias dayanum .... **Bulbophyllum dayanum**
Trias disciflora
Trias intermedia
Trias mollis
Trias nana
Trias nasuta
‡ *Trias nummularia* ‡
Trias oblonga
Trias ovata.. **Trias oblonga**
Trias picta

* † ‡ **For explanation see page 2 and 3, point 3**
* † ‡ **Voir les explications page 10 et 11, point 3**
* † ‡ **Para mayor explicación, véase la página 18 y 19, point 3**   **95**

ALL NAMES	ACCEPTED NAMES
Trias pusilla	
Trias rolfei	Trias rosea
Trias rosea	
Trias stocksii	
Trias tothastes	
Trias vitrina	Trias nasuta
Tribrachia hirta	Bulbophyllum hirtum
Tribrachia odoratissima	Bulbophyllum odoratissimum
Tribrachia pendula	Bulbophyllum pendulum
Tribrachia purpurea	Bulbophyllum careyanum
Tribrachia racemosa	Bulbophyllum reptans
Tribrachia reptans	Bulbophyllum reptans
Vesicisepalum folliculiferum	Bulbophyllum folliculiferum
Xiphizusa chloroptera	Bulbophyllum chloropterum
Xiphizusa gladiata	Bulbophyllum meridense
Xiphizusa weddelii	Bulbophyllum weddelii
Zygoglossum umbellatum	Bulbophyllum longiflorum

Part II: Accepted Names / Noms Acceptés / Nombres Aceptado

Part II: Accepted Names / Noms Acceptés / Nombres Aceptado

PART II: ACCEPTED NAMES IN CURRENT USE
Ordered alphabetically on Accepted Names for the genera:

Acrochaene, Bulbophyllum, Chaseella, Codonosiphon, Drymoda, Monomeria, Monosepalum, Pedilochilus, Saccoglossum, Sunipia and *Trias*

DEUXIÈME PARTIE: NOMS ACCEPTES D'USAGE COURANT
Par ordre alphabétique des noms acceptés pour les genre:

Acrochaene, Bulbophyllum, Chaseella, Codonosiphon, Drymoda, Monomeria, Monosepalum, Pedilochilus, Saccoglossum, Sunipia et *Trias*

PARTE II: NOMBRES ACEPTADOS DE USO ACTUAL
Presentados por orden alfabético: nombres aceptados para el genero:

Acrochaene, Bulbophyllum, Chaseella, Codonosiphon, Drymoda, Monomeria, Monosepalum, Pedilochilus, Saccoglossum, Sunipia y *Trias*

Part II: Accepted Names / Noms Acceptés / Nombres Aceptado

ACROCHAENE BINOMIALS IN CURRENT USE

ACROCHAENE BINOMES ACTUELLEMENT EN USAGE

ACROCHAENE BINOMIALES UTILIZADOS NORMALMENTE

Acrochaene punctata Lindl.
Bulbophyllum kingii Hook. f.
Monomeria punctata (Lindl.) Schltr.
Phyllorkis kingii (Hook. f.) Kuntze

Distribution: Bhutan, India, Myanmar, Thailand

Part II: Accepted Names / Noms Acceptés / Nombres Aceptado

BULBOPHYLLUM BINOMIALS IN CURRENT USE

BULBOPHYLLUM BINOMES ACTUELLEMENT EN USAGE

BULBOPHYLLUM BINOMIALES UTILIZADOS NORMALMENTE

Bulbophyllum abbreviatum Rchb. f.
Bulbophyllum trigonopus Rchb. f.
Cirrhopetalum abbreviatum Rchb. f.
Cirrhopetalum frigonopus Rchb. f.

Distribution: Unknown

Bulbophyllum abbrevilabium Carr
Epicranthes abbrevilabia (Carr) Garay & W. Kittr.
Epicranthes annamensis Guillaumin

Distribution: Malaysia, Thailand, Viet Nam

Bulbophyllum aberrans Schltr.

Distribution: Indonesia

Bulbophyllum ablepharon Schltr.

Distribution: Papua New Guinea

Bulbophyllum absconditum J. J. Sm.
Bulbophyllum ochrochlamys Schltr. (1913)
Pelma absconditum (J. J. Sm.) Finet

Distribution: Indonesia, New Caledonia, Papua New Guinea, Philippines, Vanuatu

Bulbophyllum absconditum J. J. Sm. subsp. **hastula** J. J. Verm.

Distribution: Indonesia, Papua New Guinea

Bulbophyllum acanthoglossum Schltr.
Hapalochilus acanthoglossus (Schltr.) Garay & W. Kittr.

Distribution: Papua New Guinea

Bulbophyllum acropogon Schltr.

Distribution: Papua New Guinea

Bulbophyllum acuminatifolium J. J. Sm.

Distribution: Indonesia

Part II: Accepted Names / Noms Acceptés / Nombres Aceptado

Bulbophyllum acuminatum (Ridl.) Ridl.
Bulbophyllum semibifidum (Ridl.) Ridl.
Bulbophyllum stenophyllum Ridl.
Cirrhopetalum acuminatum Ridl.
Cirrhopetalum linearifolium Ridl.
Cirrhopetalum semibifidum Ridl.

Distribution: Indonesia, Malaysia, Singapore, Thailand

Bulbophyllum acutebracteatum De Wild.
Bulbophyllum platirachis De Wild.

Distribution: Cameroon, Democratic Republic of the Congo, Gabon, Liberia, Sierra Leone

Bulbophyllum acutebracteatum De Wild. var. **rubrobrunneopapillosum** (De Wild.) J. J. Verm.
Bulbophyllum fuscoides J. B. Petersen
Bulbophyllum rubrobrunneopapillosum De Wild.

Distribution: Cameroon, Democratic Republic of the Congo, Equatorial Guinea, Gabon, South Africa

Bulbophyllum acutiflorum A. Rich.
Cirrhopetalum acutiflorum (A. Rich.) Hook. f.
Phyllorkis acutiflora (A. Rich.) Kuntze

Distribution: India

Bulbophyllum acutilingue J. J. Sm.

Distribution: Indonesia

Bulbophyllum acutispicatum H. Perrier

Distribution: Madagascar

Bulbophyllum adangense Seidenf.
Epicranthes adangensis (Seidenf.) Garay & W. Kittr.

Distribution: Thailand

Bulbophyllum adelphidium J. J. Verm.

Distribution: Indonesia

Bulbophyllum adenoblepharon Schltr.

Distribution: Papua New Guinea

104

Part II: Accepted Names / Noms Acceptés / Nombres Aceptado

Bulbophyllum adiamantinum Brade

Distribution: Brazil

Bulbophyllum adjungens Seidenf.

Distribution: Thailand

Bulbophyllum adolphi Schltr.

Distribution: Papua New Guinea

Bulbophyllum aechmophorum J. J. Verm.
  Peltopus aechmophorus (J. J. Verm.) Szlach. & Marg.

Distribution: Indonesia, Papua New Guinea

Bulbophyllum aemulum Schltr.

Distribution: Indonesia, Papua New Guinea

Bulbophyllum aeolium Ames

Distribution: Philippines

Bulbophyllum aestivale Ames

Distribution: Philippines

Bulbophyllum affine Lindl.
  Bulbophyllum kusukusense Hayata
  Phyllorkis affinis (Lindl.) Kuntze
  Sarcopodium affine (Lindl.) Lindl. & Paxt.

Distribution: Bhutan, China, India, Japan, Lao People's Democratic Republic, Thailand, Viet Nam

Bulbophyllum afzelii Schltr.

Distribution: Madagascar

Bulbophyllum afzelii Schltr. var. **microdoron** (Schltr.) Bosser
  Bulbophyllum lichenophylax Schltr. var. *microdoron* (Schltr.) H. Perrier
  Bulbophyllum microdoron Schltr.

Distribution: Madagascar

Part II: Accepted Names / Noms Acceptés / Nombres Aceptado

Bulbophyllum agapethoides Schltr.

Distribution: Indonesia

Bulbophyllum agastor Garay, Hamer & Siegerist

Distribution: Papua New Guinea

Bulbophyllum aggregatum Bosser

Distribution: Madagascar

Bulbophyllum aithorhachis J. J. Verm.

Distribution: Brunei Darussalam, Indonesia

Bulbophyllum alabastraceus P. Royen

Distribution: Papua New Guinea

Bulbophyllum alagense Ames

Distribution: Philippines

Bulbophyllum alatum J. J. Verm.

Distribution: Indonesia

Bulbophyllum albibracteum Seidenf.

Distribution: Thailand

Bulbophyllum albidostylidium Seidenf.

Distribution: Thailand

Bulbophyllum albidum (Wight) Hook. f.
  Cirrhopetalum albidum Wight
  Phyllorkis albida (Wight) Kuntze

Distribution: India

Bulbophyllum albociliatum (Liu & Su) Nakajima
  Bulbophyllum albociliatum (Liu & Su) Seidenf.
  Bulbophyllum taichungianum S. S. Ying
  Cirrhopetalum albociliatum Liu & Su

Distribution: Taiwan, Province of China

106

Part II: Accepted Names / Noms Acceptés / Nombres Aceptado

Bulbophyllum albociliatum (Liu & Su) Nakajima var. **weiminianum** T. S. Lin & Kuo Huang

Distribution: Taiwan, Province of China

Bulbophyllum albo-roseum Ames

Distribution: Philippines

Bulbophyllum alcicorne C. S. P. Parish & Rchb. f.
Phyllorkis alcicornis (C. S. P. Parish & Rchb. f.) Kuntze

Distribution: Malaysia, Myanmar, Thailand

Bulbophyllum alexandrae Schltr.

Distribution: Madagascar

Bulbophyllum algidum Ridl.
Hapalochilus algidus (Ridl.) Garay & W. Kittredge

Distribution: Indonesia

Bulbophyllum alinae Szlach. & Olszewski

Distribution: Cameroon

Bulbophyllum alkmaarense J. J. Sm.
Hapalochilus alkmaarensis (J. J. Sm.) Garay & W. Kittr

Distribution: Indonesia

Bulbophyllum alleizettei Schltr.

Distribution: Madagascar

Bulbophyllum allenkerrii Seidenf.

Distribution: Lao People's Democratic Republic, Thailand

Bulbophyllum alliifolium J. J. Sm.

Distribution: Indonesia

Bulbophyllum alsiosum Ames

Distribution: Philippines

Part II: Accepted Names / Noms Acceptés / Nombres Aceptado

Bulbophyllum alticaule Ridl.

Distribution: Indonesia

Bulbophyllum alticola Schltr.
Hapalochilus alticola (Schltr.) Garay & W. Kittr.

Distribution: Papua New Guinea

Bulbophyllum alveatum J. J. Verm.
Peltopus alveatus (J. J. Verm.) Szlach. & Marg.

Distribution: Papua New Guinea

Bulbophyllum amazonicum L. O. Williams

Distribution: Bolivia

Bulbophyllum ambatoavense Bosser

Distribution: Madagascar

Bulbophyllum amblyacron Schltr.

Distribution: Papua New Guinea

Bulbophyllum amblyanthum Schltr.

Distribution: Papua New Guinea

Bulbophyllum amblyoglossum Schltr.

Distribution: Indonesia

Bulbophyllum ambrense H. Perrier

Distribution: Madagascar

Bulbophyllum ambrosia (Hance) Schltr.
Bulbophyllum amygdalinum Aver.
Bulbophyllum watsonianum Rchb. f.
Eria ambrosia Hance

Distribution: China, Viet Nam

Bulbophyllum ambrosia (Hance) Schltr. subsp. **nepalensis** J. J. Wood

Distribution: Nepal

108

Part II: Accepted Names / Noms Acceptés / Nombres Aceptado

Bulbophyllum amoenum Bosser

Distribution: Madagascar

Bulbophyllum amphorimorphum H. Perrier

Distribution: Madagascar

Bulbophyllum amplebracteatum Teijsm. & Binn.

Distribution: Indonesia

Bulbophyllum amplifolium (Rolfe) Balakr. & Sud.Chowdhury
  Cirrhopetalum amplifolium Rolfe

Distribution: Bhutan, India

Bulbophyllum amplistigmaticum Kores

 Distribution: Fiji

Bulbophyllum anaclastum J. J. Verm.

Distribution: Malaysia

Bulbophyllum anakbaruppui J. J. Verm. & P. O'Byrne

Distribution: Indonesia

Bulbophyllum analamazoatrae Schltr.

Distribution: Madagascar

Bulbophyllum anceps Rolfe

Distribution: Indonesia

Bulbophyllum andersonii (Hook. f.) J. J. Sm.
  Bulbophyllum henryi (Rolfe) J. J. Sm.
  Cirrhopetalum andersoni Hook. f.
  Cirrhopetalum henryi Rolfe
  Cirrhopetalum rivesii Gagnep.
  Phyllorkis andersonii (Hook. f.) Kuntze

Distribution: Bhutan, China, India, Myanmar, Viet Nam

Part II: Accepted Names / Noms Acceptés / Nombres Aceptado

Bulbophyllum andohahelense H. Perrier

Distribution: Madagascar

Bulbophyllum andreeae A. D. Hawkes
Bulbophyllum acuminatum Schltr.

Distribution: Papua New Guinea

Bulbophyllum anguipes Schltr.

Distribution: Indonesia

Bulbophyllum angulatum J. J. Sm.

Distribution: Indonesia

Bulbophyllum anguliferum Ames & C. Schweinf.
Bulbophyllum muluense J. J. Wood

Distribution: Indonesia, Philippines

Bulbophyllum angusteovatum Seidenf.

Distribution: Thailand

Bulbophyllum angustifolium (Blume) Lindl.
Diphyes angustifolia Blume
Phyllorkis angustifolia (Blume) Kuntze

Distribution: Indonesia, Malaysia

Bulbophyllum anisopterum J. J. Verm. & P. O'Byrne

Distribution: Malaysia

Bulbophyllum anjozorobeense Bosser

Distribution: Madagascar

Bulbophyllum ankaizinense (Jum. & H. Perrier) Schltr.
Bulbophyllum ophiuchus Ridl. var. *ankaizinensis* Jum. & H. Perrier

Distribution: Madagascar

Part II: Accepted Names / Noms Acceptés / Nombres Aceptado

Bulbophyllum ankaratranum Schltr.

Distribution: Madagascar

Bulbophyllum ankylochele J. J. Verm.
Peltopus ankylochele (J. J. Verm.) Szlach. & Marg.

Distribution: Indonesia, Papua New Guinea

Bulbophyllum ankylorhinon J. J. Verm.

Distribution: Papua New Guinea

Bulbophyllum annamense (Garay) Sieder & Kiehn †
Cirrhopetalum annamense Garay

Distribution: Viet Nam

Bulbophyllum annandalei Ridl.

Distribution: Malaysia, Thailand

Bulbophyllum antennatum Schltr.
Bulbophyllum navigioliferum J. J. Sm.

Distribution: Indonesia, Papua New Guinea

Bulbophyllum antenniferum (Lindl.) Rchb. f.
Cirrhopetalum antenniferum Lindl.
Hyalosema antenniferum (Lindl.) Rysy
Phyllorkis antennifera (Lindl.) Kuntze

Distribution: Indonesia, Papua New Guinea, Philippines, Solomon Islands

Bulbophyllum antioquiense Kraenzl.

Distribution: Colombia

Bulbophyllum antongilense Schltr.

Distribution: Madagascar

Bulbophyllum apertum Schltr.

Distribution: Indonesia

Part II: Accepted Names / Noms Acceptés / Nombres Aceptado

Bulbophyllum aphanopetalum Schltr.
Bulbophyllum capillipes (Guillaumin) N. Halle
Cirrhopetalum capillipes Guillaumin
Peltopus aphanopetalum (Schltr.) Szlach. & Marg.

Distribution: Fiji, New Caledonia, Papua New Guinea

Bulbophyllum apheles J. J. Verm.

Distribution: Indonesia

Bulbophyllum apiculatum Schltr.

Distribution: Papua New Guinea

Bulbophyllum apiferum Carr
Bulbophyllum holttumii A. D. Hawkes

Distribution: Malaysia

Bulbophyllum apodum Hook. f.
Bulbophyllum ebulbum King & Pantl.
Bulbophyllum saccatum Kraenzl.
Bulbophyllum vaginulosum Carr
Bulbophyllum vidalii Tixier
Phyllorkis apoda (Hook. f.) Kuntze

Distribution: Bhutan, Indonesia, Malaysia, Philippines, Thailand, Viet Nam

Bulbophyllum apodum Hook. f. var. **lanceolatum** Ridl.

Distribution: Indonesia

Bulbophyllum apoense Schuit. & De Vogel
Bulbophyllum graciliscapum Ames & Rolfe

Distribution: Philippines

Bulbophyllum appendiculatum (Rolfe) J. J. Sm.
Cirrhopetalum appendiculatum Rolfe

Distribution: Bhutan

Bulbophyllum appressicaule Ridl.

Distribution: Indonesia

Bulbophyllum appressum Schltr.

Distribution: Papua New Guinea

112

Part II: Accepted Names / Noms Acceptés / Nombres Aceptado

Bulbophyllum approximatum Ridl.

Distribution: Madagascar

Bulbophyllum arachnidium Ridl.

Distribution: Indonesia

Bulbophyllum arachnoideum Schltr.
Bulbophyllum arcaniflorum Ridl.

Distribution: Papua New Guinea

Bulbophyllum arcutilabium Aver.

Distribution: Viet Nam

Bulbophyllum ardjunense J. J. Sm.

Distribution: Indonesia

Bulbophyllum arfakense J. J. Sm.
Hapalochilus arfakensis (J. J. Sm.) Garay & W. Kittr.

Distribution: Indonesia

Bulbophyllum arfakianum Kraenzl.
Hyalosema arfakianum (Kraenzl.) Rysy

Distribution: Indonesia, Papua New Guinea

Bulbophyllum argyropus (Endl.) Rchb. f.
Adelopetalum argyropum (Endl.) D. L. Jones et M. A. Clem.
Bulbophyllum corythium N. Halle
Phyllorkis argyropus (Endl.) Kuntze
Thelychiton argyropus Endl.

Distribution: Australia, New Caledonia, New Zealand

Bulbophyllum arianeae Fraga & E.C.Smidt

Distribution: Brazil

Bulbophyllum ariel Ridl.

Distribution: Indonesia

Part II: Accepted Names / Noms Acceptés / Nombres Aceptado

Bulbophyllum aristatum (Rchb. f.) Hemsl.
Bolbophyllaria aristata Rchb. f.
Phyllorkis aristata (Rchb. f.) Kuntze

Distribution: Belize, Costa Rica, Guatemala, Honduras, Mexico, Nicaragua, Panama, Venezuela

Bulbophyllum aristilabre J. J. Sm.
Hapalochilus aristilabris (J. J. Sm.) Garay & W. Kittr.

Distribution: Indonesia

Bulbophyllum aristopetalum Kores

Distribution: Fiji

Bulbophyllum armeniacum J. J. Sm.

Distribution: Indonesia, Malaysia

Bulbophyllum arminii Sieder & Kiehn †
Hapalochilus bandischii Garay, Hamer & Siegerist

Distribution: Papua New Guinea

Bulbophyllum arrectum Kraenzl.

Distribution: Philippines

Bulbophyllum arsoanum J. J. Sm.

Distribution: Indonesia

Bulbophyllum artostigma J. J. Verm.
Peltopus artostigma (J. J. Verm.) Szlach. & Marg.

Distribution: Papua New Guinea

Bulbophyllum ascochiloides J. J. Sm.

Distribution: Indonesia

Bulbophyllum asperilingue Schltr.

Distribution: Papua New Guinea

Part II: Accepted Names / Noms Acceptés / Nombres Aceptado

Bulbophyllum aspersum J. J. Sm.

Distribution: Indonesia

Bulbophyllum asperulum J. J. Sm.
  Cirrhopetalum asperulum (J. J. Sm.) Garay, Hamer & Siegerist

Distribution: Indonesia

Bulbophyllum astelidum Aver.

Distribution: Viet Nam

Bulbophyllum atratum J. J. Sm.

Distribution: Indonesia, Malaysia

Bulbophyllum atrolabium Schltr.

Distribution: Papua New Guinea

Bulbophyllum atropurpureum Barb. Rodr.
  Didactyle atropurpurea Barb. Rodr.

Distribution: Brazil

Bulbophyllum atrorubens Schltr.

Distribution: New Caledonia, Samoa, Vanuatu

Bulbophyllum atrosanguineum Aver.

Distribution: Viet Nam

Bulbophyllum aubrevillei Bosser

Distribution: Madagascar

Bulbophyllum aundense Ormerod
  Bulbophyllum barbatum P. Royen

Distribution: Papua New Guinea

Bulbophyllum auratum (Lindl.) Rchb. f.
  Bulbophyllum borneense (Schltr.) J. J. Sm.
  Bulbophyllum campanulatum Rolfe
  Bulbophyllum campanulatum Rolfe var. *inconspicum* J. J. Sm.
  Cirrhopetalum auratum Lindl.
  Cirrhopetalum campanulatum (Rolfe) Rolfe

Cirrhopetalum borneense Schltr.
Phyllorkis aurata (Lindl.) Kuntze

Distribution: India, Indonesia, Malaysia, Philippines, Thailand

Bulbophyllum aureoapex Schltr.
Hapalochilus aureoapex (Schltr.) Garay & W. Kittr.

Distribution: Papua New Guinea

Bulbophyllum aureobrunneum Schltr.

Distribution: Papua New Guinea

Bulbophyllum aureolabellum T. P. Lin
Bulbophyllum gracillimum Hayata

Distribution: Taiwan, Province of China

Bulbophyllum aureum (Hook. f.) J. J. Sm.
Cirrhopetalum aureum Hook. f.
Phyllorkis aurea (Hook. f.) Kuntze

Distribution: India

Bulbophyllum auricomum Lindl.
Bulbophyllum foenisecii C. S. P. Parish & Rchb. f.
Bulbophyllum tripetaloides (Roxb.) Schltr.
Dendrobium tripetaloides Roxb.
Phyllorkis auricoma (Lindl.) Kuntze

Distribution: Indonesia, Myanmar, Thailand

Bulbophyllum auriculatum J. J. Verm. & P. O'Byrne

Distribution: Indonesia

Bulbophyllum auriflorum H. Perrier

Distribution: Madagascar

Bulbophyllum auroreum J. J. Sm.
Bulbophyllum auroreum J. J. Sm. var. *grandiflorum* J. J. Sm.

Distribution: Indonesia

Bulbophyllum averyanovii Seidenf.

Distribution: Viet Nam

116

Part II: Accepted Names / Noms Acceptés / Nombres Aceptado

Bulbophyllum bacilliferum J. J. Sm.

Distribution: Indonesia

Bulbophyllum baculiferum Ridl.

Distribution: Indonesia, Papua New Guinea

Bulbophyllum baileyi F. Muell.
Bulbophyllum caryophyllum J. J. Sm.
Bulbophyllum punctatum Fitzg.
Carparomorchis baileyi (F. Muell.) M. A. Clem. & D. L. Jones
Phyllorkis baileyi (F. Muell.) Kuntze
Phyllorkis punctata (Fitzg.) Kuntze

Distribution: Australia

Bulbophyllum bakhuizenii van Steenis
Bulbophyllum multiflorum (Breda) Kraenzl.
Odontostyles minor Breda
Odontostyles multiflora Breda

Distribution: Indonesia, Malaysia

Bulbophyllum baladeanum J. J. Sm.
Bulbophyllum leratiae (Kranzlin) Seidenf.
Cirrhopetalum leratii Kraenzl.
Cirrhopetalum uniflorum Schltr.

Distribution: New Caledonia

Bulbophyllum balapiuense J. J. Sm.

Distribution: Indonesia

Bulbophyllum ballii P. J. Cribb

Distribution: Zimbabwe

Bulbophyllum bandischii Garay, Hamer & Siegerist
Hyalosema bandischii (Garay, Hamer & Siegerist) Rysy

Distribution: Papua New Guinea

Bulbophyllum barbatum Barb. Rodr.

Distribution: Brazil, Papua New Guinea

Part II: Accepted Names / Noms Acceptés / Nombres Aceptado

Bulbophyllum barbigerum Lindl.
  Phyllorkis barbigera (Lindl.) Kuntze

Distribution: Cameroon, Central African Republic, Congo, Côte d'Ivoire, Democratic Republic of the Congo, Gabon, Liberia, Nigeria, Sierra Leone

Bulbophyllum bariense Gagnep.

Distribution: Viet Nam

Bulbophyllum baronii Ridl.
  Phyllorkis baronii (Ridl.) Kuntze

Distribution: Madagascar

Bulbophyllum basisetum J. J. Sm.

Distribution: Philippines

Bulbophyllum bataanense Ames

Distribution: Philippines

Bulbophyllum bathieanum Schltr.

Distribution: Madagascar

Bulbophyllum batukauense J. J. Sm.

Distribution: Indonesia

Bulbophyllum bavonis J. J. Verm.

Distribution: Malawi, United Republic of Tanzania

Bulbophyllum beccarii Rchb. f.
  Phyllorkis beccarii (Rchb. f.) Kuntze

Distribution: Indonesia

Bulbophyllum berenicis Rchb. f.

Distribution: Unknown

Bulbophyllum betchei F. Muell.
  Bulbophyllum atroviolaceum Fleischm. & Rech.
  Bulbophyllum finetianum Schltr.

118

Part II: Accepted Names / Noms Acceptés / Nombres Aceptado

Bulbophyllum ponapense Schltr.

Distribution: Fiji, Indonesia, Micronesia, Federated States of, New Caledonia, Palau, Samoa, Solomon Islands, Vanuatu

Bulbophyllum biantennatum Schltr.
Hyalosema biantennatum (Schltr.) Rolfe

Distribution: Papua New Guinea

Bulbophyllum bicaudatum Schltr.

Distribution: Papua New Guinea

Bulbophyllum bicolor Lindl.
Cirrhopetalum bicolor (Lindl.) Rolfe
Phyllorkis bicolor (Lindl.) Kuntze

Distribution: China

Bulbophyllum bicoloratum Schltr.
Bulbophyllum bicolor Jum. & H. Perrier
Bulbophyllum coeruleum H. Perrier
Bulbophyllum theiochlamys Schltr.

Distribution: Madagascar

Bulbophyllum bidentatum (Barb. Rodr.) Cogn.
Didactyle bidentata Barb. Rodr.

Distribution: Brazil

Bulbophyllum bidenticulatum J. J. Verm.

Distribution: Cameroon, Côte d'Ivoire, Guinea, Liberia, Sierra Leone

Bulbophyllum bidenticulatum J. J. Verm. var. **joyceae** J. J. Verm.

Distribution: Kenya

Bulbophyllum bidi Govaerts
Bulbophyllum pumilio Ridl.

Distribution: Indonesia

Part II: Accepted Names / Noms Acceptés / Nombres Aceptado

Bulbophyllum bifarium Hook. f.
Bulbophyllum pallescens Kraenzl.
Phyllorkis bifaria (Hook. f.) Kuntze

Distribution: Cameroon

Bulbophyllum biflorum Teijsm. & Binn.
Bulbophyllum geminatum Carr
Cirrhopetalum biflorum (Teijsm. & Binn.) J. J. Sm.
Phyllorkis biflora (Teijsm. & Binn.) Kuntze

Distribution: Indonesia, Malaysia, Philippines, Thailand

Bulbophyllum bigibbosum J. J. Sm.

Distribution: Indonesia

Bulbophyllum bigibbum Schltr.

Distribution: Indonesia, Papua New Guinea

Bulbophyllum binnendijkii J. J. Sm.
Bulbophyllum ericssoni Kraenzl.
Cirrhopetalum leopardinum Teijsm. & Binn.

Distribution: Indonesia

Bulbophyllum birmense Schltr.

Distribution: Myanmar

Bulbophyllum birugatum J. J. Sm.

Distribution: Indonesia

Bulbophyllum bisepalum Schltr.

Distribution: Papua New Guinea

Bulbophyllum biseriale Carr

Distribution: Malaysia, Thailand

Bulbophyllum bisetoides Seidenf.

Distribution: Thailand, Viet Nam

Bulbophyllum bisetum Lindl.
Bolbophyllaria biseta (Lindl.) Rchb. f.

120

Part II: Accepted Names / Noms Acceptés / Nombres Aceptado

Bulbophyllum cirrhopetaloides Griff.
Phyllorkis biseta (Lindl.) Kuntze

Distribution: Bhutan, India, Thailand

Bulbophyllum bismarckense Schltr.

Distribution: Papua New Guinea

Bulbophyllum bittnerianum Schltr.

Distribution: Thailand

Bulbophyllum bivalve J. J. Sm.

Distribution: Indonesia

Bulbophyllum blepharicardium Schltr.

Distribution: Papua New Guinea

Bulbophyllum blephariglossum Schltr.

Distribution: Papua New Guinea

Bulbophyllum blepharistes Rchb. f.
Bulbophyllum malayanum J. J. Sm.
Cirrhopetalum blepharistes (Rchb. f.) Hook. f.
Cirrhopetalum distans Rchb. f.
Cirrhopetalum longescapum Teijsm. & Binn.
Cirrhopetalum spicatum Gagnep.
Phyllorkis blepharistes (Rchb. f.) Kuntze
Phyllorkis longiscapa (Teijsm. & Binn.) Kuntze

Distribution: India, Lao People's Democratic Republic, Malaysia, Myanmar, Thailand, Viet Nam

Bulbophyllum blepharochilum Garay

Distribution: Cameroon

Bulbophyllum blepharopetalum Schltr.

Distribution: Papua New Guinea

Bulbophyllum bliteum J. J. Verm.
Peltopus bliteus (J. J. Verm.) Szlach. & Marg.

Distribution: Papua New Guinea

Bulbophyllum blumei (Lindl.) J. J. Sm.
Bulbophyllum cuspidilingue Rchb. f.
Bulbophyllum ephippium Ridl.
Bulbophyllum masdevalliaceum Kraenzl.
Cirrhopetalum blumii Lindl.
Ephippium ciliatum Blume
Ephippium masdevalliaceum (Kraenzl.) M. A. Clem. & D. L. Jones
Phyllorkis blumei (Lindl.) Kuntze
Phyllorkis maxillaris (Lindl.) Kuntze

Distribution: Indonesia, Malaysia, Papua New Guinea, Philippines, Solomon Islands

Bulbophyllum boiteaui H. Perrier

Distribution: Madagascar

Bulbophyllum bolivianum Schltr.

Distribution: Bolivia

Bulbophyllum bolsteri Ames

Distribution: Philippines

Bulbophyllum bomiensis Z. H. Tsi

Distribution: China

Bulbophyllum boninense (Schltr.) J. J. Sm.
Cirrhopetalum boninense Schltr.

Distribution: China

Bulbophyllum bontocense Ames

Distribution: Philippines

Bulbophyllum boonjee B. Gray & D. L. Jones
Adelopetalum boonjee (B. Gray & D. L. Jones) D. L. Jones & M. A. Clem.

Distribution: Australia

Bulbophyllum bootanense C. S. P. Parish & Rchb. f.
Rhytionanthos bootanense (C. S. P. Parish & Rchb. f.) Garay, Hamer & Siegerist

Distribution: Myanmar

Bulbophyllum botryophorum Ridl.

Distribution: Indonesia, Malaysia, Singapore

122

Part II: Accepted Names / Noms Acceptés / Nombres Aceptado

Bulbophyllum boudetiana Fraga

Distribution: Brazil

Bulbophyllum boulbetii Tixier

Distribution: Viet Nam

Bulbophyllum bowkettae F. M. Bailey
Bulbophyllum waughense Rupp
Phyllorkis bowkettae (F. M. Bailey) Kuntze
Serpenticaulis bowkettiae (F. M. Bailey) M. A. Clem. & D. L. Jones

Distribution: Australia

Bulbophyllum brachychilum Schltr.

Distribution: Papua New Guinea

Bulbophyllum brachypetalum Schltr.
Peltopus brachypetalus (Schltr.) Szlach. & Marg.

Distribution: Papua New Guinea

Bulbophyllum brachyphyton Schltr.

Distribution: Madagascar

Bulbophyllum brachystachyum Schltr.
Bulbophyllum pseudonutans H. Perrier

Distribution: Madagascar

Bulbophyllum bracteatum F. M. Bailey
Adelopetalum bracteatum Fitzg.

Distribution: Australia

Bulbophyllum bracteolatum Lindl.
Bolbophyllaria bracteolata (Lindl.) Rchb. f.
Phyllorkis bracteolata (Lindl.) Kuntze

Distribution: Bolivia, Brazil, Guyana, Suriname, Venezuela

Bulbophyllum bractescens Rolfe ex Kerr

Distribution: Thailand

Bulbophyllum brassii J. J. Verm.
Peltopus brassii (J. J. Verm.) Szlach. & Marg.

Distribution: Indonesia, Papua New Guinea

Bulbophyllum brastagiense Carr
Ferruminaria brastagiense (Carr) Garay, Hamer & Siegerist

Distribution: Indonesia

Bulbophyllum breve Schltr.
Hapalochilus brevis (Schltr.) Garay & W. Kittr.

Distribution: Papua New Guinea

Bulbophyllum brevibrachiatum (Schltr.) J. J. Sm.
Cirrhopetalum brevibrachiatum Schltr.

Distribution: Indonesia

Bulbophyllum brevicolumna J. J. Verm.

Distribution: Indonesia

Bulbophyllum breviflorum Ridl. ex Stapf

Distribution: Indonesia, Philippines

Bulbophyllum brevilabium Schltr.

Distribution: Indonesia, Papua New Guinea

Bulbophyllum brevipes Ridl.
Phyllorkis brevipes (Ridl.) Kuntze

Distribution: Malaysia

Bulbophyllum brevipetalum H. Perrier
Bulbophyllum brevipetalum H. Perrier var. *majus* H. Perrier
Bulbophyllum brevipetalum H. Perrier var. *speculiferum* H. Perrier

Distribution: Madagascar

Bulbophyllum brevispicatum Z. H. Tsi & H. C. Chen

Distribution: China

Bulbophyllum brevistylidium Seidenf.

Distribution: Thailand

Bulbophyllum brienianum (Rolfe) Ames
Bulbophyllum adenophorum (Schltr.) J. J. Sm.
Bulbophyllum brienianum (Rolfe) J. J. Sm.
Bulbophyllum brunnescens (Ridl.) J. J. Sm
Bulbophyllum makoyanum (Rchb. f.) Ridl. var. *brienianum* (Ridl.) Ridl.
Bulbophyllum obrienianum Rolfe
Cirrhopetalum adenophorum Schltr.
Cirrhopetalum brienianum Rolfe
Cirrhopetalum brunnescens Ridl.
Cirrhopetalum makoyanum Rchb. f. var. *brienianum* Ridl.

Distribution: Indonesia, Malaysia

Bulbophyllum bryoides Guillaumin

Distribution: Viet Nam

Bulbophyllum bulhartii Sieder & Kiehn †
Bulbophyllum pulchrum Schltr.
Hapalochilus striatus Garay & W. Kittr.

Distribution: Papua New Guinea

Bulbophyllum bulliferum J. J. Sm.
Bulbophyllum barbilabium Schltr.
Bulbophyllum verrucibracteum J. J. Sm.

Distribution: Indonesia, Papua New Guinea, Solomon Islands

Bulbophyllum burfordiense Hort. ex Garay, Hamer & Siegerist
Hyalosema burfordiense (Garay, Hamer & Siegerist) Rysy

Distribution: Papua New Guinea

Bulbophyllum burkilli Gage

Distribution: Myanmar

Bulbophyllum burttii Summerh.

Distribution: Democratic Republic of the Congo, Rwanda

Bulbophyllum cadetioides Schltr.

Distribution: Papua New Guinea

Part II: Accepted Names / Noms Acceptés / Nombres Aceptado

Bulbophyllum caecilii J. J. Sm.

Distribution: Indonesia

Bulbophyllum caecum J. J. Sm.

Distribution: Indonesia

Bulbophyllum caespitosum Thouars
  Phyllorkis caespitosa Thouars

Distribution: Mauritius, Réunion

Bulbophyllum calceilabium J. J. Sm.

Distribution: Indonesia

Bulbophyllum calceolus J. J. Verm.

Distribution: Indonesia

Bulbophyllum caldericola G. F. Walsh

Distribution: Australia

Bulbophyllum callichroma Schltr.
  Bulbophyllum calothyrsus Schltr.
  Bulbophyllum manifestans J. J. Sm.

Distribution: Indonesia, Papua New Guinea

Bulbophyllum callipes J. J. Sm.
  Hapalochilus callipes (J. J. Sm.) Garay & W. Kittr.

Distribution: Indonesia

Bulbophyllum callosum Bosser

Distribution: Madagascar

Bulbophyllum caloglossum Schltr.

Distribution: Papua New Guinea

Bulbophyllum calophyllum L. O. Williams

Distribution: Philippines

126

Part II: Accepted Names / Noms Acceptés / Nombres Aceptado

Bulbophyllum calviventer J. J. Verm.
Peltopus calviventer (J. J. Verm.) Szlach. & Marg.

Distribution: Papua New Guinea

Bulbophyllum calvum Summerh.

Distribution: Cameroon, Nigeria

Bulbophyllum calyptratum Kraenzl.
Bulbophyllum buchenavianum (Kraenzl.) De Wild.
Bulbophyllum lindleyi (Rolfe) Schltr.
Megaclinium buchenavianum Kraenzl.
Megaclinium flaccidum Hook. f.
Megaclinium lepturum Kraenzl.
Megaclinium lindleyi Rolfe

Distribution: Cameroon, Congo, Côte d'Ivoire, Democratic Republic of the Congo,
Equatorial Guinea, Gabon, Ghana, Guinea, Liberia, Nigeria, Sierra Leone

Bulbophyllum calyptratum Kraenzl. var. **graminifolium** (Summerh.) J. J.
Verm.
Bulbophyllum graminifolium Summerh.
Bulbophyllum intermedium De Wild.
Megaclinium intermedium De Wild.

Distribution: Côte d'Ivoire, Democratic Republic of the Congo, Ghana, Guinea, Liberia,
Sierra Leone

Bulbophyllum calyptratum Kraenzl. var. **lucifugum** (Summerh.) J. J. Verm.
Bulbophyllum lucifugum Summerh.

Distribution: Côte d'Ivoire, Liberia, Sierra Leone

Bulbophyllum calyptropus Schltr.

Distribution: Madagascar

Bulbophyllum cameronense Garay, Hamer & Siegerist

Distribution: Malaysia

Bulbophyllum campos-portoi Brade

Distribution: Brazil

Bulbophyllum camptochilum J. J. Verm.

Distribution: Brunei Darussalam

Bulbophyllum candidum Hook. f.
Phyllorkis candida (Hook. f.) Kuntze

Distribution: India

Bulbophyllum canlaonense Ames

Distribution: Philippines

Bulbophyllum cantagallense (Barb. Rodr.) Cogn.
Didactyle cantagellense Barb. Rodr.

Distribution: Brazil

Bulbophyllum capilligerum J. J. Sm.

Distribution: Indonesia

Bulbophyllum capillipes C. S. P. Parish & Rchb. f.
Drymoda latisepala Seidenf.
Phyllorkis capillipes (C. S. P. Parish & Rchb. f.) Kuntze

Distribution: India, Myanmar, Thailand

Bulbophyllum capitatum (Blume) Lindl.
Diphyes capitata Blume
Phyllorkis capitata (Blume) Kuntze
Phyllorkis diphyes Kuntze

Distribution: Indonesia, Malaysia

Bulbophyllum capituliflorum Rolfe

Distribution: Cameroon, Congo, Democratic Republic of the Congo, Gabon

Bulbophyllum capuronii Bosser

Distribution: Madagascar

Bulbophyllum caputgnomonis J. J. Verm.

Distribution: Papua New Guinea

Bulbophyllum cardiobulbum Bosser

Distribution: Madagascar

Part II: Accepted Names / Noms Acceptés / Nombres Aceptado

Bulbophyllum cardiophyllum J. J. Verm.

Distribution: Papua New Guinea

Bulbophyllum careyanum (Hook. f.) Sprengel
Anisopetalum careyanum Hook. f.
Bulbophyllum careyanum (Hook. f.) Sprengel var. *ochraceum* Hook. f.
Bulbophyllum cupreum Hook. f.
Phyllorkis purpurea (D. Don) Kuntze
Pleurothallis purpurea D. Don
Tribrachia purpurea (D. Don) Lindl.

Distribution: Bhutan, India, Myanmar, Philippines, Thailand, Viet Nam

Bulbophyllum carinatum (Teijsm. & Binn.) Naves
Cirrhopetalum carinatum Teijsm. & Binn.

Distribution: Philippines

Bulbophyllum cariniflorum Rchb. f.
Bulbophyllum densiflorum Rolfe

Distribution: Bhutan, India, Thailand

Bulbophyllum carinilabium J. J. Verm.

Distribution: Indonesia, Malaysia

Bulbophyllum carnosilabium Summerh.

Distribution: Cameroon, Democratic Republic of the Congo, Gabon

Bulbophyllum carnosisepalum J. J. Verm.

Distribution: Cameroon, Côte d'Ivoire, Democratic Republic of the Congo, Gabon, Uganda

Bulbophyllum carrianum J. J. Verm.

Distribution: Malaysia

Bulbophyllum carunculatum Garay, Hamer & Siegerist

Distribution: Philippines

Bulbophyllum cataractarum Schltr.

Distribution: Madagascar

Part II: Accepted Names / Noms Acceptés / Nombres Aceptado

Bulbophyllum catenarium Ridl.
Bulbophyllum carunculilabrum Carr

Distribution: Indonesia, Malaysia, Philippines, Viet Nam

Bulbophyllum catenulatum Kraenzl.

Distribution: Philippines

Bulbophyllum cateorum J. J. Verm.

Distribution: Indonesia, Papua New Guinea

Bulbophyllum catillus J. J. Verm. & P. O'Byrne

Distribution: Papua New Guinea

Bulbophyllum caudatisepalum Ames & C. Schweinf.
Bulbophyllum cuneifolium Ames & C. Schweinf.
Bulbophyllum dulitense Carr
Bulbophyllum koyanense Carr
Bulbophyllum pergracile Ames & C. Schweinf.

Distribution: Indonesia, Malaysia, Philippines

Bulbophyllum caudipetalum J. J. Sm.
Hapalochilus caudatipetalum (J. J. Sm.) Garay & W. Kittr.

Distribution: Indonesia

Bulbophyllum cauliflorum Hook. f.
Phyllorkis cauliflora (Hook.) Kuntze

Distribution: Bhutan, India, Myanmar

Bulbophyllum cauliflorum Hook. f. var. **sikkimense** N. Peace & P. J. Cribb

Distribution: Bhutan

Bulbophyllum cavibulbum J. J. Sm.

Distribution: Indonesia

Bulbophyllum cavipes J. J. Verm.

Distribution: Indonesia

Part II: Accepted Names / Noms Acceptés / Nombres Aceptado

Bulbophyllum centrosemiflorum J. J. Sm.
Tapeinoglossum centrosemiflorum (J. J. Sm.) Schltr.

Distribution: Indonesia

Bulbophyllum cephalophorum Garay, Hamer & Siegerist

Distribution: Philippines

Bulbophyllum cerambyx J. J. Sm.

Distribution: Indonesia

Bulbophyllum ceratostylis J. J. Sm.
Bulbophyllum eximium Ames & C. Schweinf.

Distribution: Indonesia

Bulbophyllum ceratostyloides Ridl.

Distribution: Indonesia

Bulbophyllum cercanthum (Garay, Hamer & Siegerist) Sieder & Kiehn †
Cirrhopetalum cercanthum Garay, Hamer & Siegerist

Distribution: Indonesia

Bulbophyllum cerebellum J. J. Verm.

Distribution: Malaysia

Bulbophyllum cerinum Schltr.

Distribution: Papua New Guinea, Solomon Islands

Bulbophyllum ceriodorum Boiteau

Distribution: Madagascar

Bulbophyllum cernuum (Blume) Lindl.
Bulbophyllum gibbilingue J. J. Sm.
Diphyes cernua Blume
Phyllorkis cernua (Blume) Kuntze

Distribution: Indonesia

Part II: Accepted Names / Noms Acceptés / Nombres Aceptado

Bulbophyllum cernuum Lindl. var. **vittata** (Teijsm. & Binn.) J. J. Sm.
  Bulbophyllum vittatum Teijsm. & Binn.

Distribution: Indonesia

Bulbophyllum chaetostroma Schltr.

Distribution: Papua New Guinea

Bulbophyllum chanii J. J. Verm. & A. L. Lamb

Distribution: Indonesia

Bulbophyllum chaunobulbon Schltr.

Distribution: Papua New Guinea

Bulbophyllum chaunobulbon Schltr. var. **ctenopetalum** Schltr.

Distribution: Papua New Guinea

Bulbophyllum cheiri Lindl.
  Bulbophyllum megalanthum Griff.
  Phyllorkis cheiri (Lindl.) Kuntze
  Phyllorkis megalantha (Griff.) Kuntze
  Sarcopodium cheiri (Lindl.) Lindl. & Paxt.
  Sarcopodium megalanthum (Griff.) Lindl.

Distribution: Indonesia, Malaysia, Philippines

Bulbophyllum cheiropetalum Ridl.
  Bulbophyllum manipetalum J. J. Sm.
  Epicranthes cheiropetalum (Ridl.) Garay & W. Kittredge

Distribution: Indonesia, Malaysia, New Caledonia

Bulbophyllum chimaera Schltr.

Distribution: Papua New Guinea

Bulbophyllum chinense (Lindl.) Rchb. f.
  Cirrhopetalum chinense Lindl.
  Phyllorkis chinensis (Lindl.) Kuntze

Distribution: China

Bulbophyllum chloranthum Schltr.
  Bulbophyllum arcuatum Schltr.
  Bulbophyllum hedyothyrsus Schltr.
  Bulbophyllum macgregorii Schltr.

132

Part II: Accepted Names / Noms Acceptés / Nombres Aceptado

Bulbophyllum solutisepalum J. J. Sm.
Bulbophyllum squamiferum J. J. Sm.
Cirrhopetalum macgregorii (Schltr.) Schltr.

Distribution: Indonesia, Papua New Guinea, Solomon Islands

Bulbophyllum chlorascens J. J. Sm.

Distribution: Indonesia

Bulbophyllum chloroglossum Rchb. f.
Didactyle galeata Barb. Rodr.

Distribution: Brazil

Bulbophyllum chloropterum Rchb. f.
Phyllorkis chloroptera (Rchb. f.) Kuntze
Xiphizusa chloroptera (Rchb. f.) Rchb. f.

Distribution: Brazil

Bulbophyllum chlororhopalon Schltr.
Epicranthes chlororhopalon (Schltr.) Garay & W. Kittr.

Distribution: Papua New Guinea

Bulbophyllum chondriophorum (Gagnep.) Seidenf.
Cirrhopetalum chondriophorum Gagnep.

Distribution: China

Bulbophyllum chrysendetum Ames

Distribution: Philippines

Bulbophyllum chryseum (Kraenzl.) Ames
Bulbophyllum chryseum (Kraenzl.) J. J. Sm.
Cirrhopetalum chryseum Kraenzl.

Distribution: Philippines

Bulbophyllum chrysocephalum Schltr.

Distribution: Unknown

Part II: Accepted Names / Noms Acceptés / Nombres Aceptado

Bulbophyllum chrysochilum Schltr.
Hapalochilus chrysochilus (Schltr.) Garay & W. Kittr.

Distribution: Papua New Guinea

Bulbophyllum chrysoglossum Schltr.
Hapalochilus chrysoglossus (Schltr.) Garay & W. Kittr.

Distribution: Papua New Guinea

Bulbophyllum chrysotes Schltr.

Distribution: Papua New Guinea

Bulbophyllum ciliatilabrum H. Perrier

Distribution: Madagascar

Bulbophyllum ciliatum (Blume) Lindl.
Diphyes ciliata Blume
Phyllorkis ciliata (Blume) Kuntze

Distribution: Indonesia, Malaysia

Bulbophyllum ciliipetalum Schltr.

Distribution: Papua New Guinea

Bulbophyllum ciliolatum Schltr.

Distribution: Papua New Guinea

Bulbophyllum ciluliae Bianch. & J.A.N.Bat.

Distribution: Brazil

Bulbophyllum cimicinum J. J. Verm.
Epicranthes cimicina (J. J. Verm.) Garay & W. Kittr.

Distribution: Papua New Guinea

Bulbophyllum cirrhatum Hook. f.
Phyllorkis cirrhata (Hook.) Kuntze

Distribution: India

134

Bulbophyllum cirrhoglossum H. Perrier

Distribution: Madagascar

Bulbophyllum cirrhosum L. O. Williams

Distribution: Mexico

Bulbophyllum citrellum Ridl.

Distribution: Indonesia

Bulbophyllum citricolor J. J. Sm.
  Bulbophyllum citrellum J. J. Sm.

Distribution: Indonesia

Bulbophyllum citrinilabre J. J. Sm.

Distribution: Indonesia, Papua New Guinea, Solomon Islands

Bulbophyllum clandestinum Lindl.
  Bulbophyllum bolovenense Guillaumin
  Bulbophyllum cryptanthum Cogn.
  Bulbophyllum myrianthum Schltr.
  Bulbophyllum ovalifolium (Wight) C. S. P. Parish
  Bulbophyllum sessile (J. Koenig) J. J. Sm.
  Bulbophyllum sessile Hochr.
  Bulbophyllum sparsifolium Schltr.
  Bulbophyllum trisetosum Griff.
  Epidendrum sessile J. Koenig
  Oxysepala ovalifolia Wight
  Phyllorkis sessilis Kuntze

Distribution: Indonesia, Malaysia, Myanmar, Philippines, Viet Nam

Bulbophyllum clausseni Rchb. f.
  Didactyle clausseni Lindl.

Distribution: Brazil

Bulbophyllum clavatum Thouars
  Bulbophyllum conicum Thouars
  Phyllorkis clavata Thouars

Distribution: Comoros, Mauritius, Réunion

Bulbophyllum cleistogamum Ridl.

Distribution: Indonesia, Malaysia, Philippines, Singapore

Part II: Accepted Names / Noms Acceptés / Nombres Aceptado

Bulbophyllum clemensiae Ames

Distribution: Philippines

Bulbophyllum clipeibulbum J. J. Verm.

Distribution: Viet Nam

Bulbophyllum coccinatum H. Perrier

Distribution: Madagascar

Bulbophyllum cochleatum Lindl.
Bulbophyllum jungwirthianum Schltr.
Bulbophyllum mannii Hook. f.
Bulbophyllum pholidotoides Kraenzl.
Bulbophyllum talbothi Rendle
Phyllorkis cochleata (Lindl.) Kuntze
Phyllorkis mannii (Hook.) Kuntze

Distribution: Cameroon, Côte d'Ivoire, Equatorial Guinea, Gabon, Guinea, Kenya, Liberia, Malawi, Nigeria, Rwanda, Sao Tome and Principe, Sierra Leone, South Africa, Sudan, Uganda, United Republic of Tanzania, Zambia

Bulbophyllum cochleatum Lindl. var. **bequaertii** (De Wild.) J. J. Verm.
Bulbophyllum bequaerti De Wild.
Megaclinium bequaerti De Wild.

Distribution: Cameroon, Democratic Republic of the Congo, Rwanda, Uganda, United Republic of Tanzania

Bulbophyllum cochleatum Lindl. var. **brachyanthum** (Summerh.) J. J. Verm.
Bulbophyllum bequaerti De Wild. var. *brachyanthum* Summerh.

Distribution: Burundi, Democratic Republic of the Congo, Kenya, Rwanda, Uganda, United Republic of Tanzania

Bulbophyllum cochleatum Lindl. var. **tenuicaule** (Lindl.) J. J. Verm.
Bulbophyllum tenuicaule Lindl.
Bulbophyllum thomense Summerh.
Phyllorkis tenuicaulis (Lindl.) Kuntze

Distribution: Cameroon, Democratic Republic of the Congo, Equatorial Guinea, Kenya, Nigeria, Rwanda, Sao Tome and Principe

Bulbophyllum cochlia Garay, Hamer & Siegerist
Bulbophyllum oculatum Teijsm. & Binn.
Bulbophyllum violaceum (Blume) Rchb. f.
Cochlia violacea Blume

Distribution: Indonesia

Part II: Accepted Names / Noms Acceptés / Nombres Aceptado

Bulbophyllum cochlioides J. J. Sm.

Distribution: Indonesia

Bulbophyllum cocoinum Bateman ex Lindl.
Bulbophyllum andongense Rchb. f.
Bulbophyllum brevidenticulatum De Wild.
Bulbophyllum vitiense Rolfe

Distribution: Angola, Cameroon, Côte d'Ivoire, Democratic Republic of the Congo, Gabon, Ghana, Liberia, Sierra Leone, Uganda

Bulbophyllum coelochilum J. J. Verm.

Distribution: Indonesia

Bulbophyllum cogniauxianum (Kraenzl.) J. J. Sm.
Cirrhopetalum cogniauxianum Kraenzl.

Distribution: Brazil

Bulbophyllum collettii King & Pantl.

Distribution: India

Bulbophyllum colliferum J. J. Sm.
Bulbophyllum niveo-sulphureum Schltr.
Bulbophyllum papulilabium Schltr.

Distribution: Indonesia, Papua New Guinea

Bulbophyllum collinum Schltr.
Hapalochilus collinus (Schltr.) Garay & W. Kittr.

Distribution: Papua New Guinea

Bulbophyllum colomaculosum Z. H. Tsi & H. C. Chen

Distribution: China

Bulbophyllum coloratum J. J. Sm.
Hapalochilus coloratus (J. J. Sm.) Garay & W. Kittr.

Distribution: Indonesia

Bulbophyllum colubrimodum Ames

Distribution: Philippines

Bulbophyllum colubrinum (Rchb. f.) Rchb. f.
Bulbophyllum decipiens Schltr.
Bulbophyllum gabunense Schltr.
Bulbophyllum imschootianum (Rolfe) De Wild.
Bulbophyllum inaequale Rchb. f.
Bulbophyllum makakense J. B. Hansen
Megaclinium colubrinum Rchb. f.
Megaclinium imschootianum Rolfe
Megaclinium inaequale Rchb. f.
Phyllorkis colubrina (Rchb. f.) Kuntze

Distribution: Angola, Cameroon, Congo, Côte d'Ivoire, Democratic Republic of the Congo, Gabon, Ghana, Nigeria, Sierra Leone

Bulbophyllum comatum Lindl.
Bulbophyllum hirsutissimum Kraenzl. (1914)
Phyllorkis comata (Lindl.) Kuntze

Distribution: Equatorial Guinea, Gabon, Nigeria

Bulbophyllum comatum Lindl. var. **inflatum** (Rolfe) J. J. Verm.
Bulbophyllum inflatum Rolfe

Distribution: Côte d'Ivoire, Gabon, Liberia, Rwanda, Sierra Leone

Bulbophyllum comberi J. J. Verm.

Distribution: Indonesia, Malaysia

Bulbophyllum comberipictum J. J. Verm.

Distribution: Indonesia, Malaysia

Bulbophyllum cominsii Rolfe
Hyalosema cominsii (Rolfe) Rolfe

Distribution: Papua New Guinea

Bulbophyllum commersonii Thouars
Phyllorkis commersonii Thouars

Distribution: Mauritius, Réunion

Bulbophyllum commissibulbum J. J. Sm.

Distribution: Indonesia

Bulbophyllum comorianum H. Perrier

Distribution: Comoros, Madagascar

138

Part II: Accepted Names / Noms Acceptés / Nombres Aceptado

Bulbophyllum comosum Collett & Hemsl.
Phyllorkis comosa (Collet & Hemsl.) Kuntze

Distribution: Myanmar, Thailand

Bulbophyllum complanatum H. Perrier
Bulbophyllum sigilliforme H. Perrier

Distribution: Madagascar

Bulbophyllum compressilabellatum P. Royen

Distribution: Papua New Guinea

Bulbophyllum compressum Teijsm. & Binn.

Distribution: Indonesia

Bulbophyllum comptonii Rendle

Distribution: New Caledonia

Bulbophyllum concatenatum P. J. Cribb & P. Taylor

Distribution: United Republic of Tanzania

Bulbophyllum concavibasalis P. Royen
Hapalochilus concavibasalis (van Royen) Garay & W. Kittr.

Distribution: Indonesia

Bulbophyllum conchidioides Ridl.
Bulbophyllum pleurothalloides Schltr.

Distribution: Madagascar

Bulbophyllum conchophyllum J. J. Sm.
Epicranthes conchophylla (J. J. Sm.) Garay & W. Kittr.

Distribution: Indonesia

Bulbophyllum concinnum Hook. f.
Phyllorkis concinna (Hook. f.) Kuntze

Distribution: Indonesia, Malaysia, Thailand, Viet Nam

Part II: Accepted Names / Noms Acceptés / Nombres Aceptado

Bulbophyllum concolor J. J. Sm.
  Hapalochilus concolor (J. J. Sm.) Garay & W. Kittr.

Distribution: Indonesia

Bulbophyllum confusum (Garay, Hamer & Siegerist) Sieder & Kiehn †
  Cirrhopetalum confusum Garay, Hamer & Siegerist

Distribution: India

Bulbophyllum congestiflorum Ridl.

Distribution: Indonesia

Bulbophyllum coniferum Ridl.
  Bulbophyllum musciferum Ridl.
  Bulbophyllum obscurum J. J. Sm.
  Diphyes crassifolia Blume

Distribution: Indonesia, Malaysia, Philippines

Bulbophyllum connatum Carr

Distribution: Indonesia

Bulbophyllum conspersum J. J. Sm.

Distribution: Indonesia

Bulbophyllum contortisepalum J. J. Sm.

Distribution: Indonesia

Bulbophyllum cootesii Clements

Distribution: Philippines

Bulbophyllum copelandii Ames

Distribution: Philippines

Bulbophyllum corallinum Tixier & Guillaumin

Distribution: China, Thailand, Viet Nam

140

Part II: Accepted Names / Noms Acceptés / Nombres Aceptado

Bulbophyllum cordemoyi Frapp. ex Cordem.
Bulbophyllum jacobi Frapp.

Distribution: Réunion

Bulbophyllum coriaceum Ridl.
Bulbophyllum kinabaluense Rolfe
Bulbophyllum venustum Ames & C. Schweinf.

Distribution: Indonesia, Malaysia, Philippines

Bulbophyllum coriophorum Ridl.
Bulbophyllum compactum Kraenzl.
Bulbophyllum crenulatum Rolfe
Bulbophyllum mandrakanum Schltr.
Bulbophyllum robustum Rolfe

Distribution: Comoros, Madagascar

Bulbophyllum coriscense Rchb. f.
Phyllorkis coriscensis (Rchb. f.) Kuntze

Distribution: Gabon

Bulbophyllum cornu-cervi King & Pantl.

Distribution: Bhutan, India

Bulbophyllum cornutum (Blume) Rchb. f.
Bulbophyllum concavum Ames & C. Schweinf.
Ephippium cornutum Blume

Distribution: Indonesia, Philippines

Bulbophyllum corolliferum J. J. Sm.
Bulbophyllum corolliferum J. J. Sm. var. *atropurpureum* J. J. Sm.
Bulbophyllum curtisii (Hook. f.) J. J. Sm.
Bulbophyllum curtisii (Hook. f.) Ridl. var. *purpureum* J. J. Sm.
Bulbophyllum pulchellum Ridl. var. *purpureum* Ridl.
Cirrhopetalum concinnum Hook. f. var. *purpureum* Ridl.
Cirrhopetalum curtisii Hook. f.
Cirrhopetalum curtisii Hook. f. var. *lutescens* Garay
Cirrhopetalum curtisii Hook. f. var. *purpureum* Garay

Distribution: Indonesia, Malaysia, Thailand

Bulbophyllum correae Pabst

Distribution: Brazil

Part II: Accepted Names / Noms Acceptés / Nombres Aceptado

Bulbophyllum costatum Ames

Distribution: Philippines

Bulbophyllum coweniorum J. J. Verm. & P. O'Byrne

Distribution: Lao People's Democratic Republic

Bulbophyllum crassifolium Thwaites ex Trimen
  Phyllorkis crassifolia (Thwaites ex Trimen) Kuntze
  Trias crassifolia (Thwaites ex Trimen) C. S. Kumar

Distribution: Sri Lanka

Bulbophyllum crassinervium J. J. Sm.

Distribution: Indonesia

Bulbophyllum crassipes Hook. f.
  Bulbophyllum careyanum (Hook. f.) Sprengel var. *crassipes* (Hook.) Pradhan
  Phyllorkis crassipes (Hook. f.) Kuntze

Distribution: Bhutan, India, Malaysia, Myanmar, Thailand

Bulbophyllum crassipetalum H. Perrier

Distribution: Madagascar

Bulbophyllum crassissimum J. J. Sm.

Distribution: Indonesia

Bulbophyllum crassiusculifolium Aver.

Distribution: Viet Nam

Bulbophyllum crenilabium W. Kittr.
  Bulbophyllum pictum Schltr.

Distribution: Papua New Guinea

Bulbophyllum crepidiferum J. J. Sm.

Distribution: Indonesia

142

Bulbophyllum cribbianum Toscano
  Bulbophyllum micropetalum Barb. Rodr.
  Didactyle micropetala Barb. Rodr.

Distribution: Brazil

Bulbophyllum crispatisepalum P. Royen

Distribution: Papua New Guinea

Bulbophyllum croceum (Blume) Lindl.
  Bulbophyllum medusella Ridl.
  Diphyes crocea Blume
  Phyllorkis crocea (Blume) Kuntze

Distribution: Indonesia

Bulbophyllum crocodilus J. J. Sm.

Distribution: Indonesia

Bulbophyllum cruciatum J. J. Sm.
  Bulbophyllum immobile Schltr.
  Hapalochilus cruciatus (J. J. Sm.) Garay & W. Kittr.
  Hapalochilus immobilis (Schltr.) Garay, Hamer & Siegerist

Distribution: Indonesia, Papua New Guinea

Bulbophyllum cruciferum J. J. Sm.

Distribution: Indonesia

Bulbophyllum cruentum Garay, Hamer & Siegerist

Distribution: Papua New Guinea

Bulbophyllum cruttwellii J. J. Verm.

Distribution: Indonesia, Papua New Guinea

Bulbophyllum cryptanthoides J. J. Sm.

Distribution: Papua New Guinea

Bulbophyllum cryptanthum Schltr.

Distribution: Indonesia, Papua New Guinea

Bulbophyllum cryptophoranthus Garay
Bulbophyllum cryptophoranthoides Garay

Distribution: Philippines

Bulbophyllum cryptostachyum Schltr.

Distribution: Madagascar

Bulbophyllum cubicum Ames

Distribution: Philippines

Bulbophyllum cucullatum Schltr.
Hapalochilus cucullatus (Schltr.) Garay & W. Kittr.

Distribution: Papua New Guinea

Bulbophyllum culex Ridl.

Distribution: Indonesia

Bulbophyllum cumingii (Lindl.) Rchb. f.
Bulbophyllum amesianum (Rolfe) J. J. Sm.
Bulbophyllum stramineum Ames
Cirrhopetalum amesianum Rolfe
Cirrhopetalum cumingii Lindl.
Phyllorkis cumingii (Lindl.) Kuntze

Distribution: Philippines

Bulbophyllum cuneatum Rolfe

Distribution: Philippines

Bulbophyllum cuniculiforme J. J. Sm.
Hapalochilus cuniculiformis (J. J. Sm.) Garay & W. Kittr.

Distribution: Indonesia

Bulbophyllum cupreum Lindl.
Bulbophyllum pechei W.Bull
Phyllorkis cuprea (Lindl.) Kuntze

Distribution: India, Malaysia, Myanmar, Philippines, Thailand

Bulbophyllum curranii Ames

Distribution: Philippines

144

Part II: Accepted Names / Noms Acceptés / Nombres Aceptado

Bulbophyllum curvibulbum Frapp. ex Cordem.

Distribution: Réunion

Bulbophyllum curvicaule Schltr.

Distribution: Papua New Guinea

Bulbophyllum curvifolium Schltr.

Distribution: Madagascar

Bulbophyllum curvimentatum J. J. Verm.

Distribution: Equatorial Guinea, Sao Tome and Principe

Bulbophyllum cuspidipetalum J. J. Sm.

Distribution: Indonesia, Malaysia, Papua New Guinea

Bulbophyllum cyanotriche J. J. Verm.

Distribution: Malaysia

Bulbophyllum cyclanthum Schltr.

Distribution: Madagascar

Bulbophyllum cycloglossum Schltr.
  Peltopus cycloglossum (Schltr.) Szlach. & Marg.

Distribution: Indonesia, Papua New Guinea

Bulbophyllum cyclopense J. J. Sm.

Distribution: Indonesia

Bulbophyllum cyclophoroides J. J. Sm.

Distribution: Indonesia

Bulbophyllum cyclophyllum Schltr.

Distribution: Papua New Guinea

Part II: Accepted Names / Noms Acceptés / Nombres Aceptado

Bulbophyllum cyclosepalon Carr
Cirrhopetalum cyclosepalon (Carr) Garay, Hamer & Siegerist

Distribution: Malaysia

Bulbophyllum cylindraceum Lindl.
Bulbophyllum imbricatum Griff.
Phyllorkis cylindracea (Lindl.) Kuntze

Distribution: Bhutan, India, Indonesia

Bulbophyllum cylindricum King & Pantl.

Distribution: India

Bulbophyllum cylindrobulbum Schltr.
Bulbophyllum acutibrachium J. J. Sm.
Bulbophyllum angiense J. J. Sm.
Bulbophyllum constrictilabre J. J. Sm.
Bulbophyllum cordilabium P. Royen
Bulbophyllum disjunctibulbum J. J. Sm.
Bulbophyllum equivestigium Gilli
Bulbophyllum ferruginescens Schltr.
Bulbophyllum govidjoae Schltr.
Bulbophyllum imitans Schltr.
Bulbophyllum kempterianum Schltr.
Bulbophyllum longiserpens Schltr.
Bulbophyllum microcharis Schltr. (1905)
Bulbophyllum pallidiflavum Schltr.
Bulbophyllum perlongum Schltr.
Bulbophyllum remotum J. J. Sm.
Bulbophyllum sculptum J. J. Sm.
Bulbophyllum subalpinum P. Royen
Bulbophyllum uduense Schltr.

Distribution: Indonesia, Papua New Guinea, Solomon Islands

Bulbophyllum cylindrocarpum Frapp ex Cordem.

Distribution: Madagascar, Réunion

Bulbophyllum cylindrocarpum Frapp. ex Cordem. var. **andringitrense** Bosser

Distribution: Madagascar

Bulbophyllum cylindrocarpum Frapp ex Cordem. var. **aurantiacum** Frapp. ex Cordem.

Distribution: Réunion

Part II: Accepted Names / Noms Acceptés / Nombres Aceptado

Bulbophyllum cylindrocarpum Frapp ex Cordem. var. **olivacea** Frapp. ex Cordem.

Distribution: Réunion

Bulbophyllum dagamense Ames

Distribution: Philippines

Bulbophyllum dalatense Gagnep.

Distribution: Viet Nam

Bulbophyllum danii Pérez-Vera

Distribution: Côte d'Ivoire

Bulbophyllum dasypetalum Rolfe ex Ames
Bulbophyllum vanoverberghii Ames

Distribution: Philippines

Bulbophyllum dasyphyllum Schltr.

Distribution: Papua New Guinea

Bulbophyllum dawongense J. J. Sm.

Distribution: Indonesia

Bulbophyllum dayanum Rchb. f.
Bulbophyllum dyphoniae Tixier
Bulbophyllum hispidum Ridl.
Phyllorkis dayana (Rchb. f.) Kuntze
Trias dayanum Grant

Distribution: Cambodia, Malaysia, Myanmar, Thailand, Viet Nam

Bulbophyllum dearei (Hort.) Rchb.f.
Bulbophyllum dearei Veitch
Bulbophyllum godseffianum Weathers
Bulbophyllum goebelianum Kraenzl.
Bulbophyllum punctatum Ridl.
Bulbophyllum reticosum Ridl.
Phyllorkis dearei (Rchb. f.) Kuntze
Sarcopodium dearei Hort.

Distribution: Indonesia, Malaysia, Philippines

Part II: Accepted Names / Noms Acceptés / Nombres Aceptado

Bulbophyllum debile Bosser

Distribution: Madagascar

Bulbophyllum debrincatiae J. J. Verm.

Distribution: Philippines

Bulbophyllum debruynii J. J. Sm.

Distribution: Indonesia

Bulbophyllum decarhopalon Schltr.
  Epicranthes decarhopalon (Schltr.) Garay & W. Kittr.

Distribution: Papua New Guinea

Bulbophyllum decaryanum H. Perrier

Distribution: Madagascar

Bulbophyllum decatriche J. J. Verm.

Distribution: Indonesia

Bulbophyllum deceptum Ames

Distribution: Philippines

Bulbophyllum decumbens Schltr.

Distribution: Papua New Guinea

Bulbophyllum decurrentilobum J. J. Verm. & P. O'Byrne

Distribution: Indonesia

Bulbophyllum decurviscapum J. J. Sm.

Distribution: Indonesia

Bulbophyllum decurvulum Schltr.
  Hapalochilus decurvulus (Schltr.) Garay & W. Kittr.

Distribution: Papua New Guinea

Part II: Accepted Names / Noms Acceptés / Nombres Aceptado

Bulbophyllum dekockii J. J. Sm.
  Bulbophyllum jugicola P. Royen

Distribution: Indonesia, Papua New Guinea

Bulbophyllum delicatulum Schltr.

Distribution: Indonesia

Bulbophyllum delitescens Hance
  Cirrhopetalum delitescens (Hance) Rolfe
  Cirrhopetalum mirificum Gagnep.

Distribution: China, India, Viet Nam

Bulbophyllum deltoideum Ames & C. Schweinf.
  Bulbophyllum angustatifolium J. J. Sm.

Distribution: Indonesia, Philippines

Bulbophyllum deminutum J. J. Sm.

Distribution: Indonesia

Bulbophyllum dempoense J. J. Sm.

Distribution: Indonesia

Bulbophyllum dendrobioides J. J. Sm.

Distribution: Indonesia

Bulbophyllum dendrochiloides Schltr.

Distribution: Indonesia, Papua New Guinea

Bulbophyllum dennisii J. J. Wood
  Hyalosema dennisii (J. J. Wood) Rysy

Distribution: Solomon Islands

Bulbophyllum densibulbum W. Kittr.
  Bulbophyllum cylindrocarpum Schltr.

Distribution: Papua New Guinea

Part II: Accepted Names / Noms Acceptés / Nombres Aceptado

Bulbophyllum densifolium Schltr.

Distribution: Papua New Guinea

Bulbophyllum densum Thouars
  Phyllorkis densa Thouars

Distribution: Mauritius, Réunion

Bulbophyllum denticulatum Rolfe

Distribution: Côte d'Ivoire, Liberia, Sierra Leone

Bulbophyllum dentiferum Ridl.
  Cirrhopetalum dentiferum (Ridl.) Garay, Hamer & Siegerist

Distribution: Malaysia, Thailand

Bulbophyllum dependens Schltr.

Distribution: Papua New Guinea

Bulbophyllum depressum King & Pantl.
  Bulbophyllum acutum J. J. Sm.
  Bulbophyllum hastatum T. Tang & F. T. Wang

Distribution: China, India, Indonesia, Thailand

Bulbophyllum desmanthum Tuyama

Distribution: Palau

Bulbophyllum desmotrichoides Schltr.
  Bulbophyllum breviscapum J. J. Sm.
  Bulbophyllum planifolium W. Kittr.

Distribution: Indonesia, Papua New Guinea

Bulbophyllum devium J. B. Comber

Distribution: Indonesia

Bulbophyllum devogelii J. J. Verm.

Distribution: Indonesia

Part II: Accepted Names / Noms Acceptés / Nombres Aceptado

Bulbophyllum dewildei J. J. Verm.

Distribution: Indonesia

Bulbophyllum dhaninivatii Seidenf.

Distribution: Thailand

Bulbophyllum dianthum Schltr.

Distribution: Indonesia

Bulbophyllum dibothron J. J. Verm. & A. L. Lamb

Distribution: Malaysia

Bulbophyllum dichaeoides Schltr.

Distribution: Papua New Guinea

Bulbophyllum dichilus Schltr.

Distribution: Papua New Guinea

Bulbophyllum dichotomum J. J. Sm.

Distribution: Indonesia, Papua New Guinea, Solomon Islands, Vanuatu

Bulbophyllum dickasonii Seidenf.

Distribution: Myanmar, Thailand

Bulbophyllum dictyoneuron Schltr.

Distribution: Papua New Guinea

Bulbophyllum didymotropis Seidenf.

Distribution: Thailand

Bulbophyllum digoelense J. J. Sm.

Distribution: Indonesia, Papua New Guinea

Bulbophyllum diplantherum Carr

Distribution: Malaysia

Part II: Accepted Names / Noms Acceptés / Nombres Aceptado

Bulbophyllum diploncos Schltr.

Distribution: Indonesia

Bulbophyllum dischidiifolium J. J. Sm.

Distribution: Indonesia

Bulbophyllum dischorense Schltr.

Distribution: Papua New Guinea

Bulbophyllum discilabium H. Perrier

Distribution: Madagascar

Bulbophyllum discolor Schltr.
  Bulbophyllum commocardium Garay, Hamer & Siegerist
  Peltopus discolor (Schltr.) Szlach. & Marg.

Distribution: Indonesia, Papua New Guinea

Bulbophyllum discolor Schltr. var. **cubitale** J. J. Verm.
  Peltopus cubitalis (J. J. Verm.) Szlach. & Marg.

Distribution: Papua New Guinea

Bulbophyllum disjunctum Ames & C. Schweinf.

Distribution: Indonesia, Malaysia, Philippines

Bulbophyllum dissitiflorum Seidenf.

Distribution: Lao People's Democratic Republic, Thailand

Bulbophyllum dissolutum Ames

Distribution: Philippines

Bulbophyllum distichobulbum P. J. Cribb

Distribution: Samoa

Bulbophyllum distichum Schltr.

Distribution: Papua New Guinea

152

Part II: Accepted Names / Noms Acceptés / Nombres Aceptado

Bulbophyllum divaricatum H. Perrier

Distribution: Madagascar

Bulbophyllum djamuense Schltr.

Distribution: Papua New Guinea

Bulbophyllum dolabriforme J. J. Verm.

Distribution: Cameroon, Nigeria

Bulbophyllum dolichoblepharon (Schltr.) J. J. Sm.
  Cirrhopetalum dolichoblepharon Schltr.

Distribution: Indonesia

Bulbophyllum dolichoglottis Schltr.
  Hapalochilus dolichoglottis (Schltr.) Garay & W. Kittr.

Distribution: Papua New Guinea

Bulbophyllum doryphoroide Ames

Distribution: Philippines

Bulbophyllum dracunculus J. J. Verm.

Distribution: Malaysia

Bulbophyllum dransfieldii J. J. Verm.

Distribution: Indonesia

Bulbophyllum drepanosepalum J. J. Verm.

Distribution: Papua New Guinea

Bulbophyllum dryadum Schltr.

Distribution: Papua New Guinea

Bulbophyllum dryas Ridl.

Distribution: Indonesia, Malaysia

Part II: Accepted Names / Noms Acceptés / Nombres Aceptado

Bulbophyllum drymoglossum Maxim.
 Bulbophyllum somai Hayata

Distribution: China, Japan, Republic of Korea, Taiwan, Province of China

Bulbophyllum dschischungarense Schltr.

Distribution: Papua New Guinea

Bulbophyllum dubium J. J. Sm.

Distribution: Indonesia

Bulbophyllum dunstervillei Garay

 Distribution: Venezuela

Bulbophyllum dusenii Kraenzl.

Distribution: Brazil

Bulbophyllum ebracteolatum Kraenzl.

Distribution: Philippines

Bulbophyllum ebulbe Schltr.
 Bulbophyllum nigroscapum Ames
 Bulbophyllum polypodioides Schltr.

Distribution: Fiji, India, Indonesia, New Caledonia, Papua New Guinea, Samoa, Solomon Islands, Vanuatu

Bulbophyllum echinochilum Kraenzl.

Distribution: Philippines

Bulbophyllum echinolabium J. J. Sm.

Distribution: Indonesia

Bulbophyllum echinulus Seidenf.

Distribution: Thailand

Bulbophyllum eciliatum Schltr.

Distribution: Papua New Guinea

Bulbophyllum ecornutum (J. J. Sm.) J. J. Sm.
Bulbophyllum cornutum (Blume) Rchb. f. var. *ecornutum* J. J. Sm.
Bulbophyllum ecornutum J. J. Sm. var. *daliense* J. J. Sm.
Bulbophyllum ecornutum J. J. Sm. var. *teloense* J. J. Sm.

Distribution: Indonesia, Thailand

Bulbophyllum edentatum H. Perrier

Distribution: Madagascar

Bulbophyllum elachanthe J. J. Verm.

Distribution: Indonesia, Malaysia

Bulbophyllum elaphoglossum Schltr.

Distribution: Indonesia

Bulbophyllum elasmatopus Schltr.

Distribution: Papua New Guinea

Bulbophyllum elassoglossum Siegerist

Distribution: Philippines

Bulbophyllum elassonotum Summerh.

Distribution: India, Thailand, Viet Nam

Bulbophyllum elatum (Hook. f.) J. J. Sm.
Cirrhopetalum elatum Hook. f.
Phyllorkis elata (Hook. f.) Kuntze

Distribution: Bhutan, India, Nepal, Viet Nam

Bulbophyllum elbertii J. J. Sm.
Hyalosema elbertii (J. J. Sm.) Rolfe

Distribution: Indonesia

Bulbophyllum electrinum Seidenf.
Bulbophyllum hirundinis (Gagnep.) Seidenf. var. *electrinum* (Seidenf.) S. S. Ying
Cirrhopetalum aurantiacum W. W. Sm.
Cirrhopetalum melinanthum Schltr.

Distribution: China

Bulbophyllum elegans Gardner ex Thwaites
Bulbophyllum balaeniceps Rchb. f. †
Phyllorkis elegans (Gardn. Ex Thw.) Kuntze

Distribution: India, Sri Lanka

Bulbophyllum elegantius Schltr.

Distribution: Papua New Guinea

Bulbophyllum elegantulum (Rolfe) J. J. Sm.
Bulbophyllum pileolatum (Klinge) J. J. Sm.
Cirrhopetalum elegantulum Rolfe
Cirrhopetalum pileolatum Klinge

Distribution: India

Bulbophyllum elephantinum J. J. Sm.
Hyalosema elephantinum (J. J. Sm.) Rolfe

Distribution: Indonesia

Bulbophyllum elevatopunctatum J. J. Sm.

Distribution: Indonesia

Bulbophyllum elizae (F. Muell.) Benth.
Adelopetalum elisae (F. Muell.) D. L. Jones & M. A. Clem.
Cirrhopetalum elisae F. Muell.
Phyllorkis elisae (F. Muell.) Kuntze

Distribution: Australia

Bulbophyllum elliae Rchb. f.
Cirrhopetalum elliae (Rchb. f.) Trimen
Cirrhopetalum roseum Jayaw.
Cirrhopetalum wightii Thwaites
Phyllorkis elliae (Rchb. f.) Kuntze

Distribution: Sri Lanka

Bulbophyllum elliottii Rolfe
Bulbophyllum malawiense B. Morris

Distribution: Democratic Republic of the Congo, Madagascar, Malawi, South Africa, United Republic of Tanzania, Zambia, Zimbabwe

Bulbophyllum ellipticifolium J. J. Sm.

Distribution: Indonesia

156

Part II: Accepted Names / Noms Acceptés / Nombres Aceptado

Bulbophyllum ellipticum Schltr.

Distribution: Papua New Guinea

Bulbophyllum elmeri Ames

Distribution: Philippines

Bulbophyllum elodeiflorum J. J. Sm.

Distribution: Indonesia

Bulbophyllum elongatum (Blume) Hassk.
  Bulbophyllum gigas Ridl.
  Bulbophyllum sceptrum Rchb. f.
  Cirrhopetalum elongatum (Blume) Lindl.
  Ephippium elongatum Blume
  Phyllorkis elongata (Blume) Kuntze
  Phyllorkis sceptrum (Rchb. f.) Kuntze

Distribution: Brunei Darussalam, Indonesia, Malaysia, Papua New Guinea, Philippines

Bulbophyllum emarginatum (Finet) J. J. Sm.
  Bulbophyllum brachypodium A. S. Rao & Balakr. var. *geei* A. S. Rao & Balakr.
  Bulbophyllum yoksunense J. J. Sm. var. *geei* (Rao & Balak.) S. S. R. Bennet
  Cirrhopetalum emarginatum Finet

Distribution: Bhutan, China, India, Myanmar, Viet Nam

Bulbophyllum emiliorum Ames & Quisumb.

Distribution: Philippines

Bulbophyllum encephalodes Summerh.

Distribution: Burundi, Cameroon, Democratic Republic of the Congo, Kenya, Malawi, Uganda, United Republic of Tanzania, Zambia, Zimbabwe

Bulbophyllum endotrachys Schltr.

Distribution: Papua New Guinea

Bulbophyllum ensiculiferum J. J. Sm.

Distribution: Indonesia

Bulbophyllum entomonopsis J. J. Verm.

Distribution: Papua New Guinea

Part II: Accepted Names / Noms Acceptés / Nombres Aceptado

Bulbophyllum epapillosum Schltr.

Distribution: Papua New Guinea

Bulbophyllum epibulbon Schltr.

Distribution: Papua New Guinea, Solomon Islands

Bulbophyllum epicrianthes Hook. f.
Bulbophyllum javanicum (Blume) J. J. Sm.
Epicranthes javanica Blume
Phyllorkis javanica (Blume) Kuntze

Distribution: Indonesia, Malaysia

Bulbophyllum epicrianthes Hook. f. var. **sumatranum** (J. J. Sm.) J. J. Verm.
Bulbophyllum javanicum (Blume) J. J. Sm. var. *sumatranum* J. J. Sm.

Distribution: Indonesia

Bulbophyllum epiphytum Barb. Rodr.

Distribution: Brazil

Bulbophyllum erectum Thouars
Bulbophyllum calamarioides Schltr.
Bulbophyllum lobulatum Schltr.
Phyllorkis erecta Thouars

Distribution: Madagascar, Mauritius

Bulbophyllum erinaceum Schltr.

Distribution: Indonesia, Papua New Guinea

Bulbophyllum erioides Schltr.

Distribution: Papua New Guinea

Bulbophyllum erosipetalum C. Schweinf.
Osyricera erosipetala (C. Schweinf.) Garay, Hamer & Siegerist

Distribution: Philippines

Bulbophyllum erratum Ames

Distribution: Philippines

158

Part II: Accepted Names / Noms Acceptés / Nombres Aceptado

Bulbophyllum erythroglossum Bosser

Distribution: Madagascar

Bulbophyllum erythrostachyum Rolfe

Distribution: Madagascar

Bulbophyllum erythrostictum Ormerod
  Bulbophyllum polystictum Schltr.

Distribution: Papua New Guinea

Bulbophyllum escritorii Ames

Distribution: Philippines

Bulbophyllum euplepharum Rchb. f.
  Phyllorkis eublephara (Rchb. f.) Kuntze

Distribution: Bhutan, India, Papua New Guinea

Bulbophyllum evansii M. R. Henderson

Distribution: Malaysia

Bulbophyllum evasum Hunt & Rupp
  Karorchis evasa (T. E. Hunt & Rupp) M. A. Clem. & D. L. Jones
  Kaurorchis evasa (T. E. Hunt & Rupp) M. A. Clem. & D. L. Jones

Distribution: Australia

Bulbophyllum evrardii Gagnep.

Distribution: Viet Nam

Bulbophyllum exaltatum Lindl.
  Didactyle exaltata (Lindl.) Lindl.
  Phyllorkis exaltata (Lindl.) Kuntze

Distribution: Bolivia, Brazil, Guyana, Venezuela

Bulbophyllum exasperatum Schltr.

Distribution: Indonesia, Papua New Guinea

Bulbophyllum exiguiflorum Schltr.

Distribution: Papua New Guinea

Bulbophyllum exiguum F. Muell.
Adelopetalum exiguum (F. Muell.) D. L. Jones & M. A. Clem.
Dendrobium caleyi A. Cunn.
Dendrobium exiguum (F. Muell.) Kuntze
Phyllorkis exigua (F. Muell.) Kuntze

Distribution: Australia

Bulbophyllum exile Ames

Distribution: Philippines

Bulbophyllum exilipes Schltr.

Distribution: Papua New Guinea

Bulbophyllum expallidum J. J. Verm.

Distribution: Democratic Republic of the Congo, Malawi, Rwanda, United Republic of Tanzania, Zambia

Bulbophyllum exquisitum Ames
Bulbophyllum macgregorii Ames

Distribution: Philippines

Bulbophyllum facetum Garay, Hamer & Siegerist

Distribution: Philippines

Bulbophyllum falcatocaudatum J. J. Sm.

Distribution: Indonesia

Bulbophyllum falcatum (Lindl.) Rchb. f.
Bulbophyllum dahlemense Schltr.
Bulbophyllum hemirhachis (Pfitzer) De Wild
Bulbophyllum leptorrhachis Schltr.
Bulbophyllum oxyodon Rchb. f.
Bulbophyllum ugandae (Rolfe) De Wild.
Megaclinium endotrachys Kraenzl.
Megaclinium falcatum Lindl.
Megaclinium hemirhachis Pfitzer
Megaclinium oxyodon Rchb. f.
Megaclinium ugandae Rolfe
Phyllorkis falcata (Lindl.) Kuntze

Distribution: Cameroon, Central African Republic, Côte d'Ivoire, Equatorial Guinea, Ghana, Guinea, Liberia, Nigeria, Sierra Leone, Togo, Uganda

Bulbophyllum falcatum (Lindl.) Rchb. f. var. **bufo** (Lindl.) Govaerts
  Bulbophyllum bakossorum Schltr.
  Bulbophyllum bufo (Lindl.) Rchb. f.
  Bulbophyllum deistelianum (Kraenzl.) Schltr.
  Bulbophyllum falcatum (Lindl.) Rchb. f. var. *bufo* (Lindl.) J. J. Verm.
  Bulbophyllum longibulbum Schltr.
  Bulbophyllum lubiense De Wild.
  Bulbophyllum sereti De Wild.
  Megaclinium bufo Lindl.
  Megaclinium deistelianum Kraenzl.
  Megaclinium gentilii De Wild.
  Megaclinium sereti De Wild.
  Phyllorkis bufo (Lindl.) Kuntze

Distribution: Cameroon, Côte d'Ivoire, Democratic Republic of the Congo, Ghana, Guinea, Liberia, Nigeria, Sierra Leone

Bulbophyllum falcatum (Lindl.) Rchb. f. var. **velutinum** (Lindl.) J. J. Verm.
  Bulbophyllum arnoldianum (De Wild.) De Wild.
  Bulbophyllum brixhei De Wild.
  Bulbophyllum fractiflexum Kraenzl.
  Bulbophyllum kewense Schltr.
  Bulbophyllum kewense Schltr. var. *purpureum* De Wild.
  Bulbophyllum lanuriense De Wild.
  Bulbophyllum melanorrhachis (Rchb. f.) De Wild.
  Bulbophyllum millenii (Rolfe) Schltr.
  Bulbophyllum minutum (Rolfe) Engler
  Bulbophyllum minutum (Rolfe) Engler var. *purpureum* (De Wild.) De Wild.
  Bulbophyllum rhizophorae Lindl.
  Bulbophyllum simoni Summerh.
  Bulbophyllum solheidi De Wild.
  Bulbophyllum velutinum (Lindl.) Rchb. f.
  Megaclinium angustum Rolfe
  Megaclinium arnoldianum De Wild.
  Megaclinium brixhei De Wild.
  Megaclinium lanuriense De Wild.
  Megaclinium lasianthum Kraenzl.
  Megaclinium melanorrhachis Rchb. f.
  Megaclinium millenii Rolfe
  Megaclinium minutum Rolfe
  Megaclinium minutum Rolfe var. *purpureum* De Wild.
  Megaclinium solheidi De Wild.
  Megaclinium velutinum Lindl.
  Phyllorkis rhizophorae (Lindl.) Kuntze
  Phyllorkis velutina (Lindl.) Kuntze

Distribution: Cameroon, Côte d'Ivoire, Democratic Republic of the Congo, Equatorial Guinea, Gabon, Ghana, Liberia, Nigeria, Sao Tome and Principe, Sierra Leone

Bulbophyllum falcibracteum Schltr.
  Bulbophyllum streptosepalum Schltr.

Distribution: Papua New Guinea

Part II: Accepted Names / Noms Acceptés / Nombres Aceptado

Bulbophyllum falciferum J. J. Sm.

Distribution: Indonesia

Bulbophyllum falcifolium Schltr.

Distribution: Papua New Guinea

Bulbophyllum falcipetalum Lindl.
Bulbophyllum brauni Kraenzl.
Bulbophyllum lutescens (Rolfe) De Wild.
Megaclinium lutescens Rolfe
Phyllorkis falcipetala (Lindl.) Kuntze

Distribution: Cameroon, Côte d'Ivoire, Gabon, Ghana, Nigeria

Bulbophyllum falculicorne J. J. Sm.

Distribution: Indonesia

Bulbophyllum fallax Rolfe

Distribution: India

Bulbophyllum farinulentum J. J. Sm.
Bulbophyllum noeanum Kerr

Distribution: Indonesia, Malaysia, Thailand

Bulbophyllum farinulentum J. J. Sm. subsp. **densissimum** (Carr) J. J. Verm.
Bulbophyllum densissimum Carr

Distribution: Malaysia

Bulbophyllum farreri (W. W. Sm.) Seidenf.
Cirrhopetalum farreri W. W. Sm.

Distribution: Myanmar, Viet Nam

Bulbophyllum fasciatum Schltr.
Hapalochilus fasciatus (Schltr.) Garay & W. Kittr.

Distribution: Papua New Guinea

Bulbophyllum fasciculatum Schltr.

Distribution: Indonesia, Papua New Guinea

Bulbophyllum fasciculiferum Schltr.

Distribution: Papua New Guinea

Bulbophyllum faunula Ridl.

Distribution: Indonesia

Bulbophyllum fayi J. J. Verm.

Distribution: Cameroon

Bulbophyllum fenixii Ames

Distribution: Philippines

Bulbophyllum ferkoanum Schltr.

Distribution: Madagascar

Bulbophyllum fibratum (Gagnep.) Seidenf.
  Bulbophyllum fibratum (Gagnep.) Bân & D. H. Duong
  Cirrhopetalum fibratum Gagnep.
  Cirrhopetalum wallichii J. Graham

Distribution: Viet Nam

Bulbophyllum fibrinum J. J. Sm.
  Hapalochilus fibrinus (J. J. Sm.) Garay & W. Kittr.

Distribution: Indonesia

Bulbophyllum filamentosum Schltr.

Distribution: Papua New Guinea

Bulbophyllum filicaule J. J. Sm.

Distribution: Indonesia

Bulbophyllum filicoides Ames

Distribution: Philippines

Bulbophyllum filifolium Borba & E.C.Smidt

Distribution: Brazil

Bulbophyllum filovagans Carr

Distribution: Indonesia

Bulbophyllum fimbriatum (Lindl.) Rchb. f.
Cirrhopetalum fimbriatum Lindl.
Phyllorkis fimbriata (Lindl.) Kuntze

Distribution: India, Madagascar

Bulbophyllum finetii Szlach. & Olszewski

Distribution: Cameroon, Gabon

Bulbophyllum finisterrae Schltr.

Distribution: Papua New Guinea

Bulbophyllum fischeri Seidenf.
Bulbophyllum gamblei (Hook. f.) J. J. Sm.
Bulbophyllum thomsonii (Hook. f.) J. J. Sm.
Cirrhopetalum gamblei Hook. f.
Cirrhopetalum thomsonii Hook. f.
Phyllorkis gamblei (Hook. f.) Kuntze
Phyllorkis hookeri Kuntze

Distribution: India, Sri Lanka, Viet Nam

Bulbophyllum fissibrachium J. J. Sm.#

Distribution: Indonesia

Bulbophyllum fissipetalum Schltr.

Distribution: Papua New Guinea

Bulbophyllum flabellum-veneris (J. Koenig) Seidenf. & Ormerod ex Aver.
Bulbophyllum andersonii Kurz
Bulbophyllum gamosepalum (Griff.) J. J. Sm.
Bulbophyllum griffithianum C. S. P. Parish & Rchb. f.
Bulbophyllum lepidum (Blume) J. J. Sm.
Bulbophyllum lepidum (Blume) J. J. Sm. var. *insigne* J. J. Sm.
Bulbophyllum stramineum Ames var. *purpuratum* Guillaumin
Bulbophyllum viscidum J. J. Sm.
Cirrhopetalum andersonii Kurz
Cirrhopetalum ciliatum Klinge
Cirrhopetalum flabellum-veneris (J. Koenig) Seidenf. & Ormerod
Cirrhopetalum gagnepainii Guillaumin
Cirrhopetalum gamosepalum Griff.
Cirrhopetalum gamosepalum Griff. var. *angustum* Ridl.
Cirrhopetalum lepidum (Blume) J. J. Sm. var. *angustum* Ridl.
Cirrhopetalum lepidum (Blume) Schltr.

Cirrhopetalum siamense Rolfe ex Downie
Cirrhopetalum stramineum Teijsm. & Binn. var. *purpureum* Gagnep.
Cirrhopetalum tenuicaule Rolfe
Cirrhopetalum viscidum (J. J. Sm.) Garay, Hamer & Siegerist
Ephippium lepidum Blume
Epidendrum flabellum-veneris. J. Koenig
Phyllorkis gamosepala (Griff.) Kuntze

Distribution: Viet Nam

Bulbophyllum flagellare Schltr.

Distribution: Papua New Guinea

Bulbophyllum flammuliferum Ridl.

Distribution: Indonesia, Malaysia

Bulbophyllum flavescens (Blume) Lindl.
Bulbophyllum adenopetalum Lindl.
Bulbophyllum barrinum Ridl.
Bulbophyllum exiliscapum J. J. Sm.
Bulbophyllum flavescens (Blume) Lindl. var. *temelenense* J. J. Sm.
Bulbophyllum flavescens (Blume) Lindl. var. *triflorum* J. J. Sm.
Bulbophyllum lanceolatum Ames & C. Schweinf.
Bulbophyllum montigenum Ridl.
Bulbophyllum puberulum Ridl.
Bulbophyllum ramosii Ames
Bulbophyllum semperflorens J. J. Sm.
Bulbophyllum simulacrum Ames
Diphyes flavescens Blume
Phyllorkis adenopetala (Lindl.) Kuntze
Phyllorkis flavescens (Blume) Kuntze

Distribution: Indonesia, Malaysia, Philippines

Bulbophyllum flavicolor J. J. Sm.

Distribution: Indonesia

Bulbophyllum flavidiflorum Carr
Bulbophyllum obtusum (Blume) Lindl. var. *robustum* J. J. Sm.

Distribution: Indonesia

Bulbophyllum flaviflorum (Liu & Su) Seidenf.
Cirrhopetalum flaviflorum Liu & Su

Distribution: Taiwan, Province of China, Viet Nam

Bulbophyllum flavofimbriatum J. J. Sm.
  Epicranthes flavofimbriata (J. J. Sm.) Garay & W. Kittr.

Distribution: Indonesia

Bulbophyllum flavorubellum J. J. Verm. & P. O'Byrne

Distribution: Malaysia

Bulbophyllum flavum Schltr.

Distribution: Papua New Guinea

Bulbophyllum fletcherianum Rolfe
  Bulbophyllum fletcherianum Hort.
  Bulbophyllum fletcherianum Pearson
  Cirrhopetalum fletcheranum (Pearson) Rolfe

Distribution: Papua New Guinea

Bulbophyllum flexuosum Schltr.

Distribution: Papua New Guinea

Bulbophyllum floribundum J. J. Sm.

Distribution: Indonesia

Bulbophyllum florulentum Schltr.

Distribution: Madagascar

Bulbophyllum foetidilabrum Ormerod

Distribution: Papua New Guinea

Bulbophyllum foetidolens Carr

Distribution: Indonesia

Bulbophyllum foetidum Schltr.

Distribution: Papua New Guinea

Bulbophyllum foetidum Schltr. var. **grandiflorum** J. J. Sm.

Distribution: Indonesia

166

Part II: Accepted Names / Noms Acceptés / Nombres Aceptado

Bulbophyllum folliculiferum J. J. Sm.
  Vesicisepalum folliculiferum (J. J. Sm.) Garay, Hamer & Siegerist

Distribution: Indonesia

Bulbophyllum fonsflorum J. J. Verm.

Distribution: Papua New Guinea

Bulbophyllum foraminiferum J. J. Verm.

Distribution: Malaysia

Bulbophyllum forbesii Schltr.
  Bulbophyllum cornutum Ridl.

Distribution: Papua New Guinea

Bulbophyllum fordii (Rolfe) J. J. Sm.
  Cirrhopetalum fordii Rolfe

Distribution: Papua New Guinea

Bulbophyllum formosanum (Rolfe) Nakajima
  Bulbophyllum formosanum (Rolfe) S. S. Ying
  Bulbophyllum formosanum (Rolfe) Seidenf.
  Cirrhopetalum formosanum Rolfe

Distribution: Taiwan, Province of China

Bulbophyllum formosum Schltr.
  Hapalochilus formosus (Schltr.) Garay & W. Kittr.

Distribution: Papua New Guinea

Bulbophyllum forresti Seidenf.
  Cirrhopetalum aemulum W. W. Sm.
  Rhytionanthos aemulum (W. W. Sm.) Garay, Hamer & Siegerist

Distribution: China, India, Thailand

Bulbophyllum forsythianum Kraenzl.

Distribution: Madagascar

Bulbophyllum fractiflexum J. J. Sm.
Bulbophyllum effusum Schltr.
Bulbophyllum fractiflexoides Schltr.
Bulbophyllum genybrachyum Schltr.
Bulbophyllum lamprobulbon Schltr.
Bulbophyllum linearipetalum J. J. Sm.

Distribution: Indonesia, Papua New Guinea, Solomon Islands

Bulbophyllum fractiflexum J. J. Sm. subsp. **salomonense** J. J. Verm. & A. L. Lamb

Distribution: Papua New Guinea, Solomon Islands

Bulbophyllum francoisii H. Perrier

Distribution: Madagascar

Bulbophyllum francoisii H. Perrier var. **andrangense** (H. Perrier) Bosser
Bulbophyllum andrangense H. Perrier

Distribution: Madagascar

Bulbophyllum frappieri Schltr.
Bulbophyllum compressum Frapp. ex Cordem.
Bulbophyllum frappieri A. D. Hawkes

Distribution: Réunion

Bulbophyllum fraudulentum Garay, Hamer & Siegerist
Hyalosema fraudulentum (Garay, Hamer & Siegerist) Rysy

Distribution: Indonesia

Bulbophyllum fritillariiflorum J. J. Sm.
Hyalosema fritillariflorum (J. J. Sm.) Rolfe

Distribution: Indonesia

Bulbophyllum frostii Summerh.
Bulbophyllum bootanoides (Guill.) Seidenf.
Cirrhopetalum bootanoides Guillaumin
Cirrhopetalum frostii (Summerh.) Garay, Hamer & Siegerist

Distribution: Viet Nam

Bulbophyllum frustrans J. J. Sm.
Hapalochilus frustrans (J. J. Sm.) Garay & W. Kittr.

Distribution: Indonesia

168

Bulbophyllum fruticicola Schltr.
Fruticicola albopunctata (Schltr.) M. A. Clem. & D. L. Jones

Distribution: Papua New Guinea

Bulbophyllum fukuyamae Tuyama

Distribution:Palau

Bulbophyllum fulgens J. J. Verm.

Distribution: Indonesia

Bulbophyllum fulvibulbum J. J. Verm.

Distribution: Indonesia, Malaysia

Bulbophyllum funingense Z. H. Tsi & H. C. Chen

Distribution: China, Viet Nam

Bulbophyllum furcatum Aver.

Distribution: Viet Nam

Bulbophyllum furcillatum J. J. Verm. & P. O'Byrne

Distribution: Indonesia

Bulbophyllum fuscatum Schltr.

Distribution: Papua New Guinea

Bulbophyllum fusciflorum Schltr.

Distribution: Papua New Guinea

Bulbophyllum fusco-purpureum Wight
Phyllorkis fuscopurpurea (Wight) Kuntze

Distribution: India

Bulbophyllum fuscum Lindl.
Bulbophyllum ogoouense Guillaumin
Phyllorkis fusca (Lindl.) Kuntze

Distribution: Angola, Cameroon, Central African Republic, Côte d'Ivoire, Democratic Republic of the Congo, Gabon, Guinea, Liberia, Nigeria, Sierra Leone

Part II: Accepted Names / Noms Acceptés / Nombres Aceptado

Bulbophyllum fuscum Lindl. var. **melinostachyum** (Schltr) J. J. Verm.
Bulbophyllum melinostachyum Schltr.
Bulbophyllum obanense Rendle

Distribution: Cameroon, Côte d'Ivoire, Democratic Republic of the Congo, Equatorial Guinea, Gabon, Liberia, Malawi, Mozambique, Nigeria, Sierra Leone, Uganda, United Republic of Tanzania, Zambia, Zimbabwe

Bulbophyllum futile J. J. Sm.

Distribution: Indonesia

Bulbophyllum gadgarrense Rupp
Oxysepala gadgarrensis (Rupp) D. L. Jones & M. A. Clem.

Distribution: Australia

Bulbophyllum gajoense J. J. Sm.

Distribution: Indonesia

Bulbophyllum galactanthum Schltr.

Distribution: Papua New Guinea

Bulbophyllum galliaheneum P. Royen

Distribution: Indonesia

Bulbophyllum gautierense J. J. Sm.

Distribution: Indonesia

Bulbophyllum gemma-reginae J. J. Verm.

Distribution: Indonesia, Malaysia

Bulbophyllum geniculiferum J. J. Sm.
Hapalochilus geniculifer (J. J. Sm.) Garay & W. Kittr.

Distribution: Indonesia

Bulbophyllum geraense Rchb. f.
Bulbophyllum antenniferum (Lindl.) Rchb. f.
Didactyle antennifera Lindl.
Phyllorkis geraensis (Rchb. f.) Kuntze

Distribution: Brazil, Guyana, Venezuela

Part II: Accepted Names / Noms Acceptés / Nombres Aceptado

Bulbophyllum gerlandianum Kraenzl.

Distribution: Indonesia

Bulbophyllum gibbolabium Seidenf.

Distribution: Thailand

Bulbophyllum gibbosum (Blume) Lindl.
Bulbophyllum igneocentrum J. J. Sm.
Bulbophyllum igneocentrum J. J. Sm. var. *lativaginatum* J. J. Sm.
Bulbophyllum magnivaginatum Ames & C. Schweinf.
Bulbophyllum pangerangi Rchb. f.
Bulbophyllum selangorense Ridl.
Diphyes gibbosa Blume
Phyllorkis gibbosa (Blume) Kuntze

Distribution: Indonesia, Malaysia

Bulbophyllum gibbsiae Rolfe
Bulbophyllum minutiflorum Ames & C. Schweinf.

Distribution: Indonesia, Philippines

Bulbophyllum gilgianum Kraenzl.

Distribution: United Republic of Tanzania

Bulbophyllum gilvum J. J. Verm. & A. L. Lamb

Distribution: Malaysia

Bulbophyllum gimagaanense Ames

Distribution: Philippines

Bulbophyllum giriwoensc J. J. Sm.

Distribution: Indonesia

Bulbophyllum gjellerupii J. J. Sm.

Distribution: Indonesia

Bulbophyllum glabrum Schltr.

Distribution: Papua New Guinea

Bulbophyllum gladiatum Lindl.
 Phyllorkis gladiata (Lindl.) Kuntze

Distribution: Brazil

Bulbophyllum glanduliferum Schltr.

Distribution: Papua New Guinea

Bulbophyllum glandulosum Ames

Distribution: Philippines

Bulbophyllum glaucifolium J. J. Verm.

Distribution: Indonesia

Bulbophyllum glaucum Schltr.

Distribution: Papua New Guinea

Bulbophyllum globiceps Schltr.

Distribution: Indonesia, Papua New Guinea

Bulbophyllum globiceps Schltr. var. **boloboense** Schltr.

Distribution: Papua New Guinea

Bulbophyllum globuliforme Nicholls
 Oncophyllum globuliforme (Nicholls) D. L. Jones & M. A. Clem.

Distribution: Australia

Bulbophyllum globulosum (Ridl.) Schuit. & de Vogel
 Phreatia globulosa Ridl.

Distribution: Indonesia

Bulbophyllum globulus Hook. f.
 Phyllorkis globulus (Hook. f.) Kuntze

Distribution: Malaysia

Bulbophyllum glutinosum (Barb. Rodr.) Cogn.
 Didactyle glutinosa Barb. Rodr.

Distribution: Brazil

172

Part II: Accepted Names / Noms Acceptés / Nombres Aceptado

Bulbophyllum gnomoniferum Ames

Distribution: Philippines

Bulbophyllum gobiense Schltr.
Hapalochilus gobiensis (Schltr.) Garay & W. Kittr.

Distribution: Papua New Guinea

Bulbophyllum goliathense J. J. Sm.

Distribution: Indonesia

Bulbophyllum gomesii Fraga

Distribution: Brazil

Bulbophyllum gomphreniflorum J. J. Sm.

Distribution: Indonesia

Bulbophyllum gongshanense Z. H. Tsi
Cirrhopetalum gonshanense (Z. H. Tsi) Garay, Hamer & Siegerist

Distribution: China

Bulbophyllum gracile Thouars
Bulbophyllum thouarsii Steud.
Phyllorkis gracilis Thouars

Distribution: Madagascar, Mauritius

Bulbophyllum gracilicaule W. Kittr.
Bulbophyllum erectum Ridl.

Distribution: Indonesia

Bulbophyllum gracilipes King & Pantl.

Distribution: Bhutan, India

Bulbophyllum graciliscapum Schltr.
Bulbophyllum setipes Schltr.

Distribution: Papua New Guinea, Solomon Islands, Vanuatu

Bulbophyllum gracillimum (Rolfe) Rolfe
Bulbophyllum leratii (Schltr.) J. J. Sm.
Bulbophyllum psittacoides (Ridl.) J. J. Sm.
Cirrhopetalum gracillimum Rolfe
Cirrhopetalum leratii Schltr.
Cirrhopetalum psittacoides Ridl.
Cirrhopetalum warianum Schltr.

Distribution: Australia, Fiji, Indonesia, Malaysia, Myanmar, New Caledonia, Papua New Guinea, Solomon Islands, Thailand

Bulbophyllum gramineum Ridl.

Distribution: Indonesia

Bulbophyllum grammopoma J. J. Verm.

Distribution: Indonesia, Papua New Guinea

Bulbophyllum grandiflorum Blume
Bulbophyllum burfordiense Hook. f.
Ephippium grandiflorum Blume
Hyalosema grandiflorum (Blume) Rolfe
Phyllorkis grandiflora (Blume) Kuntze
Sarcopodium grandiflorum (Blume) Lindl.

Distribution: Indonesia, Papua New Guinea, Solomon Islands

Bulbophyllum grandifolium Schltr.

Distribution: Papua New Guinea

Bulbophyllum grandilabre Carr

Distribution: Indonesia

Bulbophyllum grandimesense B. Gray & D. L. Jones
Oxysepala grandimesensis (B. Gray & D. L. Jones) D. L. Jones & M. A. Clem.

Distribution: Australia

Bulbophyllum granulosum Barb. Rodr.
Didactyle granulosa Barb. Rodr.

Distribution: Brazil

Bulbophyllum graveolens (F. M. Baill.) J. J. Sm.
Bulbophyllum pachybulbum (Schltr.) Seidenfaden
Cirrhopetalum graveolens F. M. Bailey
Cirrhopetalum pachybulbum Schltr.

Cirrhopetalum robustum Rolfe

Distribution: Indonesia

Bulbophyllum gravidum Lindl.
Bulbophyllum cochleatum Lindl. var. *gravidum* (Lindl.) J. J. Verm.
Bulbophyllum monticolum Hook. f.
Phyllorkis gravida (Lindl.) Kuntze
Phyllorkis monticola (Hook. f.) Kuntze

Distribution: Democratic Republic of the Congo, Zambia

Bulbophyllum griffithii (Lindl.) Rchb. f.
Bulbophyllum calodictyon Schltr.
Dendrobium bolbophylli Griff.
Phyllorkis bulbophylli (Griff.) Kuntze
Sarcopodium griffithii Lindl.

Distribution: Bhutan, India

Bulbophyllum groeneveldtii J. J. Sm.

Distribution: Indonesia, Malaysia

Bulbophyllum grotianum J. J. Verm.

Distribution: Indonesia

Bulbophyllum grudense J. J. Sm.
Bulbophyllum oeneum Burkill ex Ridl.

Distribution: Indonesia, Malaysia

Bulbophyllum guamense Ames

Distribution: Guam, Dependent Territory of the United States of America

Bulbophyllum gusdorfii J. J. Sm.
Bulbophyllum gusdorfii J. J. Sm. var. *johorense* Holttum
Cirrhopetalum gusdorfii (J. J. Sm.) Garay, Hamer & Siegerist
Cirrhopetalum gusdorfii (J. J. Sm.) Garay, Hamer & Siegerist var. *johorense* (Holtt)
Garay, Hamer & Siegerist

Distribution: Indonesia, Malaysia

Bulbophyllum guttatum Schltr.

Distribution: Papua New Guinea

Bulbophyllum guttifilum Seidenf.

Distribution: Thailand

Bulbophyllum guttulatum (Hook. f.) Balakr.
Bulbophyllum umbellatum Lindl. var. *bergemanni* Regel
Cirrhopetalum guttulatum Hook. f.
Phyllorkis guttulata (Hook. f.) Kuntze

Distribution: Bhutan, China, India, Myanmar, Viet Nam

Bulbophyllum gyaloglossum J. J. Verm.

Distribution: Papua New Guinea

Bulbophyllum gymnopus Hook. f.
Drymoda gymnopus (Hook. f.) Garay, Hamer & Siegerist
Monomeria gymnopus (Hook. f.) Aver.
Phyllorkis gymnopus (Hook. f.) Kuntze

Distribution: Bhutan, India, Thailand

Bulbophyllum gyrochilum Seidenf.

Distribution: India, Thailand

Bulbophyllum habbemense P. Royen

Distribution: Indonesia

Bulbophyllum habrotinum J. J. Verm. & A. L. Lamb
Cirrhopetalum habrotinum (J. J. Verm. & A. L. Lamb) Garay, Hamer & Siegerist

Distribution: Indonesia, Malaysia

Bulbophyllum hahlianum Schltr.
Bulbophyllum macranthum Lindl. var. *albescens* J. J. Sm.

Distribution: Indonesia, Papua New Guinea, Solomon Islands

Bulbophyllum hainanense Z. H. Tsi

Distribution: China

Bulbophyllum halconense Ames

Distribution: Philippines

Part II: Accepted Names / Noms Acceptés / Nombres Aceptado

Bulbophyllum hamadryas Schltr.

Distribution: Papua New Guinea

Bulbophyllum hamadryas Schltr. var. **orientale** Schltr.

Distribution: Papua New Guinea

Bulbophyllum hamatipes J. J. Sm.
Bulbophyllum winckelii J. J. Sm.

Distribution: Indonesia

Bulbophyllum hamelini W. Watson
Bulbophyllum hamelinii Hort. ex Rolfe

Distribution: Madagascar

Bulbophyllum haniffii Carr
Epicranthes haniffii (Carr) Garay & W. Kittr.

Distribution: Lao People's Democratic Republic, Malaysia, Myanmar, Thailand

Bulbophyllum hans-meyeri J. J. Wood

Distribution: Papua New Guinea

Bulbophyllum hapalanthos Garay

Distribution: Madagascar

Bulbophyllum harposepalum Schltr.

Distribution: Papua New Guinea

Bulbophyllum hassalli Kores

Distribution: Fiji, Solomon Islands

Bulbophyllum hastiferum Schltr.

Distribution: Indonesia

Bulbophyllum hatusimanum Tuyama

Distribution: Palau

Part II: Accepted Names / Noms Acceptés / Nombres Aceptado

Bulbophyllum heldiorum J. J. Verm.
  Synarmosepalum heldiorum (J. J. Verm.) Garay, Hamer & Siegerist

Distribution: Indonesia

Bulbophyllum helenae (Kuntze) J. J. Sm.
  Cirrhopetalum cornutum Lindl.
  Phyllorkis cornuta (Lindl.) Kuntze
  Phyllorkis helenae Kuntze
  Rhytionanthos cornutum (Lindl.) Garay, Hamer & Siegerist

Distribution: Bhutan, China, India, Thailand

Bulbophyllum heliophilum J. J. Sm.

Distribution: Indonesia

Bulbophyllum helix Schltr.

Distribution: Papua New Guinea

Bulbophyllum hellwigianum Kraenzl. ex Warb.

Distribution: Papua New Guinea

Bulbophyllum hemiprionotum J. J. Verm. & A. L. Lamb

Distribution: Malaysia

Bulbophyllum henanense J. L. Lu

Distribution: China

Bulbophyllum henrici Schltr.

Distribution: Madagascar

Bulbophyllum henrici Schltr. var. **rectangulare** H. Perrier

Distribution: Madagascar

Bulbophyllum herbula Frapp. ex Cordem.

Distribution: Réunion

Bulbophyllum heteroblepharon Schltr.

Distribution: Papua New Guinea

178

Bulbophyllum heterorhopalon Schltr.
Epicranthes heterorhopalon (Schltr.) Garay & W. Kittr.

Distribution: Papua New Guinea

Bulbophyllum heterosepalum Schltr.

Distribution: Papua New Guinea

Bulbophyllum hexarhopalos Schltr.
Epicranthes hexarhopalon (Schltr.) Garay & W. Kittr.

Distribution: Fiji, New Caledonia, Papua New Guinea

Bulbophyllum hexurum Schltr.

Distribution: Papua New Guinea

Bulbophyllum hians Schltr.

Distribution: Papua New Guinea

Bulbophyllum hians Schltr. var. **alticola** Schltr.

Distribution: Papua New Guinea

Bulbophyllum hiepii Aver.

Distribution: Viet Nam

Bulbophyllum hildebrandtii Rchb. f.
Bulbophyllum johannum H. Perrier
Bulbophyllum maculatum Jum. & H. Perrier
Bulbophyllum madagascariense Schltr.
Bulbophyllum melanopogon Schltr.
Phyllorkis hildebrandtii (Rchb. f.) Kuntze

Distribution: Madagascar

Bulbophyllum hiljeae J. J. Verm.
Peltopus hiljeae (J. J. Verm.) Szlach. & Marg.

Distribution: Papua New Guinea

Bulbophyllum hirsutissimum Kraenzl. (1912)
Bulbophyllum kraenzlinianum De Wild.

Distribution: Cameroon

179

Bulbophyllum hirsutiusculum H. Perrier

Distribution: Madagascar

Bulbophyllum hirsutum (Blume) Lindl.
  Diphyes hirsuta Blume
  Phyllorkis hirsuta (Blume) Kuntze

Distribution: Indonesia

Bulbophyllum hirtulum Ridl.
  Bulbophyllum cincinnatum Ridl.
  Bulbophyllum trichoglottis Ridl.
  Bulbophyllum trichoglottis Ridl. var. *sumatranum* J. J. Sm.

Distribution: India, Indonesia, Malaysia, Nepal, Thailand, Viet Nam

Bulbophyllum hirtum (J. E. Sm.) Lindl.
  Bulbophyllum kerri Rolfe
  Bulbophyllum suave Griff.
  Phyllorkis hirta (J. E. Sm.) Kuntze
  Stelis hirta J. E. Sm.
  Tribrachia hirta (J. E. Sm.) Lindl.

Distribution: Bhutan, India, Myanmar, Thailand, Viet Nam

Bulbophyllum hirudiniferum J. J. Verm.
  Epicranthes hirudinifera (J. J. Sm.) Garay & W. Kittr.

Distribution: Papua New Guinea

Bulbophyllum hirundinis (Gagnep.) Seidenf.
  Bulbophyllum remotifolium (Fukuyama) Nakajima
  Bulbophyllum remotifolium (Fukuyama) S. S. Ying
  Cirrhopetalum hirundinis Gagnep.
  Cirrhopetalum remotifolium Fukuyama

Distribution: China, Taiwan, Province of China, Viet Nam

Bulbophyllum hodgsoni M. R. Henderson

Distribution: Malaysia

Bulbophyllum hollandianum J. J. Sm.

Distribution: Indonesia

Bulbophyllum holochilum J. J. Sm.
  Hapalochilus holochilus (J. J. Sm.) Garay & W. Kittr.

Distribution: Indonesia

180

Part II: Accepted Names / Noms Acceptés / Nombres Aceptado

Bulbophyllum holochilum J. J. Sm. var. **aurantiacum** J. J. Sm.

Distribution: Indonesia

Bulbophyllum holochilum J. J. Sm. var. **pubescens** J. J. Sm.

Distribution: Indonesia

Bulbophyllum hookeri (Duthie) J. J. Sm.
  Cirrhopetalum hookeri Duthie

Distribution: India

Bulbophyllum horizontale Bosser

Distribution: Madagascar

Bulbophyllum horridulum J. J. Verm.

Distribution: Democratic Republic of the Congo

Bulbophyllum hovarum Schltr.

Distribution: Madagascar

Bulbophyllum howcroftii Garay, Hamer & Siegerist

Distribution: Papua New Guinea

Bulbophyllum hoyifolium J. J. Verm.

Distribution: Papua New Guinea

Bulbophyllum humbertii Schltr.

Distribution: Madagascar

Bulbophyllum humblottii Rolfe
  Bulbophyllum album Jum. & H. Perrier
  Bulbophyllum laggiarae Schltr.
  Bulbophyllum linguiforme P. J. Cribb
  Bulbophyllum luteolabium H. Perrier

Distribution: Madagascar, Malawi, Seychelles, United Republic of Tanzania, Zimbabwe

Part II: Accepted Names / Noms Acceptés / Nombres Aceptado

Bulbophyllum humile Schltr.
  Hapalochilus humilis (Schltr.) Garay & W. Kittr.

Distribution: Papua New Guinea

Bulbophyllum humiligibbum J. J. Sm.

Distribution: Indonesia

Bulbophyllum hyalinum Schltr.

Distribution: Comoros, Madagascar

Bulbophyllum hydrophilum J. J. Sm.

Distribution: Indonesia, Papua New Guinea

Bulbophyllum hymenanthum Hook. f.
  Phyllorkis hymenantha (Hook. f.) Kuntze

Distribution: Bhutan, India, Indonesia, Thailand, Viet Nam

Bulbophyllum hymenobracteum Schltr.

Distribution: Indonesia, Papua New Guinea

Bulbophyllum hymenobracteum Schltr. var. **giriwoense** J. J. Sm.

Distribution: Indonesia

Bulbophyllum hymenochilum Kraenzl.
  Bulbophyllum rhombifolium (Carr) Masam.
  Cirrhopetalum rhombifolium Carr

Distribution: Indonesia

Bulbophyllum hystricinum Schltr.

Distribution: Indonesia, Papua New Guinea

Bulbophyllum ialibuense Ormerod

Distribution: Papua New Guinea

Bulbophyllum iboense Schltr.

Distribution: Papua New Guinea

182

Bulbophyllum icteranthum Schltr.

Distribution: Papua New Guinea

Bulbophyllum idenburgense J. J. Sm.

Distribution: Indonesia

Bulbophyllum igneum J. J. Sm.

Distribution: Indonesia

Bulbophyllum ignevenosum Carr

Distribution: Malaysia, Viet Nam

Bulbophyllum ignobile J. J. Sm.

Distribution: Indonesia

Bulbophyllum ikongoense H. Perrier

Distribution: Madagascar

Bulbophyllum illecebrum J. J. Verm. & P. O'Byrne

Distribution: Indonesia

Bulbophyllum illudens Ridl.

Distribution: Indonesia

Bulbophyllum imbricans J. J. Sm.

Distribution: Indonesia

Bulbophyllum imbricatum Lindl.
Bulbophyllum congolense (De Wild.) De Wild.
Bulbophyllum gilleti (De Wild.) De Wild.
Bulbophyllum kamerunense Schltr.
Bulbophyllum laurentianum Kraenzl.
Bulbophyllum ledermanni (Kraenzl.) De Wild.
Bulbophyllum leucorhachis (Rolfe) Schltr.
Bulbophyllum linderi Summerh.
Bulbophyllum stenorhachis Kraenzl.
Bulbophyllum strobiliferum Kraenzl.
Bulbophyllum triste (Rolfe) Schltr.
Megaclinium congolense De Wild.
Megaclinium gilletii De Wild.
Megaclinium hebetatum Kraenzl.
Megaclinium imbricatum (Lindl.) Rolfe

Part II: Accepted Names / Noms Acceptés / Nombres Aceptado

Megaclinium laurentianum (Kraenzl.) De Wild.
Megaclinium ledermannii Kraenzl.
Megaclinium leucorhachis Rolfe
Megaclinium strobiliferum (Kraenzl.) Rolfe
Megaclinium triste Rolfe
Phyllorkis imbricata (Lindl.) Kuntze

Distribution: Cameroon, Central African Republic, Congo, Côte d'Ivoire, Democratic Republic of the Congo, Equatorial Guinea, Gabon, Ghana, Liberia, Nigeria, Sao Tome and Principe, Sierra Leone, United Republic of Tanzania

Bulbophyllum imerinense Schltr.

Distribution: Madagascar

Bulbophyllum imitator J. J. Verm.

Distribution: Papua New Guinea

Bulbophyllum impar Ridl.

Distribution: Indonesia

Bulbophyllum inaequale (Blume) Lindl.
Diphyes inaequalis Blume
Phyllorkis inaequalis (Blume) Kuntze

Distribution: Indonesia

Bulbophyllum inaequale (Blume) Lindl. var. **angustifolium** J. J. Sm.

Distribution: Indonesia

Bulbophyllum inaequisepalum Schltr.

Distribution: Papua New Guinea

Bulbophyllum inauditum Schltr. (1913)

Distribution: Papua New Guinea

Bulbophyllum incarum Kraenzl.

Distribution: Peru

Bulbophyllum inciferum J. J. Verm.
Peltopus inciferus (J. J. Verm.) Szlach. & Marg.

Distribution: Papua New Guinea

Part II: Accepted Names / Noms Acceptés / Nombres Aceptado

Bulbophyllum incisilabrum J. J. Verm. & P. O'Byrne

Distribution: Indonesia

Bulbophyllum inclinatum J. J. Sm.
Hapalochilus inclinatus (J. J. Sm.) Garay & W. Kittr.

Distribution: Indonesia

Bulbophyllum incommodum Kores

Distribution: Fiji

Bulbophyllum inconspicuum Maxim.

Distribution: Japan

Bulbophyllum incumbens Schltr.

Distribution: Papua New Guinea

Bulbophyllum incurvum Thouars
Phyllorkis incurva Thouars

Distribution: Mauritius, Réunion

Bulbophyllum iners Rchb. f.
Phyllorkis iners (Rchb. f.) Kuntze

Distribution: India

Bulbophyllum infundibuliforme J. J. Sm.
Bulbophyllum garupinum Schltr.

Distribution: Indonesia

Bulbophyllum injoloense De Wild.
Megaclinium injoloense (De Wild.) De Wild.

Distribution: Democratic Republic of the Congo

Bulbophyllum injoloense De Wild. subsp. **pseudoxypterum** (J. J. Verm.) J. J. Verm.
Bulbophyllum injoloense De Wild. var. *pseudoxypterum* J. J. Verm.

Distribution: Democratic Republic of the Congo, Zambia

Part II: Accepted Names / Noms Acceptés / Nombres Aceptado

Bulbophyllum inops Rchb. f.

Distribution: Unknown

Bulbophyllum inornatum J. J. Verm.

Distribution: United Republic of Tanzania

Bulbophyllum inquirendum J. J. Verm.

Distribution: Indonesia, Papua New Guinea

Bulbophyllum insectiferum Barb. Rodr.

Distribution: Brazil

Bulbophyllum insolitum Bosser

Distribution: Madagascar

Bulbophyllum insulsum (Gagnep.) Seidenf.
  Bulbophyllum insulsoides Seidenf.
  Bulbophyllum racemosum Hayata
  Cirrhopetalum insulsum Gagnep.
  Cirrhopetalum racemosum (Hayata) Hayata

Distribution: Viet Nam

Bulbophyllum intermedium F. M. Bailey
  Bulbophyllum shephardi (F. Muell.) F. Muell. var. *intermedium* (F. M. Bailey) Nicholls
  Oxysepala intermedia (F. M. Bailey) D. L. Jones & M. A. Clem.

Distribution: Australia

Bulbophyllum intersitum J. J. Verm.
  Peltopus intersitus (J. J. Verm.) Szlach. & Marg.

Distribution: Papua New Guinea

Bulbophyllum intertextum Lindl.
  Bulbophyllum amauryae Rendle
  Bulbophyllum intertextum Lindl. var. *parvilabium* G. Will.
  Bulbophyllum pertenue Kraenzl.
  Bulbophyllum quintasii Rolfe
  Bulbophyllum seychellarum Rchb. f.
  Bulbophyllum triaristellum Kraenzl. & Schltr.
  Bulbophyllum usambarae Kraenzl.
  Bulbophyllum viride Rolfe
  Phyllorkis intertexta (Lindl.) Kuntze

Part II: Accepted Names / Noms Acceptés / Nombres Aceptado

Phyllorkis seychellarum (Rchb. f.) Kuntze

Distribution: Angola, Cameroon, Côte d'Ivoire, Democratic Republic of the Congo, Equatorial Guinea, Ethiopia, Gabon, Guinea, Kenya, Liberia, Malawi, Nigeria, Sao Tome and Principe, Seychelles, Sierra Leone, United Republic of Tanzania, Zambia, Zimbabwe

Bulbophyllum intricatum Seidenf.

Distribution: Thailand

Bulbophyllum inunctum J. J. Sm.
  Bulbophyllum longiflorum Ridl.

Distribution: Indonesia, Malaysia

Bulbophyllum inversum Schltr.

Distribution: Papua New Guinea

Bulbophyllum invisum Ames

Distribution: Philippines

Bulbophyllum involutum L. E. Borba & J. Semir

Distribution: Brazil

Bulbophyllum ionophyllum J. J. Verm.

Distribution: Indonesia

Bulbophyllum ipanemense Hoehne

Distribution: Brazil

Bulbophyllum ischnopus Schltr.

Distribution: Papua New Guinea

Bulbophyllum ischnopus Schltr. var. **rhodoneuron** Schltr.

Distribution: Papua New Guinea

Bulbophyllum iterans J. J. Verm. & P. O'Byrne

Distribution: Indonesia

Part II: Accepted Names / Noms Acceptés / Nombres Aceptado

Bulbophyllum ivorense P. J. Cribb & Perez-Vera
Bulbophyllum elongatum (De Wild.) De Wild.
Bulbophyllum flavidum Lindl. var. *elongatum* De Wild.

Distribution: Cameroon, Côte d'Ivoire, Democratic Republic of the Congo, Gabon, Liberia, Nigeria

Bulbophyllum jaapii Szlach. & Olszewski

Distribution: Cameroon

Bulbophyllum jadunae Schltr.
Hapalochilus jadunae (Schltr.) Garay & W. Kittr.

Distribution: Papua New Guinea

Bulbophyllum jaguariahyvae Kraenzl.

Distribution: Brazil

Bulbophyllum jamaicense Cogn.

Distribution: Jamaica

Bulbophyllum janus J. J. Verm.
Bulbophyllum platyrrhachis Ridl.

Distribution: Indonesia

Bulbophyllum japonicum (Makino) Makino
Bulbophyllum inabai Hayata
Bulbophyllum japonicum (Makino) J. J. Sm.
Cirrhopetalum inabai (Hayata) Hayata
Cirrhopetalum japonicum Makino

Distribution: China, Japan, Taiwan, Province of China

Bulbophyllum jensenii J. J. Sm.
Hapalochilus jensenii (J. J. Sm.) Garay & W. Kittr.

Distribution: Indonesia

Bulbophyllum jiewhoei J. J. Verm & P. O'Byrne

Distribution: Unknown

Bulbophyllum johannis Kraenzl.

Distribution: Madagascar

Part II: Accepted Names / Noms Acceptés / Nombres Aceptado

Bulbophyllum johannulii J. J. Verm.
Epicranthes johannulii (J. J. Verm.) Garay & W. Kittr.

Distribution: Papua New Guinea

Bulbophyllum johnsonii T. E. Hunt
Serpenticaulis johnsonii (T. E. Hunt) M. A. Clem. & D. L. Jones

Distribution: Australia

Bulbophyllum jolandae J. J. Verm.

Distribution: Indonesia, Malaysia

Bulbophyllum josephi (Kuntze) Summerh.
Bulbophyllum amanicum Kraenzl.
Bulbophyllum aurantiacum Hook. f.
Bulbophyllum gustavii Schltr.
Bulbophyllum schlechteri De Wild.
Bulbophyllum sennii Chiov.
Bulbophyllum winkleri Schltr.(1914)
Phyllorkis josephi Kuntze

Distribution: Burundi, Cameroon, Côte d'Ivoire, Democratic Republic of the Congo, Ethiopia, Kenya, Malawi, Mozambique, Rwanda, Uganda, United Republic of Tanzania, Zimbabwe

Bulbophyllum josephi (Kuntze) Summerh. var. **mahonii** (Rolfe) J. J. Verm.
Bulbophyllum mahoni Rolfe
Bulbophyllum modicum Summerh.

Distribution: Cameroon, Côte d'Ivoire, Democratic Republic of the Congo, Equatorial Guinea, Guinea, Liberia, Malawi, Nigeria, Zambia

Bulbophyllum jumellanum Schltr.

Distribution: Madagascar

Bulbophyllum kainochiloides H. Perrier

Distribution: Madagascar

Bulbophyllum kaitiense Rchb. f.
Cirrhopetalum neilgherrense Wight
Phyllorkis kaitiensis (Rchb.f.) Kuntze

Distribution: India

Part II: Accepted Names / Noms Acceptés / Nombres Aceptado

Bulbophyllum kanburiense Seidenf.

Distribution: Myanmar, Thailand, Viet Nam

Bulbophyllum kaniense Schltr.
Bulbophyllum dispersum Schltr.
Bulbophyllum dispersum Schltr. var. *roseans* Schltr.
Bulbophyllum extensum Schltr.

Distribution: Papua New Guinea

Bulbophyllum kauloense Schltr.

Distribution: Papua New Guinea

Bulbophyllum kautskyi Toscano

Distribution: Brazil

Bulbophyllum keekee N. Halle

Distribution: New Caledonia (French)

Bulbophyllum kegelii Hamer & Garay

Distribution: Brazil, Suriname, Trinidad and Tobago

Bulbophyllum kelelense Schltr.
Hapalochilus kelelensis (Schltr.) Garay & W. Kittr.

Distribution: Papua New Guinea

Bulbophyllum kempfii Schltr.

Distribution: Papua New Guinea

Bulbophyllum kemulense J. J. Sm.

Distribution: Indonesia

Bulbophyllum kenae J. J. Verm.
Peltopus kenae (J. J. Verm.) Szlach. & Marg.

Distribution: Papua New Guinea

Bulbophyllum kenejianum Schltr.

Distribution: Papua New Guinea

190

Part II: Accepted Names / Noms Acceptés / Nombres Aceptado

Bulbophyllum kenejiense W. Kittr.
Bulbophyllum hydrophilum Schltr.

Distribution: Papua New Guinea

Bulbophyllum keralensis Muktesh & Stephen

Distribution: India

Bulbophyllum kermesinum Ridl.
Hapalochilus kermesinus (Ridl.) Garay & W. Kittredge

Distribution: Indonesia

Bulbophyllum kestron J. J. Verm. & A. L. Lamb

Distribution: Indonesia

Bulbophyllum kettridgei (Garay, Hamer & Siegerist) J. J. Verm.
Synarmosepalum kittredgei Garay, Hamer & Siegerist

Distribution: Philippines

Bulbophyllum khaoyaiense Seidenf.
Bulbophyllum tripudians C. S. P. Parish & Rchb. f. var. *pumilum* Seidenf. & Smitinand

Distribution: Thailand

Bulbophyllum khasyanum Griff.
Bulbophyllum bowringianum Rchb. f.
Bulbophyllum conchiferum Rchb. f.
Bulbophyllum gibsonii Lindl. ex Rchb.f.
Bulbophyllum cylindraceum Lindl. var. *khasynaum* (Griff.) Hook. f.
Phyllorkis conchifera (Rchb. f.) Kuntze

Distribution: Bhutan, China, India, Malaysia, Myanmar, Thailand, Viet Nam

Bulbophyllum kieneri Bosser

Distribution: Madagascar

Bulbophyllum kirkwoodae T. E. Hunt
Serpenticaulis kirkwoodiae (T. E. Hunt) M. A. Clem. & D. L. Jones

Distribution: Australia

Bulbophyllum kirroanthum Schltr.

Distribution: Indonesia

Part II: Accepted Names / Noms Acceptés / Nombres Aceptado

Bulbophyllum kivuense J. J. Verm.

Distribution: Rwanda

Bulbophyllum kjellbergii J. J. Sm.

Distribution: Indonesia

Bulbophyllum klabatense Schltr.

Distribution: Indonesia

Bulbophyllum klossii Ridl.
Hyalosema klossii (Ridl.) Rolfe

Distribution: Indonesia

Bulbophyllum kontumense Gagnep.

Distribution: Viet Nam

Bulbophyllum korimense J. J. Sm.

Distribution: Indonesia

Bulbophyllum korinchense Ridl.

Distribution: Indonesia

Bulbophyllum korinchense Ridl. var. **grandflorum** Ridl.

Distribution: Indonesia

Bulbophyllum korinchense Ridl. var. **parviflorum** Ridl.

Distribution: Indonesia

Bulbophyllum korthalsii Schltr.
Bulbophyllum arachnites Ridl.

Distribution: Indonesia, Malaysia

Bulbophyllum kupense P. J. Cribb & B. J. Pollard

Distribution: Cameroon

Part II: Accepted Names / Noms Acceptés / Nombres Aceptado

Bulbophyllum kusaiense Tuyama
Hapalochilus kusaiense (Tuyama) Garay, Hamer & Siegerist

Distribution: Palau

Bulbophyllum kwangtungense Schltr.

Distribution: China

Bulbophyllum labatii Bosser

Distribution: Madagascar

Bulbophyllum laciniatum (Barb. Rodr.) Cogn.
Bulbophyllum laciniatum (Barb. Rodr.) Cogn. var. *janeirense* Cogn.
Didactyle laciniata Barb. Rodr.
Didactyle laciniata Barb. Rodr. var. *janeirense* Cogn.

Distribution: Brazil

Bulbophyllum lacinulosum J. J. Sm.

Distribution: Indonesia

Bulbophyllum laetum J. J. Verm.

Distribution: Malaysia

Bulbophyllum lageniforme F. M. Bailey
Adelopetalum lageniforme (F. M. Bailey) D. L. Jones & M. A. Clem.
Bulbophyllum adenocarpum Schltr.

Distribution: Australia

Bulbophyllum lakatoense Bosser

Distribution: Madagascar

Bulbophyllum lambii J. J. Verm.

Distribution: Indonesia

Bulbophyllum lamelluliferum J. J. Sm.

Distribution: Indonesia

Bulbophyllum lamii J. J. Sm.

Distribution: Indonesia

Bulbophyllum lamingtonense D. L. Jones
  Oxysepala lamingtonensis (D. L. Jones) D. L. Jones & M. A. Clem.

Distribution: Australia

Bulbophyllum lancifolium Ames

Distribution: Philippines

Bulbophyllum lancilabium Ames

Distribution: Philippines

Bulbophyllum lancipetalum Ames

Distribution: Philippines

Bulbophyllum lancisepalum H. Perrier

Distribution: Madagascar

Bulbophyllum languidum J. J. Sm.

Distribution: Indonesia

Bulbophyllum lanuginosum J. J. Verm.

Distribution: Thailand

Bulbophyllum laoticum Gagnep.

Distribution: Lao People's Democratic Republic

Bulbophyllum lasianthum Lindl.
  Anisopetalum lasianthum Kuhl ex Hook. f.
  Phyllorkis lasiantha (Lindl.) Kuntze

Distribution: Indonesia, Malaysia, Philippines

Bulbophyllum lasiochilum C. S. P. Parish & Rchb. f.
  Bulbophyllum breviscapum (Rolfe) J. J. Sm.
  Bulbophyllum breviscapum (Rolfe) Ridl.
  Cirrhopetalum breviscapum Rolfe
  Cirrhopetalum lasiochilum (C. S. P. Parish & Rchb. f.) Hook. f.

194

Part II: Accepted Names / Noms Acceptés / Nombres Aceptado

Phyllorkis lasiochila (C. S. P. Parish & Rchb. f.) Kuntze

Distribution: Malaysia, Myanmar, Thailand

Bulbophyllum lasioglossum Rolfe ex Ames

Distribution: Philippines

Bulbophyllum lasiopetalum Kraenzl.

Distribution: Philippines

Bulbophyllum latibrachiatum J. J. Sm.

Distribution: Indonesia

Bulbophyllum latibrachiatum J. J. Sm. var. **epilosum** J. J. Sm.

Distribution: Indonesia

Bulbophyllum latipes J. J. Sm.
   Dactylorhynchus flavescens Schltr.

Distribution: Indonesia, Papua New Guinea

Bulbophyllum latipetalum H. Perrier

Distribution: Madagascar

Bulbophyllum latisepalum Ames & C. Schweinf.

Distribution: Indonesia, Philippines

Bulbophyllum laxiflorum (Blume) Lindl.
   Bulbophyllum luzonense Ames
   Bulbophyllum pedicellatum Ridl.
   Bulbophyllum radiatum Lindl.
   Bulbophyllum syllectum Kraenzl.
   Diphyes laxiflora Blume
   Phyllorkis laxiflora (Blume) Kuntze
   Phyllorkis radiata (Lindl.) Kuntze

Distribution: Indonesia, Lao People's Democratic Republic, Malaysia, Myanmar, Philippines, Thailand, Viet Nam

Part II: Accepted Names / Noms Acceptés / Nombres Aceptado

Bulbophyllum laxiflorum (Blume) Lindl. var. **celebicum** Schltr.

Distribution: Indonesia

Bulbophyllum laxum Schltr.
  Bulbophyllum trichopus Schltr.

Distribution: Papua New Guinea

Bulbophyllum leandrianum H. Perrier

Distribution: Madagascar

Bulbophyllum lecouflei Bosser

Distribution: Madagascar

Bulbophyllum ledungense T. Tang & F. T. Wang

Distribution: China

Bulbophyllum lehmannianum Kraenzl.

Distribution: Colombia, Venezuela

Bulbophyllum leibergii Ames & Rolfe

Distribution: Philippines

Bulbophyllum lemnifolium Schltr.

Distribution: Papua New Guinea

Bulbophyllum lemniscatoides Rolfe

Distribution: Indonesia, Thailand, Viet Nam

Bulbophyllum lemniscatoides Rolfe var. **exappendiculatum** J. J. Sm.

Distribution: Indonesia

Bulbophyllum lemniscatum C. S. P. Parish ex Hook. f.
  Phyllorkis lemniscata (C. S. P. Parish ex Hook. f.) Kuntze

Distribution: Myanmar, Thailand

Part II: Accepted Names / Noms Acceptés / Nombres Aceptado

Bulbophyllum lemuraeoides H. Perrier

Distribution: Madagascar

Bulbophyllum lemurense Bosser & P. J. Cribb
  Bulbophyllum clavigerum H. Perrier

Distribution: Madagascar

Bulbophyllum leniae J. J. Verm.

Distribution: Indonesia

Bulbophyllum leoni Kraenzl.
  Bulbophyllum humblotianum Kraenzl.

Distribution: Comoros, Madagascar

Bulbophyllum leontoglossum Schltr.
  Hapalochilus leontoglossus (Schltr.) Garay & W. Kittr.

Distribution: Papua New Guinea

Bulbophyllum leopardinum (Wall.) Lindl.
  Dendrobium leopardinum Wall.
  Phyllorkis leopardina (Wall.) Kuntze
  Sarcopodium leopardinum (Wall.) Lindl. & Paxt.

Distribution: Bhutan, India, Thailand

Bulbophyllum leopardinum (Wall.) Lindl. var. **tuberculatum** N. P.
Balakrishnan & S. Chowdhury

Distribution: Bhutan, India

Bulbophyllum lepantense Ames

Distribution: Philippines

Bulbophyllum lepanthiflorum Schltr.
  Bulbophyllum lepanthiflorum Schltr. var. *rivulare* Schltr.

Distribution: Indonesia, Papua New Guinea

Bulbophyllum leproglossum J. J. Verm. & A. L. Lamb

Distribution: Indonesia

Bulbophyllum leptanthum Hook. f.
  Phyllorkis leptantha (Hook. f.) Kuntze

Distribution: Bhutan, India

Bulbophyllum leptobulbon J. J. Verm.

Distribution: Papua New Guinea

Bulbophyllum leptocaulon Kraenzl.

Distribution: Philippines

Bulbophyllum leptochlamys Schltr.

Distribution: Madagascar

Bulbophyllum leptoleucum Schltr.

Distribution: Indonesia, Papua New Guinea

Bulbophyllum leptophyllum W. Kittr.
  Bulbophyllum lonchophyllum Schltr. (1919)

Distribution: Indonesia, Papua New Guinea

Bulbophyllum leptopus Schltr.

Distribution: Papua New Guinea

Bulbophyllum leptosepalum Hook. f.
  Phyllorkis leptosepala (Hook. f.) Kuntze

Distribution: Malaysia

Bulbophyllum leptostachyum Schltr.

Distribution: Madagascar

Bulbophyllum leucorhodum Schltr.
  Hapalochilus leucorhodus (Schltr.) Garay & W. Kittr.

Distribution: Papua New Guinea

Bulbophyllum leucothyrsus Schltr.

Distribution: Papua New Guinea

198

Part II: Accepted Names / Noms Acceptés / Nombres Aceptado

Bulbophyllum levanae Ames

Distribution: Philippines

Bulbophyllum levanae Ames var. **giganteum** Quis. & C. Schweinf.

Distribution: Philippines

Bulbophyllum levatii Kraenzl.

Distribution: Solomon Islands, Vanuatu

Bulbophyllum levatii Kraenzl. var. **mischanthum** J. J. Verm.

Distribution: Papua New Guinea

Bulbophyllum leve Schltr.

Distribution: Papua New Guinea

Bulbophyllum levidense J. J. Sm.

Distribution: Indonesia

Bulbophyllum levinei Schltr.

Distribution: China

Bulbophyllum levyae Garay, Hamer & Siegerist

Distribution: Papua New Guinea

Bulbophyllum lewisense B. Gray & D. L. Jones
   Oxysepala lewisense (B. Gray & D. L. Jones) D. L. Jones & M. A. Clem.

Distribution: Australia

Bulbophyllum leysenianum Burb.
   Hyalosema leysenianum (Burb.) Rolfe

Distribution: Indonesia

Bulbophyllum leytense Ames

Distribution: Philippines

Part II: Accepted Names / Noms Acceptés / Nombres Aceptado

Bulbophyllum lichenoides Schltr.

Distribution: Papua New Guinea

Bulbophyllum lichenophylax Schltr.
Bulbophyllum quinquecornutum H. Perrier

Distribution: Madagascar

Bulbophyllum ligulatum W. Kittr.
Bulbophyllum eublepharum Schltr.

Distribution: Papua New Guinea

Bulbophyllum ligulifolium J. J. Sm.

Distribution: Indonesia

Bulbophyllum liliacinum Ridl.
Bulbophyllum liliacinum Ridl. var. *sorocianum* M. Ahmed, Pasha & Aziz Khan

Distribution: Bangladesh, India, Malaysia, Thailand

Bulbophyllum lilianae Rendle
Adelopetalum lilianae (Rendle) D. L. Jones & M. A. Clem.
Bulbophyllum revolutum Dockrill & St. Cloud

Distribution: Australia

Bulbophyllum limbatum Lindl.
Bulbophyllum blepharosepalum Schltr.
Phyllorkis limbata (Lindl.) Kuntze

Distribution: Brunei Darussalam, Indonesia, Malaysia, Myanmar, Singapore, Thailand

Bulbophyllum lindleyanum Griff.
Bulbophyllum caesariatum Ridl.
Bulbophyllum ringens Rchb. f.
Phyllorkis lindleyana (Griff.) Kuntze
Phyllorkis ringens (Rchb.f.) Kuntze

Distribution: Myanmar, Thailand

Bulbophyllum lineare Frapp. ex Cordem.

Distribution: Réunion

Part II: Accepted Names / Noms Acceptés / Nombres Aceptado

Bulbophyllum lineariflorum J. J. Sm.

Distribution: Indonesia

Bulbophyllum linearifolium King & Pantl.

Distribution: Malaysia

Bulbophyllum linearilabium J. J. Sm.

Distribution: Indonesia

Bulbophyllum lineariligulatum Schltr.

Distribution: Madagascar

Bulbophyllum lineatum (Teijsm. & Binn.) J. J. Sm.
  Bulbophyllum merguense C. S. P. Parish & Rchb. f.
  Cirrhopetalum lineatum Teijsm. & Binn.
  Cirrhopetalum merguense (C. S. P. Parish & Rchb. f.) Hook. f.
  Phyllorkis merguensis (C. S. P. Parish & Rchb. f.) Kuntze

Distribution: Indonesia, Myanmar

Bulbophyllum lineolatum Schltr.

Distribution: Papua New Guinea

Bulbophyllum linggense J. J. Sm.

Distribution: Indonesia

Bulbophyllum lingulatum Rendle
  Bulbophyllum lingulatum Rendle *f. microphyton* (Guillaumin) N. Halle
  Bulbophyllum microphyton Guillaumin
  Bulbophyllum trilobum Schltr.

Distribution: New Caledonia

Bulbophyllum liparidioides Schltr.

Distribution: Madagascar

Bulbophyllum lipense Ames

Distribution: Philippines

Bulbophyllum lissoglossum J. J. Verm.

Distribution: Indonesia, Malaysia

Bulbophyllum lizae J. J. Verm.

Distribution: Sao Tome and Principe

Bulbophyllum lobbii Lindl.
Bulbophyllum claptonense Rolfe
Bulbophyllum henshallii Lindl.
Bulbophyllum lobbii Lindl. var. *colosseum* Hort.
Bulbophyllum lobbii Lindl. var. *henshallii* (Lindl.) Henfrey
Bulbophyllum lobbii Lindl. var. *nettesiae* Cogn.
Bulbophyllum polystictum Ridl.
Phyllorkis lobbii (Lindl.) Kuntze
Sarcobodium lobbii Beer
Sarcopodium lobbii (Lindl.) Lindl. & Paxt.
Sestochilos uniflorum Breda

Distribution: Cambodia, India, Indonesia, Malaysia, Myanmar, Philippines

Bulbophyllum lockii Aver. & Averyanova

Distribution: Viet Nam

Bulbophyllum loherianum (Kraenzl.) Ames
Cirrhopetalum loherianum Kraenzl.

Distribution: Philippines

Bulbophyllum lohokii J. J. Verm. & A. L. Lamb
Ferruminaria lohokii (J. J. Verm. & A. L. Lamb) Garay & G. A. Romero
Hapalochilus lohokii (J. J. Verm. & A. L. Lamb) Garay, Hamer & Siegerist

Distribution: Malaysia

Bulbophyllum lokonense Schltr.

Distribution: Indonesia

Bulbophyllum lonchophyllum Schltr. (1913)

Distribution: Papua New Guinea

Bulbophyllum longebracteatum Seidenf.

Distribution: Lao People's Democratic Republic, Thailand

Part II: Accepted Names / Noms Acceptés / Nombres Aceptado

Bulbophyllum longerepens Ridl.

Distribution: Indonesia

Bulbophyllum longhutense J. J. Sm.

Distribution: Indonesia

Bulbophyllum longibrachiatum Z. H. Tsi
  Cirrhopetalum longibrachiatum (Z. H. Tsi) Garay, Hamer & Siegerist

Distribution: China, Viet Nam

Bulbophyllum longibracteatum Seidenf.

Distribution: Lao People's Democratic Republic, Thailand

Bulbophyllum longicaudatum (J. J. Sm.) J. J. Sm.
  Bulbophyllum blumei (Lindl.) J. J. Sm. var. *longicaudatum* J. J. Sm.

Distribution: Indonesia

Bulbophyllum longidens (Rolfe) Seidenf.
  Cirrhopetalum longidens Rolfe

Distribution: China

Bulbophyllum longiflorum Thouars
  Bulbophyllum clavigerum (Fitzg.) Dockr.
  Bulbophyllum clavigerum (Fitzg.) F. Muell.
  Bulbophyllum layardii (F. Muell. & Kraenzl.) J. J. Sm.
  Bulbophyllum trisetum Ames
  Bulbophyllum umbellatum J. J. Sm.
  Cirrhopetalum africanum Schltr.
  Cirrhopetalum clavigerum Fitzg.
  Cirrhopetalum kenejianum Schltr.
  Cirrhopetalum layardi F. Muell. & Kraenzl. ex Kraenzl.
  Cirrhopetalum longiflorum (Thouars) Schltr.
  Cirrhopetalum thouarsii Lindl.
  Cirrhopetalum thouarsii Lindl. var. *concolor* Rolfe
  Cirrhopetalum trisetum (Ames) Garay, Hamer & Siegerist
  Cirrhopetalum umbellatum (G. Forst.) Frappier ex Cordem.
  Cirrhopetalum umbellatum (G. Forst.) Hook. & Arn.
  Cirrhopetalum umbellatum Schltr.
  Cymbidium umbellatum (G. Forst.) Spreng.
  Epidendrum umbellatum G. Forst.
  Phyllorkis clavigera (Fitzg.) Kuntze
  Phyllorkis longiflora Thouars
  Zygoglossum umbellatum (G. Forst.) Reinw.

Distribution: Australia, Cook Islands, Democratic Republic of the Congo, Fiji, French Polynesia, Guam, Dependent Territory of the United States of America, Indonesia, Madagascar, Malawi, Mauritius, New Caledonia, Papua New Guinea, Philippines, Réunion,

Part II: Accepted Names / Noms Acceptés / Nombres Aceptado

Samoa, Seychelles, Solomon Islands, Uganda, United Republic of Tanzania, Vanuatu, Viet Nam, Zimbabwe

Bulbophyllum longilabre Schltr.
  Hapalochilus longilabris (Schltr.) Garay & W. Kittr.

Distribution: Papua New Guinea

Bulbophyllum longimucronatum Ames & C. Schweinf.

Distribution: Indonesia, Philippines

Bulbophyllum longipedicellatum J. J. Sm.

Distribution: Indonesia

Bulbophyllum longipedicellatum J. J. Sm. var. **gjellerupii** J. J. Sm.

Distribution: Indonesia

Bulbophyllum longipetalum Pabst

Distribution: Brazil

Bulbophyllum longipetiolatum Ames

Distribution: Philippines

Bulbophyllum longirostre Schltr.

Distribution: Papua New Guinea

Bulbophyllum longiscapum Rolfe
  Bulbophyllum macrolepis L. O. Williams
  Bulbophyllum praealtum Kraenzl.
  Macrolepis longiscapa A. Rich.

Distribution: Fiji, Niue, Samoa, Solomon Islands, Tonga, Vanuatu, Wallis and Futuna Islands (France)

Bulbophyllum longisepalum Rolfe
  Hyalosema longisepalum (Rolfe) Rolfe

Distribution: Indonesia

Bulbophyllum longispicatum Cogn.

Distribution: Brazil

204

Part II: Accepted Names / Noms Acceptés / Nombres Aceptado

Bulbophyllum longissimum (Ridl.) Ridl.
Bulbophyllum longissimum (Ridl.) J. J. Sm
Cirrhopetalum longissimum Ridl.

Distribution: Thailand

Bulbophyllum longivagans Carr

Distribution: Indonesia

Bulbophyllum longivaginans H. Perrier

Distribution: Madagascar

Bulbophyllum lophoglottis (Guillaumin) N. Halle
Canacorchis lophoglottis Guillaumin

Distribution: New Caledonia

Bulbophyllum lophoton J. J. Verm.
Peltopus lophotos (J. J. Verm.) Szlach. & Marg.

Distribution: Papua New Guinea

Bulbophyllum lordoglossum J. J. Verm. & A. L. Lamb

Distribution: Malaysia

Bulbophyllum lorentzianum J. J. Sm.

Distribution: Indonesia

Bulbophyllum loroglossum J. J. Verm.
Peltopus loroglossus (J. J. Verm.) Szlach. & Marg.

Distribution: Papua New Guinea

Bulbophyllum louisiadum Schltr.

Distribution: Papua New Guinea

Bulbophyllum loxophyllum Schltr.

Distribution: Papua New Guinea

Bulbophyllum luanii Tixier

Distribution: Thailand, Viet Nam

205

Bulbophyllum lucidum Schltr.
Bulbophyllum rictorium Schltr.

Distribution: Madagascar

Bulbophyllum luciphilum Stevart

Distribution: Sao Tome and Principe

Bulbophyllum luckraftii F. Muell.

Distribution: Solomon Islands

Bulbophyllum luederwaldtii Hoehne & Schltr.

Distribution: Brazil

Bulbophyllum lumbriciforme J. J. Sm.

Distribution: Indonesia, Malaysia

Bulbophyllum lundianum Rchb. f. & Warm.
Phyllorkis lundiana (Rchb. f. & Warm.) Kuntze

Distribution: Brazil

Bulbophyllum lupulinum Lindl.
Bulbophyllum ituriense De Wild.
Bulbophyllum urbanianum Kraenzl.
Phyllorkis lupulina (Lindl.) Kuntze

Distribution: Cameroon, Côte d'Ivoire, Democratic Republic of the Congo, Ethiopia, Guinea, Nigeria, Sierra Leone, Zambia

Bulbophyllum luteobracteatum Jum. & H. Perrier

Distribution: Madagascar

Bulbophyllum luteopurpureum J. J. Sm.

Distribution: Indonesia

Bulbophyllum lygeron J. J. Verm.

Distribution: Indonesia

Bulbophyllum lyperocephalum Schltr.

Distribution: Madagascar

Part II: Accepted Names / Noms Acceptés / Nombres Aceptado

Bulbophyllum lyperostachyum Schltr.

Distribution: Madagascar

Bulbophyllum lyriforme J. J. Verm. & P. O'Byrne

Distribution: Papua New Guinea

Bulbophyllum maboroense Schltr.

Distribution: Papua New Guinea

Bulbophyllum machupicchuense D. E. Benn. & Christenson

Distribution: Peru

Bulbophyllum macilentum J. J. Verm.

Distribution: Papua New Guinea

Bulbophyllum macneiceae Schuit. & De Vogel

Distribution: Papua New Guinea

Bulbophyllum macphersoni Rupp
  Blepharochilum macphersoni (Rupp) M. A. Clem. & D. L. Jones
  Bulbophyllum purpurascens F. M. Bailey
  Osyricera purpurascens H. Deane
  Phyllorkis purpurascens (F. M. Bailey) Kuntze

Distribution: Australia

Bulbophyllum macphersoni Rupp var. **spathulatum** Dockrill & St. Cloud
  Blepharochilum sladeanum (A. D. Hawkes) M. A. Clem. & D. L. Jones
  Bulbophyllum cochleatum Schltr.
  Bulbophyllum sladeanum A. D. Hawkes

Distribution: Australia

Bulbophyllum macraei (Lindl.) Rchb. f.
  Bulbophyllum boninense Makino
  Bulbophyllum makinoanum (Schltr.) Masam.
  Bulbophyllum tanegasimense Masam.
  Bulbophyllum uraiense Hayata
  Cirrhopetalum autumnale Fukuyama
  Cirrhopetalum boninense Makino
  Cirrhopetalum macraei Lindl.
  Cirrhopetalum makinoanum (Makino) Schltr.
  Cirrhopetalum tanegasimense (Masam.) Masam.
  Cirrhopetalum uraiense (Hayata) Hayata
  Cirrhopetalum walkerianum Wight

Phyllorkis macraei (Lindl.) Kuntze

Distribution: India, Japan, Sri Lanka, Taiwan, Province of China, Viet Nam

Bulbophyllum macraei (Lindl.) Rchb. f. var. **autumnale** (Fukuyama) S. S. Ying
Bulbophyllum autumnale (Fukuyama) S. S. Ying

Distribution: Taiwan, Province of China

Bulbophyllum macranthoides Kraenzl.
Bulbophyllum tollenoniferum J. J. Sm.

Distribution: Indonesia

Bulbophyllum macranthum Lindl.
Bulbophyllum cochinchinense Gagnep.
Bulbophyllum purpureum Naves
Carparomorchis macracantha (Lindl.) M. A. Clem. & D. L. Jones
Phyllorkis macrantha (Lindl.) Kuntze
Sarcopodium macranthum (Lindl.) Lindl. & Paxt.
Sarcopodium purpureum Rchb. f.

Distribution: India, Indonesia, Malaysia, Myanmar, Papua New Guinea, Philippines, Solomon Islands, Thailand, Viet Nam

Bulbophyllum macrobulbum J. J. Sm.
Bulbophyllum balfourianum Hort.

Distribution: Indonesia, Papua New Guinea

Bulbophyllum macrocarpum Frapp. ex Cordem.

Distribution: Réunion

Bulbophyllum macroceras Barb. Rodr.

Distribution: Brazil

Bulbophyllum macrochilum Rolfe
Bulbophyllum intervallatum J. J. Sm.
Bulbophyllum mirandum Kraenzl.
Bulbophyllum stella Ridl.
Bulbophyllum vanessa King & Pantl. †

Distribution: Indonesia, Malaysia, Singapore

Bulbophyllum macrocoleum Seidenf.

Distribution: Thailand, Viet Nam

Part II: Accepted Names / Noms Acceptés / Nombres Aceptado

Bulbophyllum macrorhopalon Schltr.
Epicranthes macrorhopalon (Schltr.) Garay & W. Kittr.

Distribution: Papua New Guinea

Bulbophyllum macrourum Schltr.
Bulbophyllum pensile Schltr.

Distribution: Papua New Guinea

Bulbophyllum maculatum Boxall ex Naves

Distribution: Philippines

Bulbophyllum maculosum Ames

Distribution: Philippines

Bulbophyllum magnibracteatum Summerh.

Distribution: Cameroon, Central African Republic, Côte d'Ivoire, Democratic Republic of the Congo, Equatorial Guinea, Gabon, Ghana, Liberia, Nigeria

Bulbophyllum mahakamense J. J. Sm.

Distribution: Indonesia

Bulbophyllum maijenense Schltr.

Distribution: Papua New Guinea

Bulbophyllum major (Ridl.) van Royen
Bulbophyllum ischnopus Schltr. var. *major* Ridl.

Distribution: Indonesia

Bulbophyllum makoyanum (Rchb. f.) Ridl.
Cirrhopetalum makoyanum Rchb. f.

Distribution: Indonesia, Malaysia, Philippines, Singapore

Bulbophyllum malachadenia Cogn.
Malachadenia clavata Lindl.

Distribution: Brazil

Part II: Accepted Names / Noms Acceptés / Nombres Aceptado

Bulbophyllum maleolens Kraenzl.

Distribution: Madagascar

Bulbophyllum malleolabrum Carr

Distribution: Indonesia, Malaysia

Bulbophyllum mamberamense J. J. Sm.

Distribution: Indonesia

Bulbophyllum mananjarense Poiss.

Distribution: Madagascar

Bulbophyllum manarae Foldats

Distribution: Venezuela

Bulbophyllum mandibulare Rchb. f.

Distribution: Indonesia

Bulbophyllum mangenotii Bosser

Distribution: Madagascar

Bulbophyllum manobulbum Schltr.

Distribution: Papua New Guinea, Solomon Islands

Bulbophyllum maquilinguense Ames & Quisumb.

Distribution: Philippines

Bulbophyllum marginatum Schltr.

Distribution: Papua New Guinea

Bulbophyllum marivelense Ames

Distribution: Philippines

Bulbophyllum marojejiense H. Perrier

Distribution: Madagascar

210

Bulbophyllum maromanganum Schltr.

Distribution: Madagascar

Bulbophyllum marovoense H. Perrier

Distribution: Madagascar

Bulbophyllum marudiense Carr

Distribution: Indonesia

Bulbophyllum masaganapense Ames

Distribution: Philippines

Bulbophyllum masarangicum Schltr.
  Bulbophyllum masarangicum Schltr. var. *nanodes* Schltr.

Distribution: Indonesia

Bulbophyllum maskeliyense Livera

Distribution: India, Sri Lanka

Bulbophyllum masoalanum Schltr.

Distribution: Madagascar

Bulbophyllum masonii (Senghas) J. J. Wood
  Cirrhopetalum masonii Senghas

Distribution: Papua New Guinea

Bulbophyllum mastersianum (Rolfe) J. J. Sm.
  Cirrhopetalum mastersianum Rolfe

Distribution: Indonesia, Viet Nam

Bulbophyllum matitanense H. Perrier

Distribution: Madagascar

Bulbophyllum matitanense H. Perrier subsp. **rostratum** H. Perrier

Distribution: Madagascar

Bulbophyllum mattesii Sieder & Kiehn †
Bulbophyllum oblanceolatum Schltr.

Distribution: Papua New Guinea

Bulbophyllum maudeae A. D. Hawkes
Bulbophyllum nigrilabium H. Perrier

Distribution: Madagascar

Bulbophyllum maxillare (Lindl.) Rchb. f.
Cirrhopetalum maxillare Lindl.

Distribution: Philippines

Bulbophyllum maxillarioides Schltr.

Distribution: Papua New Guinea

Bulbophyllum maximum (Lindl.) Rchb. f.
Bulbophyllum cyrtopetalum Schltr.
Bulbophyllum djumaense (De Wild.) De Wild.
Bulbophyllum djumaense (De Wild.) De Wild. var. *grandifolium* De Wild.
Bulbophyllum moirianum A. D. Hawkes
Bulbophyllum nyassanum Schltr.
Bulbophyllum oxypterum (Lindl.) Rchb. f. var. *mosambicense* (Finet) De Wild
Bulbophyllum platyrhachis (Rolfe) Schltr.
Bulbophyllum subcoriaceum De Wild.
Megaclinium djumaense De Wild. var. *grandifolium* De Wild.
Megaclinium djumaensis De Wild.
Megaclinium maximum Lindl.
Megaclinium oxypterum Lindl. var. *mozambicense* Finet
Megaclinium platyrhachis Rolfe
Megaclinium purpuratum Lindl.
Megaclinium subcoriaceum De Wild.
Phyllorkis maxima (Lindl.) Kuntze

Distribution: Angola, Cameroon, Central African Republic, Congo, Côte d'Ivoire, Democratic Republic of the Congo, Gabon, Ghana, Guinea, Kenya, Liberia, Malawi, Mozambique, Nigeria, Sao Tome and Principe, Sierra Leone, Uganda, United Republic of Tanzania, Zambia, Zimbabwe

Bulbophyllum maximum (Lindl.) Rchb. f. var. **oxypterum** (Lindl.) Pérez-Vera
Bulbophyllum ciliatum Schltr.
Bulbophyllum oxypterum (Lindl.) Rchb. f.
Megaclinium oxypterum Lindl.
Phyllorkis oxyptera (Lindl.) Kuntze

Distribution: Côte d'Ivoire

Bulbophyllum mayombeense Garay

Distribution: Congo

212

Part II: Accepted Names / Noms Acceptés / Nombres Aceptado

Bulbophyllum mayrii J. J. Sm.

Distribution: Indonesia

Bulbophyllum mearnsii Ames
  Bulbophyllum carinatum Ames

Distribution: Philippines

Bulbophyllum mediocre Summerh.

Distribution: Cameroon, Sao Tome and Principe

Bulbophyllum medusae (Lindl.) Rchb. f.
  Cirrhopetalum medusae Lindl.
  Phyllorkis medusae (Lindl.) Kuntze

Distribution: Indonesia, Malaysia, Solomon Islands, Thailand

Bulbophyllum megalonyx Rchb. f.
  Phyllorkis megalonyx (Rchb. f.) Kuntze

Distribution: Comoros, Madagascar

Bulbophyllum melanoglossum Hayata
  Bulbophyllum linchianum S. S. Ying
  Bulbophyllum melanoglossum Hayata var. *rubropunctatum* (S. S. Ying) S. S. Ying
  Bulbophyllum rubropunctatum S. S. Ying
  Bulbophyllum striatum (Liu & Su) Nakajima
  Cirrhopetalum linchianum (Ying) Garay, Hamer & Siegerist
  Cirrhopetalum melanoglossum (Hayata) Hayata
  Cirrhopetalum striatum Liu & Su

Distribution: China, Taiwan, Province of China

Bulbophyllum melanoxanthum J. J. Verm. & B. A. Lewis

Distribution: Papua New Guinea, Solomon Islands

Bulbophyllum melilotus J. J. Sm.

Distribution: Indonesia

Bulbophyllum melinanthum Schltr.
  Ferruminaria melinantha (Schltr.) Garay, Hamer & Siegerist

Distribution: Papua New Guinea

Part II: Accepted Names / Noms Acceptés / Nombres Aceptado

Bulbophyllum melinoglossum Schltr.
  Hapalochilus melinoglossus (Schltr.) Garay & W. Kittr.

Distribution: Papua New Guinea

Bulbophyllum meliphagirostrum P. Royen

Distribution: Indonesia

Bulbophyllum melleum H. Perrier

Distribution: Madagascar

Bulbophyllum melloi Pabst

Distribution: Brazil

Bulbophyllum membranaceum Teijsm. & Binn.
  Bulbophyllum avicella Ridl.
  Bulbophyllum ciliatoides Seidenf.
  Bulbophyllum gibbonianum Schltr.
  Bulbophyllum nuruanum Schltr.
  Phyllorkis membranacea (Teijsm. & Binn.) Kuntze

 Distribution: Fiji, Indonesia, Malaysia, Palau, Papua New Guinea, Samoa, Solomon Islands, Thailand, Tonga, Vanuatu

Bulbophyllum membranifolium Hook. f.
  Bulbophyllum badium Ridl.
  Bulbophyllum crista-galli Kraenzl.
  Bulbophyllum cryptophoranthoides Kraenzl.
  Bulbophyllum insigne Ridl.
  Bulbophyllum sanguineomaculatum Ridl.
  Bulbophyllum scandens Kraenzl.
  Phyllorkis membranifolia (Hook. f.) Kuntze

Distribution: Indonesia, Malaysia, Philippines

Bulbophyllum menghaiense Z. H. Tsi

Distribution: China

Bulbophyllum menglunense Z. H. Tsi & Y. Z. Ma

Distribution: China

Bulbophyllum mentiferum J. J. Sm.

Distribution: Indonesia

Part II: Accepted Names / Noms Acceptés / Nombres Aceptado

Bulbophyllum mentosum Barb. Rodr.

Distribution: Brazil

Bulbophyllum meridense Rchb. f.
  Didactyle gladiata Lindl.
  Didactyle meridensis (Rchb. f.) Lindl.
  Phyllorkis meridensis (Rchb. f.) Kuntze
  Xiphizusa gladiata Rchb. f.

Distribution: Colombia, Ecuador, Peru, Venezuela

Bulbophyllum meristorhachis Garay & Dunst.

Distribution: Venezuela

Bulbophyllum merrittii Ames

Distribution: Philippines

Bulbophyllum mesodon J. J. Verm.

Distribution: Indonesia

Bulbophyllum metonymon Summerh.
  Bulbophyllum schlechteri H. Perrier
  Bulbophyllum schlechteri Kraenzl.
  Bulbophyllum zaratananae Schltr.

Distribution: Madagascar

Bulbophyllum micholitzianum Kraenzl.

Distribution: Indonesia

Bulbophyllum micholitzii Rolfe
  Hyalosema micholitzii (Rolfe) Rolfe

Distribution: Indonesia

Bulbophyllum micranthum Barb. Rodr.

Distribution: Brazil

Bulbophyllum microblepharon Schltr.

Distribution: Papua New Guinea

Bulbophyllum microbulbon Schltr.

Distribution: Papua New Guinea

Bulbophyllum microcala P. F. Hunt
  Bulbophyllum microcharis Schltr. (1923)

Distribution: Papua New Guinea

Bulbophyllum microdendron Schltr.

Distribution: Papua New Guinea

Bulbophyllum microglossum Ridl.

Distribution: Indonesia, Malaysia, Thailand

Bulbophyllum microlabium W. Kittr.
  Bulbophyllum parvilabium Schltr. (1919)

Distribution: Indonesia

Bulbophyllum micronesiacum Schltr.

Distribution: Kiribati, Palau

Bulbophyllum micropetaliforme Leite

Distribution: Brazil

Bulbophyllum micropetalum Rchb. f.

Distribution: Myanmar, Thailand

Bulbophyllum microrhombos Schltr.
  Hapalochilus microrhombos (Schltr.) Garay & W. Kittr.

Distribution: Papua New Guinea, Solomon Islands, Vanuatu

Bulbophyllum microsphaerum Schltr.

Distribution: Papua New Guinea

Bulbophyllum microtepalum Rchb. f.
  Phyllorkis microtepala (Rchb. f.) Kuntze

Distribution: Thailand

216

Part II: Accepted Names / Noms Acceptés / Nombres Aceptado

Bulbophyllum microtes Schltr.

Distribution: Papua New Guinea

Bulbophyllum microthamnus Schltr.

Distribution: Papua New Guinea

Bulbophyllum mimiense Schltr.

Distribution: Papua New Guinea

Bulbophyllum minahassae Schltr.
  Hyalosema minahassae (Schltr.) Rolfe

Distribution: Indonesia

Bulbophyllum minax Schltr.

Distribution: Madagascar

Bulbophyllum mindanaense Ames

Distribution: Philippines

Bulbophyllum mindorense Ames

Distribution: Philippines

Bulbophyllum minutibulbum W. Kittr.
  Bulbophyllum serpens Schltr.

Distribution: Papua New Guinea

Bulbophyllum minutilabrum H. Perrier

Distribution: Madagascar

Bulbophyllum minutipetalum Schltr.
  Bulbophyllum blepharadenium Schltr.
  Bulbophyllum squamipetalum Schltr.
  Peltopus minutipetalus (Schltr.) Szlach. & Marg.

Distribution: Papua New Guinea, Vanuatu

Bulbophyllum minutissimum (F. Muell.) F. Muell.
  Bulbophyllum moniliforme F. Muell.
  Bulbophyllum moniliforme R. King
  Dendrobium minutissimum F. Muell.
  Oncophyllum minutissimum (F. Muell.) D. L. Jones & M. A. Clem.

Part II: Accepted Names / Noms Acceptés / Nombres Aceptado

Phyllorkis minutissima (F. Muell.) Kuntze

Distribution: Australia

Bulbophyllum minutulum Ridl.

Distribution: Indonesia, Malaysia

Bulbophyllum minutum Thouars
Bulbophyllum implexum Jum. & H. Perrier
Phyllorkis minuta Thouars

Distribution: Madagascar

Bulbophyllum mirabile Hallier f.

Distribution: Brunei Darussalam, Indonesia, Malaysia

Bulbophyllum mirandaianum Hoehne

Distribution: Brazil

Bulbophyllum mirificum Schltr.

Distribution: Madagascar

Bulbophyllum mirum J. J. Sm.
Cirrhopetalum mirum (J. J. Sm.) Schltr.
Rhytionanthos mirum (J. J. Sm.) Garay, Hamer & Siegerist

Distribution: Indonesia, Singapore

Bulbophyllum mischobulbon Schltr.

Distribution: Papua New Guinea

Bulbophyllum mobilifilum Carr
Epicranthes mobilifilum (Carr) Garay & W. Kittr.

Distribution: Malaysia

Bulbophyllum moldenkeanum A. D. Hawkes
Bulbophyllum microglossum H. Perrier

Distribution: Madagascar

Part II: Accepted Names / Noms Acceptés / Nombres Aceptado

Bulbophyllum molossus Rchb. f.
Bulbophyllum sessiliflorum Kraenzl.

Distribution: Madagascar

Bulbophyllum mona-lisae Sieder & Kiehn †
Bulbophyllum melanoglossum Kraenzl.

Distribution: Philippines

Bulbophyllum monanthos Ridl.

Distribution: Malaysia, Thailand

Bulbophyllum monanthum (Kuntze) J. J. Sm.
Bulbophyllum devangiriense N. P. Balakr.
Bulbophyllum pteroglossum Schltr.
Bulbophyllum tiagii A. S. Chauhan
Bulbophyllum uniflorum Griff.
Phyllorkis monantha Kuntze
Sarcopodium uniflorum (Griff.) Lindl.

Distribution: Bhutan, Viet Nam

Bulbophyllum moniliforme C. S. P. Parish & Rchb. f.
Phyllorkis moniliformis (C. S. P. Parish & Rchb. f.) Kuntze

Distribution: Cambodia, India, Lao People's Democratic Republic, Malaysia, Myanmar, Thailand, Viet Nam

Bulbophyllum monosema Schltr.
Hapalochilus monosema (Schltr.) Garay & W. Kittr.

Distribution: Papua New Guinea

Bulbophyllum monstrabile Ames

Distribution: Philippines

Bulbophyllum montanum Schltr.

Distribution: Papua New Guinea

Bulbophyllum montense Ridl.
Bulbophyllum vinculibulbum Ames & C. Schweinf.

Distribution: Indonesia, Philippines

Bulbophyllum moratii Bosser

Distribution: Madagascar

Part II: Accepted Names / Noms Acceptés / Nombres Aceptado

Bulbophyllum morenoi Dodson & Vasquez

Distribution: Bolivia, Venezuela

Bulbophyllum moroides J. J. Sm.

Distribution: Indonesia

Bulbophyllum morotaiense J. J. Sm.
Cirrhopetalum morotaiense (J. J. Sm.) Garay, Hamer & Siegerist

Distribution: Indonesia

Bulbophyllum morphologorum Kraenzl.
Bulbophyllum dixoni Rolfe

Distribution: Thailand, Viet Nam

Bulbophyllum mucronatum (Blume) Lindl.
Diphyes mucronata Blume
Phyllorkis mucronata (Blume) Kuntze

Distribution: Indonesia

Bulbophyllum mucronifolium Rchb. f. & Warm.
Phyllorkis mucronifolia (Rchb. f. & Warm.) Kuntze

Distribution: Brazil

Bulbophyllum mulderae J. J. Verm.

Distribution: Indonesia, Papua New Guinea

Bulbophyllum multiflexum J. J. Sm.

Distribution: Indonesia

Bulbophyllum multiflorum Ridl.
Bulbophyllum ridleyi Kraenzl.
Phyllorkis multiflora (Ridl.) Kuntze

Distribution: Madagascar

Bulbophyllum multiligulatum H. Perrier

Distribution: Madagascar

Part II: Accepted Names / Noms Acceptés / Nombres Aceptado

Bulbophyllum multivaginatum Jum. & H. Perrier

Distribution: Madagascar

Bulbophyllum muriceum Schltr.

Distribution: Papua New Guinea

Bulbophyllum murkelense J. J. Sm.

Distribution: Indonesia

Bulbophyllum muscarirubrum Seidenf.

Distribution: Thailand

Bulbophyllum muscicola Schltr. (1913)

Distribution: Madagascar

Bulbophyllum muscohaerens J. J. Verm. & A. L. Lamb

Distribution: Malaysia

Bulbophyllum mutabile (Blume) Lindl.
 Bulbophyllum altispex Ridl. ex Stapf
 Bulbophyllum ceratostyloides (Schltr.) Schltr.
 Bulbophyllum mutabile (Blume) Lindl. var. *ceratostyloides* Schltr.
 Bulbophyllum pauciflorum Ames
 Bulbophyllum pokapindjangense J. J. Sm.
 Bulbophyllum semipellucidum J. J. Sm.
 Callista brachypetala (Lindl.) Kuntze
 Dendrobium brachypetalum Lindl.
 Diphyes mutabilis Blume
 Phyllorkis mutabilis (Blume) Kuntze

Distribution: Indonesia, Malaysia, Philippines, Thailand

Bulbophyllum mutabile (Blume) Lindl. var. **obesum** J. J. Verm.

Distribution: Indonesia

Bulbophyllum mutatum J. J. Sm.
 Hapalochilus mutatus (J. J. Sm.) Garay & W. Kittr.

Distribution: Indonesia

Bulbophyllum myolaense Garay, Hamer & Siegerist

Distribution: Papua New Guinea

Part II: Accepted Names / Noms Acceptés / Nombres Aceptado

Bulbophyllum myon J. J. Verm.

Distribution: Indonesia, Papua New Guinea

Bulbophyllum myrmecochilum Schltr.

Distribution: Madagascar

Bulbophyllum myrtillus Schltr.

Distribution: Papua New Guinea

Bulbophyllum mysorense (Rolfe) J. J. Sm.
 Cirrhopetalum mysorense Rolf

Distribution: India

Bulbophyllum mystax Schuit. & De Vogel

Distribution: Indonesia

Bulbophyllum mystrochilum Schltr.
 Hapalochilus mystrochilus (Schltr.) Garay & W. Kittr.

Distribution: Papua New Guinea

Bulbophyllum mystrophyllum Schltr.

Distribution: Papua New Guinea

Bulbophyllum nabawanense J. J. Wood & A. L. Lamb

Distribution: Indonesia

Bulbophyllum nagelii L. O. Williams

Distribution: Mexico

Bulbophyllum namoronae Bosser

Distribution: Madagascar

Bulbophyllum nannodes Schltr.
 Tapeinoglossum nannodes (Schltr.) Schltr.

Distribution: Papua New Guinea

222

Part II: Accepted Names / Noms Acceptés / Nombres Aceptado

Bulbophyllum nanopetalum Seidenf.

Distribution: Thailand

Bulbophyllum napelli Lindl.
 Bulbophyllum monosepalum Barb. Rodr.
 Phyllorkis napelli (Lindl.) Kuntze

Distribution: Brazil

Bulbophyllum napelloides Kraenzl.

Distribution: Brazil

Bulbophyllum nasica Schltr.
 Bulbophyllum blumei (Lindl.) J. J. Sm. var. *pumilum* J. J. Sm.

Distribution: Indonesia, Papua New Guinea

Bulbophyllum nasilabium Schltr.

Distribution: Papua New Guinea

Bulbophyllum nasseri Garay

Distribution: Philippines

Bulbophyllum navicula Schltr.

Distribution: Papua New Guinea

Bulbophyllum nebularum Schltr.

Distribution: Papua New Guinea

Bulbophyllum neglectum Bosser

Distribution: Madagascar

Bulbophyllum negrosianum Ames

Distribution: Philippines

Bulbophyllum neilgherrense Wight
 Phyllorkis neilgherensis (Wight) Kuntze

Distribution: Bangladesh, India, Myanmar

Bulbophyllum nematocaulon Ridl.
 Bulbophyllum johannis-winkleri J. J. Sm.
 Bulbophyllum oreas Ridl.

Distribution: Indonesia, Malaysia

Bulbophyllum nematopodum F. Muell.
 Papulipetalum nematopodum (F. Muell.) M. A. Clem. & D. L. Jones
 Phyllorkis nematopoda (F. Muell.) Kuntze

Distribution: Australia

Bulbophyllum nematorhizis Schltr.

Distribution: Papua New Guinea

Bulbophyllum nemorale L. O. Williams

Distribution: Indonesia, Philippines

Bulbophyllum nemorosum (Barb. Rodr.) Cogn.
 Didactyle nemorosa Barb. Rodr.

Distribution: Brazil

Bulbophyllum neo-caledonicum Schltr.
 Pelma neocaledonicum (Schltr.) Finet

Distribution: New Caledonia, Vanuatu

Bulbophyllum neo-pommeranicum Schltr.

Distribution: Papua New Guinea

Bulbophyllum neoebudicus (Garay, Hamer & Siegerist) Sieder †
 Hapalochilus neoebudicus Garay, Hamer & Siegerist

 Distribution: Vanuatu

Bulbophyllum neoguinense J. J. Sm.

Distribution: Indonesia

Bulbophyllum nephropetalum Schltr.

Distribution: Papua New Guinea

Part II: Accepted Names / Noms Acceptés / Nombres Aceptado

Bulbophyllum nervulosum Frapp. ex Cordem.

Distribution: Réunion

Bulbophyllum nesiotes Seidenf.

Distribution: Thailand

Bulbophyllum newportii (F. M. Bailey) Rolfe
  Adelopetalum newportii (F. M. Bailey) D. L. Jones & M. A. Clem.
  Bulbophyllum wanjurum T. E. Hunt
  Sarcochilus newportii F. M. Bailey

Distribution: Australia

Bulbophyllum ngoclinhensis Aver.

Distribution: Viet Nam

Bulbophyllum ngoyense Schltr.

Distribution: New Caledonia

Bulbophyllum nieuwenhuisii J. J. Sm.

Distribution: Indonesia

Bulbophyllum nigericum Summerh.

Distribution: Cameroon, Côte d'Ivoire, Nigeria

Bulbophyllum nigrescens Rolfe
  Bulbophyllum anguste-ellipticum Seidenf.

Distribution: China, Myanmar, Thailand, Viet Nam

Bulbophyllum nigriflorum H. Perrier

Distribution: Madagascar

Bulbophyllum nigrilabium Schltr.

Distribution: Indonesia, Papua New Guinea

Bulbophyllum nigripetalum Rolfe

Distribution: China, Thailand

Part II: Accepted Names / Noms Acceptés / Nombres Aceptado

Bulbophyllum nigritianum Rendle
Bulbophyllum africanum A. D. Hawkes
Bulbophyllum albidum De Wild.

Distribution: Cameroon, Côte d'Ivoire, Democratic Republic of the Congo, Gabon, Ghana, Liberia, Nigeria, Sierra Leone

Bulbophyllum nigropurpureum Carr

Distribution: Malaysia, Singapore

Bulbophyllum nipondhii Seidenf.

Distribution: Thailand

Bulbophyllum nitens Jum. & H. Perrier

Distribution: Madagascar

Bulbophyllum nitens Jum. & H. Perrier var. **intermedium** H. Perrier

Distribution: Madagascar

Bulbophyllum nitens Jum. & H. Perrier var. **majus** H. Perrier

Distribution: Madagascar

Bulbophyllum nitens Jum. & H. Perrier var. **minus** H. Perrier

Distribution: Madagascar

Bulbophyllum nitens Jum. & H. Perrier var. **pulverulentum** H. Perrier

Distribution: Madagascar

Bulbophyllum nitidum Schltr.
Hapalochilus nitidus (Schltr.) Senghas

Distribution: Papua New Guinea

Bulbophyllum nodosum (Rolfe) J. J. Sm.
Cirrhopetalum nodosum Rolfe
Rhytionanthos nodosum (Rolfe) Garay, Hamer & Siegerist

Distribution: India

Bulbophyllum notabilipetalum Seidenf.

Distribution: Thailand

226

Part II: Accepted Names / Noms Acceptés / Nombres Aceptado

Bulbophyllum novaciae J. J. Verm. & P. O'Byrne

Distribution: Indonesia

Bulbophyllum novae-hiberniae Schltr.
Hapalochilus novae-hiberniae (Schltr.) Garay & W. Kittr.

Distribution: Papua New Guinea, Solomon Islands

Bulbophyllum nubigenum Schltr.

Distribution: Papua New Guinea

Bulbophyllum nubinatum J. J. Verm.

Distribution: Indonesia

Bulbophyllum nummularia (Wendl. & Kraenzl.) Rolfe
Megaclinium nummularia Wendl. & Krzl.

Distribution: Cameroon

Bulbophyllum nummularioides Schltr.

Distribution: Papua New Guinea

Bulbophyllum nutans (Thouars) Thouars
Bulbophyllum andringitranum Schltr.
Bulbophyllum chrysobulbum H. Perrier
Bulbophyllum serpens Lindl.
Bulbophyllum tsinjoarivense H. Perrier
Cymbidium reptans Sw.
Dendrobium reptans Sw.
Phyllorkis nutans Thouars

Distribution: Madagascar, Mauritius, Réunion

Bulbophyllum nutans (Thouars) Thouars var. **variifolium** (Schltr.) Bosser
Bulbophyllum ambohitrense H. Perrier
Bulbophyllum variifolium Schltr.

Distribution: Madagascar

Bulbophyllum nymphopolitanum Kraenzl.

Distribution: Philippines

Part II: Accepted Names / Noms Acceptés / Nombres Aceptado

Bulbophyllum oblanceolatum King & Pantl.

Distribution: Malaysia

Bulbophyllum obliquum Schltr.

Distribution: Indonesia

Bulbophyllum obovatifolium J. J. Sm.

Distribution: Indonesia

Bulbophyllum obscuriflorum H. Perrier

Distribution: Madagascar

Bulbophyllum obtusatum (Jum. & H. Perrier) Schltr.
  Bulbophyllum obtusum Jum. & H. Perrier

Distribution: Madagascar

Bulbophyllum obtusiangulum Z. H. Ts

Distribution: China

Bulbophyllum obtusilabium W. Kittr.
  Bulbophyllum rhizomatosum Schltr. (1924)

Distribution: Madagascar

Bulbophyllum obtusipetalum J. J. Sm.
  Bulbophyllum spinulipes J. J. Sm.

Distribution: Indonesia, Malaysia

Bulbophyllum obtusum (Blume) Lindl.
  Bulbophyllum parvilabium Schltr. (1911)
  Diphyes obtusa Blume
  Phyllorkis obtusa (Blume) Kuntze

Distribution: Indonesia, Malaysia

Bulbophyllum obyrnei Garay, Hamer & Siegerist
  Osyricera ovata F. M. Bailey

Distribution: Papua New Guinea

228

Bulbophyllum occlusum Ridl.
Phyllorkis occlusa (Ridl.) Kuntze

Distribution: Madagascar, Réunion

Bulbophyllum occultum Thouars
Dendrochilum occultum (Thouars) Lindl.
Diphyes occulta (Thouars) Kuntze
Phyllorkis occulta Thouars

Distribution: Madagascar, Mauritius, Réunion

Bulbophyllum ochraceum (Barb. Rodr.) Cogn.
Didactyle ochracea Barb. Rodr.

Distribution: Brazil

Bulbophyllum ochrochlamys Schltr. (1924)

Distribution: Madagascar

Bulbophyllum ochroleucum Schltr.
Bulbophyllum adpressiscapum J. J. Sm.
Bulbophyllum furciferum J. J. Sm.
Bulbophyllum piundensis P. Royen
Bulbophyllum ramosum Schltr.
Bulbophyllum rostratum J. J. Sm.

Distribution: Indonesia, Papua New Guinea

Bulbophyllum ochthochilum J. J. Verm.

Distribution: Papua New Guinea

Bulbophyllum ochthodes J. J. Verm.

Distribution: Malaysia

Bulbophyllum octarrhenipetalum J. J. Sm.
Bulbophyllum longipiliferum J. J. Sm.
Bulbophyllum quadrans J. J. Sm.
Peltopus octarrhenipetalus (J. J. Sm.) Szlach. & Marg.

Distribution: Indonesia

Bulbophyllum octorhopalon Seidenf.
Epicranthes octorhopalon (Seidenf.) Garay & W. Kittr.

Distribution: Malaysia

Part II: Accepted Names / Noms Acceptés / Nombres Aceptado

Bulbophyllum odoardi Pfitzer

Distribution: Indonesia

Bulbophyllum odontoglossum Schltr.

Distribution: Papua New Guinea

Bulbophyllum odontopetalum Schltr.

Distribution: Papua New Guinea

Bulbophyllum odoratissimum (J. E. Sm.) Lindl.
Bulbophyllum hyacinthiodorum W. W. Sm.
Bulbophyllum congestum Rolfe
Phyllorkis odoratissima (J. E. Sm.) Kuntze
Stelis caudata D. Don
Stelis odoratissima J. E. Sm.
Tribrachia odoratissima (J. E. Sm.) Lindl.

Distribution: Bhutan, China, India, Lao People's Democratic Republic, Myanmar, Nepal, Thailand, Viet Nam

Bulbophyllum odoratissimum (J. E. Sm.) Lindl. var. **racemosum** N. P. Balakr.

Distribution: Bhutan

Bulbophyllum odoratum (Blume) Lindl.
Bulbophyllum braccatum Rchb. f.
Bulbophyllum brookesii Ridl.
Bulbophyllum crassicaudatum Ames & C. Schweinf.
Bulbophyllum elatius Ridl.
Bulbophyllum hortense J. J. Sm.
Bulbophyllum hortensoides Ames
Bulbophyllum niveum (J. J. Sm.) J. J. Sm.
Bulbophyllum odoratum (Blume) Lindl. var. *niveum* J. J. Sm.
Bulbophyllum polyarachne Ridl.
Diphyes odorata Blume
Phyllorkis odorata (Blume) Kuntze

Distribution: Indonesia, Malaysia, Philippines

Bulbophyllum odoratum (Blume) Lindl. var. **grandiflorum** J. J. Sm. ex Bull

Distribution: Indonesia

Bulbophyllum odoratum (Blume) Lindl. var. **obtusisepalum** J. J. Sm.

Distribution: Indonesia

Part II: Accepted Names / Noms Acceptés / Nombres Aceptado

Bulbophyllum odoratum (Blume) Lindl. var. **polyarachne** J. J. Sm.

Distribution: Indonesia

Bulbophyllum oerstedii (Rchb. f.) Hemsl.
Bolbophyllaria oerstedii Rchb. f.
Phyllorkis oerstedtii (Rchb. f.) Kuntze

Distribution: Belize, Brazil, Colombia, Costa Rica, Ecuador, Guatemala, Honduras, Mexico, Nicaragua, Panama, Venezuela

Bulbophyllum oliganthum Schltr.

Distribution: Papua New Guinea

Bulbophyllum oligoblepharon Schltr.

Distribution: Indonesia

Bulbophyllum oligochaete Schltr.

Distribution: Papua New Guinea

Bulbophyllum oligoglossum Rchb. f.
Phyllorkis oligoglossa (Rchb. f.) Kuntze

Distribution: Myanmar

Bulbophyllum olivinum J. J. Sm.

Distribution: Indonesia, Papua New Guinea

Bulbophyllum olivinum J. J. Sm. subsp. **linguiferum** J. J. Verm.

Distribution: Papua New Guinea

Bulbophyllum olorinum J. J. Sm.
Hapalochilus olorinus (J. J. Sm.) Garay & W. Kittr.

Distribution: Indonesia

Bulbophyllum omerandrum Hayata
Cirrhopetalum omerandrum (Hayata) Hayata

Distribution: Taiwan, Province of China

Bulbophyllum onivense H. Perrier

Distribution: Madagascar

Bulbophyllum oobulbum Schltr.

Distribution: Papua New Guinea

Bulbophyllum ophiuchus Ridl.
  Bulbophyllum mangoroanum Schltr.

Distribution: Madagascar

Bulbophyllum ophiuchus Ridl. var. **baronianum** H. Perrier

Distribution: Madagascar

Bulbophyllum orbiculare J. J. Sm.
  Bulbophyllum glabrilabre J. J. Sm.
  Bulbophyllum habropus Schltr.
  Bulbophyllum verrucirhachis Schltr.

Distribution: Indonesia, Papua New Guinea, Solomon Islands

Bulbophyllum orbiculare J. J. Sm. subsp. **cassideum** J. J. Verm.
  Bulbophyllum cassideum J. J. Sm.
  Bulbophyllum errabundum Ridl.
  Bulbophyllum truncatisepalum J. J. Sm.

Distribution: Indonesia, Papua New Guinea

Bulbophyllum orectopetalum Garay, Hamer & Siegerist

Distribution: India, Indonesia, Thailand, Viet Nam

Bulbophyllum oreocharis Schltr.

Distribution: Papua New Guinea

Bulbophyllum oreodorum Schltr.

Distribution: Madagascar

Bulbophyllum oreodoxa Schltr.
  Bulbophyllum chaetopus Schltr.

Distribution: Papua New Guinea

Bulbophyllum oreogenum Schltr.

Distribution: Papua New Guinea

Bulbophyllum oreonastes Rchb. f.
Bulbophyllum hookerianum Kraenzl.
Bulbophyllum infundibuliflorum J. B. Petersen
Bulbophyllum planiaxe J. B. Petersen
Bulbophyllum rhopalochilum Kraenzl.
Bulbophyllum zenkerianum Kraenzl.
Phyllorkis oreonastes (Rchb. f.) Kuntze

Distribution: Cameroon, Central African Republic, Côte d'Ivoire, Democratic Republic of
the Congo, Equatorial Guinea, Gabon, Ghana, Guinea, Liberia, Mozambique, Nigeria,
Rwanda, Sierra Leone, Uganda, Zambia, Zimbabwe

Bulbophyllum orezii Sath. Kumar
Bulbophyllum josephi M.Kumar Sequiera

Distribution: India

Bulbophyllum orientale Seidenf.

Distribution: China, Thailand, Viet Nam

Bulbophyllum origami J. J. Verm.
Peltopus origami (J. J. Verm.) Szlach. & Marg.

Distribution: Indonesia, Papua New Guinea

Bulbophyllum ornatissimum (Rchb. f.) J. J. Sm.
Cirrhopetalum ornatissimum Rchb. f.
Mastigion ornatissimum (Rchb. f.) Garay, Hamer & Siegerist
Phyllorkis ornatissima (Rchb. f.) Kuntze

Distribution: India, Myanmar, Philippines

Bulbophyllum ornatum Schltr.

Distribution: Papua New Guinea

Bulbophyllum ornithorhynchum (J. J. Sm.) Garay, Hamer & Siegerist
Cirrhopetalum ornithorhynchius J. J. Sm.
Hyalosema ornithorhynchum (J. J. Sm.) Rysy

Distribution: Indonesia

Bulbophyllum orohense J. J. Sm.

Distribution: Indonesia

Bulbophyllum orsidice Ridl.

Distribution: Indonesia

Bulbophyllum ortalis J. J. Verm.
Peltopus ortalis (J. J. Verm.) Szlach. & Marg.

Distribution: Papua New Guinea

Bulbophyllum orthoglossum Kraenzl.

Distribution: Philippines

Bulbophyllum orthosepalum J. J. Verm.
Bulbophyllum hashimotoi Yukawa & Karasawa

Distribution: Papua New Guinea

Bulbophyllum osyricera Schltr.
Bulbophyllum crassifolioides Aver.
Bulbophyllum crassifolium (Blume) J. J. Sm.
Osyricera crassifolia Blume

Distribution: Indonesia

Bulbophyllum osyriceroides J. J. Sm.
Osyricera osyriceroides (J. J. Sm.) Garay, Hamer & Siegerist

Distribution: Indonesia

Bulbophyllum othonis (Kuntze) J. J. Sm.
Bulbophyllum nutans (Lindl.) Rchb. f.
Cirrhopetalum nutans Lindl.
Phyllorkis othonis Kuntze

Distribution: Philippines

Bulbophyllum otochilum J. J. Verm.

Distribution: Indonesia

Bulbophyllum otoglossum Tuyama

Distribution: Bhutan

Bulbophyllum ovale Ridl.

Distribution: Indonesia

Bulbophyllum ovalifolium (Blume) Lindl.
Bulbophyllum parvulum Lindl.
Bulbophyllum tinea Ridl.
Diphyes ovalifolia Blume
Diphyes pusilla Blume
Phyllorkis ovalifolia (Blume) Kuntze
Phyllorkis parvula (Lindl.) Kuntze

Distribution: Indonesia, Malaysia

Bulbophyllum ovalitepalum J. J. Sm.

Distribution: Indonesia

Bulbophyllum ovatilabellum Seidenf.

Distribution: Thailand

Bulbophyllum ovatolanceatum J. J. Sm.

Distribution: Indonesia

Bulbophyllum ovatum Seidenf.

Distribution: Thailand

Bulbophyllum oxyanthum Schltr.
Hapalochilus oxyanthus (Schltr.) Garay & W. Kittr.

Distribution: Papua New Guinea

Bulbophyllum oxycalyx Schltr.
Bulbophyllum rubescens Schltr. var. *meizobulbon* Schltr.

Distribution: Madagascar

Bulbophyllum oxycalyx Schltr. var. **rubescens** (Schltr.) Bosser
Bulbophyllum caeruleolineatum H. Perrier
Bulbophyllum loxodiphyllum H. Perrier
Bulbophyllum rostriferum H. Perrier
Bulbophyllum rubescens Schltr.

Distribution: Madagascar

Bulbophyllum oxychilum Schltr.
Bulbophyllum buntingii Rendle
Bulbophyllum ellipticum De Wild.

Distribution: Cameroon, Central African Republic, Côte d'Ivoire, Democratic Republic of the Congo, Gabon, Ghana, Liberia, Nigeria, Uganda

Part II: Accepted Names / Noms Acceptés / Nombres Aceptado

Bulbophyllum oxysepaloides Ridl.

Distribution: Indonesia

Bulbophyllum pabstii Garay
Bulbophyllum fractiflexum Pabst

Distribution: Brazil

Bulbophyllum pachyacris J. J. Sm.

Distribution: Indonesia

Bulbophyllum pachyanthum Schltr.

 Distribution: Fiji, New Caledonia, Samoa, Solomon Islands, Tonga

Bulbophyllum pachyglossum Schltr.

Distribution: Papua New Guinea, Solomon Islands

Bulbophyllum pachyneuron Schltr.

Distribution: Indonesia

Bulbophyllum pachypus Schltr.

Distribution: Madagascar

Bulbophyllum pachyrachis (A. Rich.) Griseb.
Bolbophyllaria pachyrhachis (A. Rich) Rchb. f.
Bulbophyllum vinosum Schltr.
Phyllorkis pachyrhachis (A. Rich.) Kuntze
Pleurothallis pachyrrhachis A. Rich.

Distribution: Belize, Bolivia, Costa Rica, Cuba, Ecuador, El Salvador, Guatemala, Jamaica, Mexico, Panama, United States of America, Venezuela

Bulbophyllum pachytelos Schltr.
Bulbophyllum geminum Schltr.
Bulbophyllum proximum Schltr.

Distribution: Indonesia, Papua New Guinea

Bulbophyllum pahudi (De Vriese) Rchb.f.
Bulbophyllum javanicum Miq.
Cirrhopetalum capitatum (Blume) Lindl.
Cirrhopetalum flagelliforme Teijsm. & Binn. ex Rchb. f.
Cirrhopetalum pahudi De Vriese

Ephippium capitatum Blume

Distribution: Indonesia

Bulbophyllum paleiferum Schltr.

Distribution: Madagascar

Bulbophyllum palilabre J. J. Sm.

Distribution: Indonesia

Bulbophyllum pallens (Jum. & H. Perrier) Schltr.
  Bulbophyllum ophiuchus Ridl. var. *pallens* Jum. & H. Perrier

Distribution: Madagascar

Bulbophyllum pallidiflorum Schltr.

Distribution: New Caledonia

Bulbophyllum pallidum Seidenf.

Distribution: Thailand

Bulbophyllum pampangense Ames

Distribution: Philippines

Bulbophyllum pan Ridl.

Distribution: Malaysia

Bulbophyllum pandanetorum Summerh.

Distribution: Gabon

Bulbophyllum pandurella Schltr.

Distribution: Madagascar

Bulbophyllum paniscus Ridl.

Distribution: Indonesia

Part II: Accepted Names / Noms Acceptés / Nombres Aceptado

Bulbophyllum pantlingii S. Z. Lucksom
Bulbophyllum flavidum S. Z. Lucksom

Distribution: Bhutan

Bulbophyllum pantoblepharon Schltr.

Distribution: Madagascar

Bulbophyllum pantoblepharon Schltr. var. **vestitum** H. Perrier

Distribution: Madagascar

Bulbophyllum papangense H. Perrier

Distribution: Madagascar

Bulbophyllum papilio J. J. Sm.

Distribution: Indonesia

Bulbophyllum papillatum J. J. Sm.
Bulbophyllum papillosum J. J. Sm.

Distribution: Indonesia

Bulbophyllum papillipetalum Ames

Distribution: Philippines

Bulbophyllum papillosefilum Carr
Epicranthes papillosofilum (Carr) Garay & W. Kittr.

Distribution: Malaysia

Bulbophyllum papuliferum Schltr.

Distribution: Indonesia

Bulbophyllum papuliglossum Schltr.

Distribution: Papua New Guinea

Bulbophyllum papulipetalum Schltr.
Papulipetalum angustifolium M. A. Clem. & D. L. Jones

Distribution: Papua New Guinea

238

Part II: Accepted Names / Noms Acceptés / Nombres Aceptado

Bulbophyllum papulosum Garay

Distribution: Philippines

Bulbophyllum parabates J. J. Verm.

Distribution: Papua New Guinea

Bulbophyllum paranaense Schltr.

Distribution: Brazil

Bulbophyllum paranaense Schltr. var. **pauloense** Hoehne & Schltr.

Distribution: Brazil

Bulbophyllum pardalinum Ridl.

Distribution: Indonesia

Bulbophyllum pardalotum Garay, Hamer & Siegerist

Distribution: Philippines

Bulbophyllum parviflorum C. S. P. Parish & Rchb. f.
 Bulbophyllum thomsoni Hook. f.
 Phyllorkis parviflora (C. S. P. Parish & Rchb. f.) Kuntze
 Phyllorkis thomsonii (Hook. f.) Kuntze

Distribution: Bhutan, India, Myanmar, Thailand, Viet Nam

Bulbophyllum parvum Summerh.

 Distribution: Sierra Leone

Bulbophyllum patella J. J. Verm.
 Peltopus patella (J. J. Verm.) Szlach. & Marg.

Distribution: Papua New Guinea

Bulbophyllum patens King ex Hook. f.
 Phyllorkis patens (King ex Hook. f.) Kuntze

Distribution: Indonesia, Malaysia, Thailand

Bulbophyllum paucisetum J. J. Sm.

Distribution: Indonesia

Bulbophyllum paululum Schltr.

Distribution: Papua New Guinea

Bulbophyllum pectenveneris (Gagnep.) Seidenf.
 Bulbophyllum tingabarinum Garay, Hamer & Siegerist
 Cirrhopetalum miniatum Rolfe
 Cirrhopetalum pectenveneris Gagnep.

Distribution: China, Lao People's Democratic Republic, Taiwan, Province of China, Viet Nam

Bulbophyllum pectinatum Finet
 Bulbophyllum pectinatum Finet var. *transarisanense* (Hayata) S. S. Ying
 Bulbophyllum spectabile Rolfe
 Bulbophyllum transarisanense Hayata
 Bulbophyllum viridiflorum Hayata

Distribution: China, India, Myanmar, Taiwan, Province of China , Thailand, Viet Nam

Bulbophyllum pelicanopsis J. J. Verm. & A. L. Lamb

Distribution: Indonesia

Bulbophyllum peltopus Schltr.
 Bulbophyllum planilabre Schltr.
 Peltopus greuterianus Szlach. & Marg.
 Peltopus planilabris (Schltr.) Szlach. & Marg.

Distribution: Indonesia, Papua New Guinea

Bulbophyllum pemae Schltr.
 Hapalochilus pemae (Schltr.) Garay & W. Kittr.

Distribution: Papua New Guinea

Bulbophyllum penduliscapum J. J. Sm.
 Bulbophyllum jarense Ames
 Bulbophyllum macrophyllum Kraenzl.

Distribution: Indonesia, Malaysia, Philippines

Bulbophyllum pendulum Thouars
 Phyllorkis pendula Thouars
 Tribrachia pendula (Thouars) Lindl.

Distribution: Mauritius, Réunion

Bulbophyllum penicillium C. S. P. Parish & Rchb. f.
 Bulbophyllum inopinatum W. W. Sm.
 Phyllorkis penicillium (C. S. P. Parish & Rchb. f.) Kuntze

240

Distribution: Bhutan, India, Myanmar, Viet Nam

Bulbophyllum peninsulare Seidenf.

Distribution: Thailand

Bulbophyllum pentaneurum Seidenf.

Distribution: Thailand

Bulbophyllum pentasticha (Pfitzer ex Kraenzl.) H. Perrier
Bolbophyllaria pentasticha Pfitzer ex Kraenzl.

Distribution: Madagascar

Bulbophyllum peperomiifolium J. J. Sm.

Distribution: Indonesia

Bulbophyllum peramoenum Ames

Distribution: Philippines

Bulbophyllum percorniculatum H. Perrier

Distribution: Madagascar

Bulbophyllum perductum J. J. Sm.

Distribution: Indonesia

Bulbophyllum perductum J. J. Sm. var. **sebesiense** J. J. Sm.

Distribution: Indonesia

Bulbophyllum perexiguum Ridl.

Distribution: Indonesia

Bulbophyllum perforans J. J. Sm.

Distribution: Indonesia

Bulbophyllum perii Schltr.

Distribution: Brazil

241

Part II: Accepted Names / Noms Acceptés / Nombres Aceptado

Bulbophyllum perparvulum Schltr.
Bulbophyllum perpusillum Ridl.

Distribution: Indonesia

Bulbophyllum perpendiculare Schltr.

Distribution: Indonesia

Bulbophyllum perpusillum Wendl. & Kraenzl.

Distribution: Madagascar

Bulbophyllum perreflexum Bosser & P. J. Cribb

Distribution: Madagascar

Bulbophyllum perrieri Schltr.

Distribution: Madagascar

Bulbophyllum pervillei Rolfe ex Elliot

Distribution: Madagascar

Bulbophyllum petiolare Thwaites
Phyllorkis petiolaris (Thwaites) Kuntze

Distribution: Sri Lanka

Bulbophyllum petiolatum J. J. Sm.

Distribution: Indonesia

Bulbophyllum peyerianum (Kranzlin) Seidenf.
Bulbophyllum klossii Ridl.
Cirrhopetalum peyerianum Kraenzl.

Distribution: Indonesia

Bulbophyllum peyrotii Bosser
Bulbophyllum fimbriatum H. Perrier
Bulbophyllum flickingerianum A. D. Hawkes
Bulbophyllum mayae A. D. Hawkes

Distribution: Madagascar

Part II: Accepted Names / Noms Acceptés / Nombres Aceptado

Bulbophyllum phaeanthum Schltr.

Distribution: Indonesia

Bulbophyllum phaeoglossum Schltr.

Distribution: Papua New Guinea

Bulbophyllum phaeoneuron Schltr.

Distribution: Indonesia

Bulbophyllum phaeorhabdos Schltr.

Distribution: Papua New Guinea

Bulbophyllum phalaenopsis J. J. Sm.

Distribution: Indonesia

Bulbophyllum phayamense Seidenf.

Distribution: Thailand

Bulbophyllum philippinense Ames

Distribution: Philippines

Bulbophyllum phillipsianum Kores

Distribution: Fiji

Bulbophyllum phormion J. J. Verm.

Distribution: Indonesia, Papua New Guinea

Bulbophyllum phreatiopse J. J. Verm.

Distribution: Papua New Guinea

Bulbophyllum phymatum J. J. Verm.
  Epicranthes phymatum (J. J. Verm.) Garay & W. Kittr.

Distribution: Papua New Guinea

Bulbophyllum physocoryphum Seidenf.

Distribution: Thailand

Bulbophyllum picturatum (Lodd.) Rchb.f.
Bulbophyllum eberhardtii (Gagnep.) Seidenf.
Cirrhopetalum eberhardtii Gagnep.
Cirrhopetalum picturatum Lodd.
Phyllorkis picturata (Lodd.) Kuntze

Distribution: India, Myanmar, Thailand

Bulbophyllum pidacanthum J. J. Verm.

Distribution: Indonesia, Papua New Guinea

Bulbophyllum piestobulbon Schltr.
Bulbophyllum bambusifolium J. J. Sm.

Distribution: Papua New Guinea, Solomon Islands

Bulbophyllum piestoglossum J. J. Verm.

Distribution: Philippines

Bulbophyllum pileatum Lindl.
Bulbophyllum fibrosum J. J. Sm.
Phyllorkis pileata (Lindl.) Kuntze
Sarcopodium pileatum (Lindl.) Lindl.

Distribution: Indonesia, Malaysia, Singapore

Bulbophyllum piliferum J. J. Sm.

Distribution: Indonesia, Papua New Guinea

Bulbophyllum pilosum J. J. Verm.

Distribution: Malaysia

Bulbophyllum piluliferum King & Pantl.

Distribution: Bhutan, India

Bulbophyllum pinicolum Gagnep.

Distribution: Cambodia, Viet Nam

Bulbophyllum pipio Rchb. f.
Bulbophyllum milesii Summerh.
Phyllorkis pipio (Rchb. f.) Kuntze

Distribution: Cameroon, Côte d'Ivoire, Ghana, Nigeria, Sierra Leone

244

Part II: Accepted Names / Noms Acceptés / Nombres Aceptado

Bulbophyllum pisibulbum J. J. Sm.

Distribution: Indonesia

Bulbophyllum placochilum J. J. Verm.

Distribution: Indonesia, Malaysia

Bulbophyllum plagiatum Ridl.

Distribution: Indonesia

Bulbophyllum plagiopetalum Schltr.

Distribution: Papua New Guinea

Bulbophyllum planibulbe (Ridl.) Ridl.
Bulbophyllum tenerum Ridl.
Cirrhopetalum planibulbe Ridl.

Distribution: Indonesia, Malaysia, Thailand

Bulbophyllum planibulbe (Ridl.) Ridl. var. **sumatranum** J. J. Sm.

Distribution: Indonesia

Bulbophyllum planitiae J. J. Sm.

Distribution: Indonesia

Bulbophyllum platypodum H. Perrier

Distribution: Madagascar

Bulbophyllum pleiopterum Schltr.

Distribution: Madagascar

Bulbophyllum pleurothallianthum Garay

Distribution: Indonesia

Bulbophyllum pleurothalloides Ames

Distribution: Philippines

Bulbophyllum pleurothallopsis Schltr.

Distribution: Madagascar

Bulbophyllum plicatum J. J. Verm.
 Peltopus plicatus (J. J. Verm.) Szlach. & Marg.

Distribution: Papua New Guinea

Bulbophyllum plumatum Ames
 Bulbophyllum jacobsonii J. J. Sm.
 Rhytionanthos plumatum (Ames) Garay, Hamer & Siegerist

Distribution: Indonesia, Malaysia, Philippines

Bulbophyllum plumosum (Barb. Rodr.) Cogn.
 Didactyle plumosa Barb. Rodr.

Distribution: Brazil

Bulbophyllum plumula Schltr.

Distribution: Papua New Guinea

Bulbophyllum pocillum J. J. Verm.

Distribution: Indonesia

Bulbophyllum poekilon Carr

Distribution: Malaysia

Bulbophyllum poilanei Gagnep.

Distribution: China, Viet Nam

Bulbophyllum polliculosum Seidenf.
 Cirrhopetalum polliculosum (Seidenf.) Garay, Hamer & Siegerist

Distribution: Thailand

Bulbophyllum polyblepharon Schltr.

Distribution: Papua New Guinea

Bulbophyllum polycyclum J. J. Verm.

Distribution: Indonesia, Malaysia

Part II: Accepted Names / Noms Acceptés / Nombres Aceptado

Bulbophyllum polygaliflorum J. J. Wood

Distribution: Indonesia, Malaysia

Bulbophyllum polyphyllum Schltr.

Distribution: Papua New Guinea

Bulbophyllum polyrhizum Lindl.
  Phyllorkis polyrhiza (Lindl.) Kuntze

Distribution: Bhutan, India, Myanmar, Thailand

Bulbophyllum popayanense Leme & Kraenzl.

Distribution: Colombia

Bulbophyllum porphyrostachys Summerh.

Distribution: Cameroon, Nigeria

Bulbophyllum porphyrotriche J. J. Verm.

Distribution: Indonesia

Bulbophyllum posticum J. J. Sm.
  Bulbophyllum bicornutum Schltr.
  Bulbophyllum diceras Schltr.

Distribution: Indonesia, Papua New Guinea

Bulbophyllum potamophila Schltr.

Distribution: Papua New Guinea

Bulbophyllum praestans Kraenzl.

Distribution: Indonesia

Bulbophyllum praetervisum J. J. Verm.

Distribution: Brunei Darussalam, Malaysia

Bulbophyllum prianganense J. J. Sm.
  Bulbophyllum hamatifolium J. J. Sm.

Distribution: Indonesia

Part II: Accepted Names / Noms Acceptés / Nombres Aceptado

Bulbophyllum prismaticum Thouars
Lyraea prismatica Lindl.
Phyllorkis prismatica Thouars

Distribution: Mauritius, Réunion

Bulbophyllum pristis J. J. Sm.

Distribution: Indonesia

Bulbophyllum proboscideum (Gagnep.) Seidenf. & Smitinand
Cirrhopetalum proboscideum Gagnep.
Mastigion proboscideum (Gagnep.) Garay, Hamer & Siegerist

Distribution: Lao People's Democratic Republic, Thailand

Bulbophyllum procerum Schltr.

Distribution: Papua New Guinea

Bulbophyllum proculcastris J. J. Verm.

Distribution: Malaysia

Bulbophyllum proencai Leite

Distribution: Brazil

Bulbophyllum profusum Ames

Distribution: Philippines

Bulbophyllum propinquum Kraenzl.
Bulbophyllum chlorostachys Schltr.

Distribution: Thailand

Bulbophyllum prorepens Summerh.

Distribution: Democratic Republic of the Congo, Rwanda

Bulbophyllum protectum H. Perrier

Distribution: Madagascar

Bulbophyllum protractum Hook. f.
Phyllorkis protracta (Hook. f.) Kuntze

Distribution: India, Myanmar, Thailand, Viet Nam

Part II: Accepted Names / Noms Acceptés / Nombres Aceptado

Bulbophyllum proudlockii (King & Pantl.) J. J. Sm.
Cirrhopetalum proudlockii King & Pantl.

Distribution: India

Bulbophyllum pseudofilicaule J. J. Sm.

Distribution: Indonesia

Bulbophyllum pseudopelma J. J. Verm. & P. O'Byrne

Distribution: Indonesia

Bulbophyllum pseudopicturatum (Garay) Sieder & Kiehn †
Cirrhopetalum pseudopicturatum Garay

Distribution: Thailand

Bulbophyllum pseudoserrulatum J. J. Sm.

Distribution: Indonesia

Bulbophyllum pseudotrias J. J. Verm.

Distribution: Papua New Guinea

Bulbophyllum psilorhopalon Schltr.
Epicranthes psilorhopalon (Schltr.) Garay & W. Kittr.

Distribution: Papua New Guinea

Bulbophyllum psittacoglossum Rchb. f.
Bulbophyllum affinoides Guillaumin
Phyllorkis psittacoglossa (Rchb. f.) Kuntze
Sarcopodium psittacoglossum (rchb.f.) Rchb. f.

Distribution: China, Myanmar, Thailand, Viet Nam

Bulbophyllum psychoon Rchb. f.
Phyllorkis psychoon (Rchb. f.) Kuntze

Distribution: India

Bulbophyllum ptiloglossum Wendl. & Kraenzl.

Distribution: Madagascar

Part II: Accepted Names / Noms Acceptés / Nombres Aceptado

Bulbophyllum ptilotes Schltr.

Distribution: Papua New Guinea

Bulbophyllum ptychantyx J. J. Verm.
Peltopus ptychantyx (J. J. Verm.) Szlach. & Marg.

Distribution: Papua New Guinea

Bulbophyllum pubiflorum Schltr.

Distribution: Indonesia

Bulbophyllum pugilanthum J. J. Wood

Distribution: Indonesia

Bulbophyllum pugioniforme J. J. Sm.
Bulbophyllum attenuatum Rolfe

Distribution: Indonesia

Bulbophyllum puguahaanense Ames
Bulbophyllum chekaense Carr
Cirrhopetalum minutiflorum Garay, Hamer & Siegerist
Cirrhopetalum puguahaanense (Ames) Garay, Hamer & Siegerist

Distribution: Indonesia, Malaysia, Philippines

Bulbophyllum pulchellum Ridl.
Cirrhopetalum concinnum Hook. f.
Phyllorkis ridleyana Kuntze

Distribution: Malaysia, Thailand

Bulbophyllum pulchrum Ridl. var. **brachysepalum** Ridl.

Distribution: Malaysia

Bulbophyllum pulchrum (N. E. Br.) J. J. Sm.
Bulbophyllum elegans (Teijsm. & Binn.) J. J. Sm.
Bulbophyllum insulare J. J. Sm.
Bulbophyllum koordersii (Rolfe) J. J. Sm.
Bulbophyllum pulchrum (N. E. Br.) J. J. Sm. var. *cliftonii* Hort.
Bulbophyllum strangularium (Rchb. f.) J. J. Sm.
Cirrhopetalum elegans Teijsm. & Binn.
Cirrhopetalum koordersii Rolfe
Cirrhopetalum pulchrum N. E. Br.
Cirrhopetalum strangularium Rchb. f.
Cirrhopetalum zygoglossum Garay, Hamer & Siegerist

Distribution: India, Indonesia

Part II: Accepted Names / Noms Acceptés / Nombres Aceptado

Bulbophyllum pulvinatum Schltr.

Distribution: Papua New Guinea

Bulbophyllum pumilio C. S. P. Parish & Rchb. f.
  Cirrhopetalum pumilio (C. S. P. Parish & Rchb. f.) Hook. f.
  Phyllorkis pumilio (C. S. P. Parish & Rchb. f.) Kuntze

Distribution: Myanmar, Thailand, Viet Nam

Bulbophyllum pumilum (Sw.) Lindl.
  Bulbophyllum calabaricum Rolfe
  Bulbophyllum dorotheae Rendle
  Bulbophyllum drallei Rchb. f.
  Bulbophyllum elachon J. J. Verm.
  Bulbophyllum flavidum Lindl.
  Bulbophyllum flavidum Lindl. var. *purpureum* Rolfe
  Bulbophyllum gabonis Lindl. & Rchb. f.
  Bulbophyllum herminiostachys (Rchb. f.) Rchb. f.
  Bulbophyllum imogeniae Hamilton
  Bulbophyllum leucopogon Kraenzl.
  Bulbophyllum moliwense Schltr.
  Bulbophyllum nanum De Wild.
  Bulbophyllum papillosum Finet
  Bulbophyllum parvimentatum Lindl.
  Bulbophyllum pavimentatum Lindl.
  Bulbophyllum porphyroglossum Kraenzl.
  Bulbophyllum recurvum Lindl.
  Bulbophyllum verecundum Summerh.
  Bulbophyllum winkleri Schltr.(1906)
  Bulbophyllum yangambiense Louis & Mullenders ex Geerinck
  Dendrobium pumilum Sw.
  Genyorchis pumila Schltr.
  Phyllorkis flavida (Lindl.) Kuntze
  Phyllorkis herminostachys (Rchb. f.) Kuntze
  Phyllorkis pavimenta (Lindl.) Kuntze
  Phyllorkis recurva (Lindl.) Kuntze
  Taurostalix herminiostachys Rchb. f.

Distribution: Cameroon, Congo, Côte d'Ivoire, Democratic Republic of the Congo, Equatorial Guinea, Gabon, Ghana, Guinea, Liberia, Nigeria, Sierra Leone

Bulbophyllum punamense Schltr.

Distribution: Papua New Guinea

Bulbophyllum punctatum Barb. Rodr.

Distribution: Brazil

Bulbophyllum pungens Schltr.

Distribution: Papua New Guinea

Part II: Accepted Names / Noms Acceptés / Nombres Aceptado

Bulbophyllum pungens Schltr. var. **pachyphyllum** Schltr.

Distribution: Papua New Guinea

Bulbophyllum puntjakense J. J. Sm.

Distribution: Indonesia

Bulbophyllum purpurascens Teijsm. & Binn.
  Bulbophyllum citrinum (Ridl.) Ridl.
  Bulbophyllum curtisii (Hook. f.) Ridl.
  Bulbophyllum perakense Ridl.
  Bulbophyllum rhizophoreti Ridl.
  Bulbophyllum tenasserimense J. J. Sm.
  Cirrhopetalum citrinum Ridl.
  Cirrhopetalum compactum Rolfe
  Cirrhopetalum lendyanum Rchb. f.
  Cirrhopetalum pallidum Schltr.

Distribution: Indonesia, Malaysia, Myanmar, Thailand

Bulbophyllum purpureifolium Aver.

Distribution: Viet Nam

Bulbophyllum purpurellum Ridl.

Distribution: Indonesia

Bulbophyllum purpureorhachis (De Wild.) Schltr.
  Megaclinium purpureorachis De Wild.

Distribution: Cameroon, Congo, Côte d'Ivoire, Democratic Republic of the Congo, Gabon

Bulbophyllum purpureum Thwaites

Distribution: Sri Lanka

Bulbophyllum pusillum Thouars
  Phyllorkis pusilla Thouars

Distribution: Mauritius

Bulbophyllum pustulatum Ridl.

Distribution: Indonesia, Malaysia

Bulbophyllum putidum (Teijsm. & Binn.) J. J. Sm.
  Bulbophyllum fascinator (Rolfe) Rolfe

252

Cirrhopetalum appendiculatum Rolfe var. *fascinator* (Rolfe) Hort.
Cirrhopetalum fascinator Rolfe
Cirrhopetalum putidum Teijsm. & Binn.
Mastigion appendiculatum (Rolfe) Garay, Hamer & Siegerist
Mastigion fascinator (Rolfe) Garay, Hamer & Siegerist
Mastigion putidum (Teijsm. & Binn.) Garay, Hamer & Siegerist

Distribution: India, Indonesia, Lao People's Democratic Republic, Malaysia, Philippines, Thailand, Viet Nam

Bulbophyllum putii Seidenf.

Distribution: Thailand

Bulbophyllum pygmaeum (Sm.) Lindl.
Bulbophyllum ichthyostomum Colenso
Dendrobium pygmaeum Sm.
Ichthyostomum pygmaeum (Sm.) D. L. Jones, M. A. Clem. & Molloy
Phyllorkis pygmaea (Sm.) Kuntze

Distribution: New Zealand

Bulbophyllum pyridion J. J. Verm.

Distribution: Indonesia

Bulbophyllum pyroglossum Schuit. & De Vogel

Distribution: Papua New Guinea

Bulbophyllum quadrangulare J. J. Sm.
Bulbophyllum adenambon Schltr.

Distribution: Indonesia

Bulbophyllum quadrangulare J. J. Sm. var. **latisepalum** J. J. Sm.

Distribution: Indonesia

Bulbophyllum quadrangulum Z. H. Tsi

Distribution: China

Bulbophyllum quadrialatum H. Perrier

Distribution: Madagascar

Part II: Accepted Names / Noms Acceptés / Nombres Aceptado

Bulbophyllum quadricarinum Kores

Distribution: Fiji

Bulbophyllum quadricaudatum J. J. Sm.
Hapalochilus quadricaudatus (J. J. Sm.) Garay & W. Kittr.

Distribution: Indonesia

Bulbophyllum quadrichaete Schltr.

Distribution: Papua New Guinea

Bulbophyllum quadricolor (Barb. Rodr.) Cogn.
Didactyle quadricolor Barb. Rodr.

Distribution: Brazil

Bulbophyllum quadrifalciculatum J. J. Sm.

Distribution: Indonesia

Bulbophyllum quadrifarium Rolfe

Distribution: Madagascar

Bulbophyllum quadrisetum Lindl.
Phyllorkis quadriseta (Lindl.) Kuntze

Distribution: Brazil

Bulbophyllum quadrisubulatum J. J. Sm.

Distribution: Indonesia

Bulbophyllum quasimodo J. J. Verm.

Distribution: Indonesia, Papua New Guinea

Bulbophyllum quinquelobum Schltr.

Distribution: Papua New Guinea

Bulbophyllum quinquelobum Schltr. var. **lancilabium** Schltr.

Distribution: Papua New Guinea

254

Bulbophyllum racemosum Rolfe

Distribution: Indonesia

Bulbophyllum radicans F. M. Bailey
Bulbophyllum cilioglossum R. S. Rogers & Nicholls
Fruticicola radicans (F. M. Bailey) M. A. Clem. & D. L. Jones

Distribution: Australia

Bulbophyllum rajanum J. J. Sm.

Distribution: Indonesia

Bulbophyllum ramulicola Schuit. & De Vogel

Distribution: Indonesia

Bulbophyllum ranomafanae Bosser & P. J. Cribb

Distribution: Madagascar

Bulbophyllum rariflorum J. J. Sm.

Distribution: Indonesia, Malaysia

Bulbophyllum rarum Schltr.

Distribution: Papua New Guinea

Bulbophyllum rauhii Toill.-Gen. & Bosser

Distribution: Madagascar

Bulbophyllum rauhii Toill.-Gen. & Bosser var. **andranobeense** Bosser

Distribution: Madagascar

Bulbophyllum raui Arora

Distribution: India

Bulbophyllum reclusum Seidenf.

Distribution: Thailand

Part II: Accepted Names / Noms Acceptés / Nombres Aceptado

Bulbophyllum rectilabre J. J. Sm.
Hapalochilus rectilabris (J. J. Sm.) Garay & W. Kittr.

Distribution: Indonesia

Bulbophyllum recurviflorum J. J. Sm.

Distribution: Indonesia

Bulbophyllum recurvilabre Garay

Distribution: Philippines

Bulbophyllum reductum J. J. Verm. & P. O'Byrne

Distribution: Indonesia

Bulbophyllum reevei J. J. Verm.
Peltopus reevei (J. J. Verm.) Szlach. & Marg.

Distribution: Indonesia, Papua New Guinea

Bulbophyllum reflexiflorum H. Perrier
Bulbophyllum bosseri Lemcke
Bulbophyllum inauditum Schltr. (1925)

Distribution: Madagascar

Bulbophyllum reflexiflorum H. Perrier subsp. **pogonochilum** (Summerh.)
Bosser
Bulbophyllum comosum H. Perrier
Bulbophyllum pogonochilum Summerh.

Distribution: Madagascar

Bulbophyllum reflexum Ames & C. Schweinf.

Distribution: Indonesia, Philippines

Bulbophyllum refractilingue J. J. Sm.

Distribution: Indonesia

Bulbophyllum refractum (Zoll.) Rchb. f.
Cirrhopetalum refractum Zoll.
Phyllorkis refracta (Zoll.) Kuntze
Phyllorkis wallichii Kuntze

Distribution: Indonesia, Myanmar, Viet Nam

256

Part II: Accepted Names / Noms Acceptés / Nombres Aceptado

Bulbophyllum regnelli Rchb. f.
Didactyle regnellii (Rchb. f.) Rodr.

Distribution: Brazil

Bulbophyllum reichenbachianum Kraenzl
Bulbophyllum mannii Rchb. f.
Bulbophyllum setiferum (Rolfe) J. J. Sm.
Cirrhopetalum mannii Rolfe
Cirrhopetalum setiferum (Rchb. f.) Mukerjee

Distribution: Unknown

Bulbophyllum reichenbachii (Kuntze) Schltr.
Bulbophyllum gracile C. S. P. Parish & Rchb. f.
Phyllorkis reichenbachii Kuntze

Distribution: Myanmar, Thailand

Bulbophyllum reifii Sieder & Kiehn †
Bulbophyllum rhizomatosum Schltr. (1923)

Distribution: Papua New Guinea

Bulbophyllum reilloi Ames

Distribution: Philippines

Bulbophyllum remiferum Carr

Distribution: Indonesia

Bulbophyllum renipetalum Schltr.

Distribution: Papua New Guinea

Bulbophyllum renkinianum (Laurent) De Wild.
Bulbophyllum cercoglossum Summerh.
Megaclinium renkinianum Laurent

Distribution: Cameroon, Democratic Republic of the Congo, Gabon

Bulbophyllum repens Griff.
Phyllorkis repens (Griff.) Kuntze

Distribution: India, Myanmar, Thailand

Bulbophyllum reptans (Lindl.) Lindl.
Bulbophyllum clarkei Rchb. f.
Bulbophyllum grandiflorum Griff.
Bulbophyllum ombrophilum Gagnep.
Bulbophyllum reptans (Lindl.) Lindl. var. *acuta* C. L. Malhotra & B. Balodi
Bulbophyllum reptans (Lindl.) Lindl. var. *subracemosa* Hook. f.
Ione racemosa (Sm.) Seidenf.
Phyllorkis reptans (Lindl.) Kuntze
Tribrachia racemosa Lindl.
Tribrachia reptans Lindl.

Distribution: Bangladesh, Bhutan, China, India, Myanmar, Thailand, Viet Nam

Bulbophyllum restrepia (Ridl.) Ridl.
Cirrhopetalum restrepia Ridl.

Distribution: Indonesia, Singapore

Bulbophyllum resupinatum Ridl.

Distribution: Cameroon, Côte d'Ivoire, Democratic Republic of the Congo, Gabon, Ghana, Nigeria, Sao Tome and Principe

Bulbophyllum resupinatum Ridl. var. **filiforme** (Kraenzl.) J. J. Verm.
Bulbophyllum daloaense P. J. Cribb & Perez-Vera
Bulbophyllum filiforme Kraenzl.
Bulbophyllum longispicatum Kraenzl. & Schltr.
Bulbophyllum macrostachyum Kraenzl.
Bulbophyllum rubroviolaceum De Wild.
Bulbophyllum victoris P. J. Cribb & Perez-Vera

Distribution: Cameroon, Côte d'Ivoire, Democratic Republic of the Congo, Gabon, Liberia, Sierra Leone

Bulbophyllum reticulatum Bateman
Bulbophyllum carinatum Cogn.
Bulbophyllum katherinae A. D. Hawkes
Phyllorkis reticulata (Bateman) Kuntze

Distribution: Indonesia

Bulbophyllum retusiusculum Rchb. f.
Bulbophyllum flavisepalum (Hayata) Masamune
Bulbophyllum flavisepalum Hayata
Bulbophyllum langbianense Seidenf. & Smitinand
Bulbophyllum micholitzii (Rolfe) Ho
Bulbophyllum micholitzii (Rolfe) J. J. Sm.
Bulbophyllum micholitzii (Rolfe) Seidenf. & Smitinand
Bulbophyllum muscicolum Rchb. f.
Bulbophyllum touranense (Gagnep.) Ho
Cirrhopetalum flavisepalum (Hayata) Hayata
Cirrhopetalum micholitzii Rolfe
Cirrhopetalum retusiusculum (Rchb. f.) Hemsl.
Cirrhopetalum touranense Gagnep.
Cirrhopetalum wallichii Lindl.

Phyllorkis retusiuscula (Rchb. f.) Kuntze

Distribution: Bhutan, China, India, Indonesia, Lao People's Democratic Republic, Malaysia, Myanmar, Nepal, Taiwan, Province of China , Thailand, Viet Nam

Bulbophyllum retusiusculum Rchb. f. var. **oreogenes** (W. W. Sm.) Z. H. Tsi
   Bulbophyllum oreogenes (W. W. Sm.) Seidenf.
   Cirrhopetalum oreogenes W. W. Sm.

Distribution: China

Bulbophyllum retusiusculum Rchb. f. var. **tigridum** (Hance) Z. H. Tsi

Distribution: China

Bulbophyllum rheedei Manilal & Sath.Kumar,
   Epidendrum sterile Lam. var. *gamma* Lam.
   Rhytionanthos rheedei (Manilal & Sath.Kumar,) Garay, Hamer & Siegerist

Distribution: India

Bulbophyllum rhizomatosum Ames & C. Schweinf.
   Bulbophyllum minimibulbum Carr

Distribution: Indonesia, Malaysia, Philippines

Bulbophyllum rhodoglossum Schltr.

Distribution: Papua New Guinea

Bulbophyllum rhodoleucum Schltr.
   Peltopus rhodoleucus (Schltr.) Szlach. & Marg.

Distribution: Indonesia, Papua New Guinea

Bulbophyllum rhodoneuron Schltr.

Distribution: Papua New Guinea

Bulbophyllum rhodosepalum Schltr.

Distribution: Indonesia

Bulbophyllum rhodostachys Schltr.

Distribution: Madagascar

Bulbophyllum rhodostictum Schltr.

Distribution: Papua New Guinea

Bulbophyllum rhomboglossum Schltr.

Distribution: Papua New Guinea, Vanuatu

Bulbophyllum rhopaloblepharon Schltr.

Distribution: Papua New Guinea

Bulbophyllum rhopalophorum Schltr.

Distribution: Papua New Guinea

Bulbophyllum rhynchoglossum Schltr. (1910)

Distribution: Indonesia

Bulbophyllum ricaldonei Leite

Distribution: Brazil

Bulbophyllum rienanense H. Perrier

Distribution: Madagascar

Bulbophyllum rigidifilum J. J. Sm.
  Epicranthes rigidifilum (J. J. Sm.) Garay & W. Kittr.

Distribution: Indonesia

Bulbophyllum rigidipes Schltr.

Distribution: Papua New Guinea

Bulbophyllum rigidum King & Pantl.

Distribution: Bhutan, India

Bulbophyllum riparium J. J. Sm.

Distribution: Indonesia

Bulbophyllum rivulare Schltr.

Distribution: Papua New Guinea

Part II: Accepted Names / Noms Acceptés / Nombres Aceptado

Bulbophyllum riyanum Fukuyama

Distribution: China, Taiwan, Province of China

Bulbophyllum rojasii L. O. Williams

Distribution: Paraguay

Bulbophyllum rolfeanum Seidenf. & Smitinand

Distribution: Thailand

Bulbophyllum rolfei (Kuntze) Seidenf.
Bulbophyllum dyeranum (King & Pantl.) Seidenf.
Bulbophyllum parvulum (Hook. f.) J. J. Sm.
Cirrhopetalum dyerianum King & Pantl.
Cirrhopetalum parvulum Hook. f.
Phyllorkis rolfei Kuntze

Distribution: Bhutan, India

Bulbophyllum romburghii J. J. Sm.

Distribution: Indonesia

Bulbophyllum roraimense Rolfe

Distribution: Brazil, Venezuela

Bulbophyllum rosemarianum C. S. Kumar, P. C. S. Kumar & Saleem

Distribution: India

Bulbophyllum roseopunctatum Schltr.

Distribution: Papua New Guinea

Bulbophyllum rostriceps Rchb. f.
Phyllorkis rostriceps (Rchb. f.) Kuntze

Distribution: Fiji

Bulbophyllum rothschildianum (O' Brien) J. J. Sm.
Cirrhopetalum rothschildianum O'Brien

Distribution: Bhutan, India

Bulbophyllum roxburghii (Lindl.) Rchb. f.
Aerides radiatum Roxb. ex Lindl.

Cirrhopetalum roxburghii Lindl.
Phyllorkis roxburghii (Lindl.) Kuntze

Distribution: Bhutan, India, Indonesia

Bulbophyllum rubiferum J. J. Sm.

Distribution: Indonesia

Bulbophyllum rubiginosum Schltr.

Distribution: Madagascar

Bulbophyllum rubipetalum P. Royen

Distribution: Papua New Guinea

Bulbophyllum rubroguttatum Seidenf.
Cirrhopetalum rubroguttatum (Seidenf.) Garay, Hamer & Siegerist

Distribution: Thailand

Bulbophyllum rubrolabellum T. P. Lin
Bulbophyllum odoratissimum (J. E. Sm.) Lindl. var. *rubrolabellum* (T. P. Lin) S. S. Ying

Distribution: Taiwan, Province of China

Bulbophyllum rubrolabium Schltr.

Distribution: Madagascar

Bulbophyllum rubrolineatum Schltr.

Distribution: Papua New Guinea

Bulbophyllum rubromaculatum W. Kittr.
Bulbophyllum nigrescens Schltr.

Distribution: Papua New Guinea

Bulbophyllum rubrum Jum. & H. Perrier
Bulbophyllum ambongense Schltr.

Distribution: Madagascar

Bulbophyllum ruficaudatum Ridl.
Bulbophyllum microbulbon (Ridl.) Ridl.
Bulbophyllum nanobulbon Seidenf.

Cirrhopetalum microbulbon Ridl.

Distribution: Indonesia, Malaysia, Singapore

Bulbophyllum rufilabrum C. S. P. Parish ex Hook. f.
  Phyllorkis rufilabra (C. S. P. Parish & Hook. f.) Kuntze

Distribution: Myanmar, Thailand

Bulbophyllum rufinum Rchb. f.
  Phyllorkis rufina (Rchb. f.) Kuntze

Distribution: Cambodia, India, Lao People's Democratic Republic, Myanmar, Thailand, Viet Nam

Bulbophyllum ruginosum H. Perrier

Distribution: Madagascar

Bulbophyllum rugosibulbum Summerh.

Distribution: Zimbabwe

Bulbophyllum rugosisepalum Seidenf.

Distribution: Thailand

Bulbophyllum rugosum Ridl.
  Bulbophyllum melliferum J. J. Sm.

Distribution: Indonesia, Malaysia, Singapore

Bulbophyllum rugulosum J. J. Sm.

Distribution: Indonesia

Bulbophyllum rupestre J. J. Sm.

Distribution: Indonesia

Bulbophyllum rupicola Barb. Rodr.

Distribution: Brazil

Bulbophyllum rutenbergianum Schltr.
  Bulbophyllum coursianum H. Perrier
  Bulbophyllum peniculus Schltr.

Part II: Accepted Names / Noms Acceptés / Nombres Aceptado

Bulbophyllum spathulifolium H. Perrier

Distribution: Madagascar

Bulbophyllum saccolabioides J. J. Sm.

Distribution: Indonesia

Bulbophyllum salaccense Rchb. f.

Distribution: Indonesia, Malaysia

Bulbophyllum salebrosum J. J. Sm.

Distribution: Indonesia

Bulbophyllum saltatorium Lindl.
  Phyllorkis saltatoria (Lindl.) Kuntze

Distribution: Cameroon, Côte d'Ivoire, Democratic Republic of the Congo, Equatorial Guinea, Gabon, Ghana, Liberia, Nigeria, Sierra Leone

Bulbophyllum saltatorium Lindl. var. **albociliatum** (Finet) J. J. Verm.
  Bulbophyllum calamarium Lindl. var. *albociliatum* Finet
  Bulbophyllum distans J. J. Sm.
  Bulbophyllum distans Lindl.
  Bulbophyllum flexiliscapum Summerh.
  Bulbophyllum graciliscapum Summerh.
  Bulbophyllum kindtianum De Wild.
  Bulbophyllum mildbraedii Kraenzl.
  Bulbophyllum miniatum Hort.
  Bulbophyllum nudiscapum Rolfe
  Phyllorkis distans (Lindl.) Kuntze

Distribution: Cameroon, Central African Republic, Congo, Côte d'Ivoire, Democratic Republic of the Congo, Equatorial Guinea, Gabon, Ghana, Liberia, Nigeria, Uganda

Bulbophyllum saltatorium Lindl. var. **calamarium** (Lindl.) J. J. Verm.
  Bulbophyllum calamarium Lindl.
  Bulbophyllum rupincola Rchb. f.
  Phyllorkis calamaria (Lindl.) Kuntze

Distribution: Cameroon, Congo, Côte d'Ivoire, Democratic Republic of the Congo, Equatorial Guinea, Gabon, Ghana, Liberia, Sierra Leone

Bulbophyllum sambiranense Jum. & H. Perrier

Distribution: Madagascar

264

Bulbophyllum sambiranense Jum. & H. Perrier var. **ankiranense** H. Perrier

Distribution: Madagascar

Bulbophyllum sambiranense Jum. & H. Perrier var. **latibracteatum** H. Perrier

Distribution: Madagascar

Bulbophyllum samoanum Schltr.
Bulbophyllum christophersenii L. O. Williams

Distribution: Fiji, New Caledonia, Samoa, Solomon Islands, Vanuatu

Bulbophyllum sanderianum Rolfe

Distribution: Brazil

Bulbophyllum sandersonii (Hook. f.) Rchb. f.
Bulbophyllum bibundiense Schltr.
Bulbophyllum melleri (Hook. f.) Rchb. f.
Bulbophyllum mooreanum Robyns & Tournay
Bulbophyllum pusillum (Rolfe) De Wild.
Bulbophyllum tentaculigerum Rchb. f.
Megaclinium melleri Hook. f.
Megaclinium pusillum Rolfe
Megaclinium sandersoni Hook. f.
Megaclinium tentaculigerum (Rchb. f.) Durand & Schinz

Distribution: Cameroon, Democratic Republic of the Congo, Gabon, Kenya, Malawi, Mozambique, Rwanda, South Africa, Uganda, United Republic of Tanzania, Zambia, Zimbabwe

Bulbophyllum sandersonii (Hook. f.) Rchb. f. subsp. **stenopetalum** (Kraenzl.) J. J. Verm.
Bulbophyllum minus (De Wild.) De Wild.
Bulbophyllum rhodopetalum Kraenzl.
Bulbophyllum stenopetalum Kraenzl.
Megaclinium minor De Wild.
Megaclinium minus De Wild.

Distribution: Cameroon, Côte d'Ivoire, Democratic Republic of the Congo, Gabon, Ghana, Liberia, Nigeria

Bulbophyllum sandrangatense Bosser

Distribution: Madagascar

Bulbophyllum sangae Schltr.

Distribution: Gabon

Part II: Accepted Names / Noms Acceptés / Nombres Aceptado

Bulbophyllum sanguineopunctatum Seidenf. & A. D. Kerr

Distribution: Lao People's Democratic Republic

Bulbophyllum sanguineum H. Perrier

Distribution: Madagascar

Bulbophyllum sanitii Seidenf.

Distribution: Thailand

Bulbophyllum santoense J. J. Verm.
  Peltopus santoensis (J. J. Verm.) Szlach. & Marg.

Distribution: Vanuatu

Bulbophyllum santosii Ames

Distribution: Philippines

Bulbophyllum sapphirinum Ames

Distribution: Philippines

Bulbophyllum sarasinorum Schltr.

Distribution: Indonesia

Bulbophyllum sarcanthiforme Ridl.

Distribution: Indonesia

Bulbophyllum sarcodanthum Schltr.

Distribution: Indonesia, Papua New Guinea

Bulbophyllum sarcophylloides Garay, Hamer & Siegerist
  Cirrhopetalum sarcophyllum King & Pantl. var. *minor* King & Pantl.

Distribution: Bhutan

Bulbophyllum sarcophyllum (King & Pantl.) J. J. Sm.
  Bulbophyllum panigraphianum S. Misra
  Cirrhopetalum panigraphianum (S. Misra) S. Misra

Cirrhopetalum sarcophyllum King & Pantl.

Distribution: Bhutan, India

Bulbophyllum sarcorhachis Schltr.

Distribution: Madagascar

Bulbophyllum sarcorhachis Schltr. var. **beforonense** (Schltr.) H. Perrier
Bulbophyllum befaonense Schltr.

Distribution: Madagascar

Bulbophyllum sarcorhachis Schltr. var. **flavemarginatum** H. Perrier

Distribution: Madagascar

Bulbophyllum sarcoscapum Teijsm. & Binn.

Distribution: Indonesia

Bulbophyllum saronae Garay
Hyalosema saronae (Garay) Rysy

Distribution: Papua New Guinea

Bulbophyllum sauguetiense Schltr.

Distribution: Papua New Guinea

Bulbophyllum saurocephalum Rchb. f.

Distribution: Philippines

Bulbophyllum savaiense Schltr.

Distribution: Fiji, Samoa, Vanuatu

Bulbophyllum savaiense Schltr. subsp. **gorumense** (Schltr.) J. J. Verm.
Bulbophyllum bolaninum Schltr.
Bulbophyllum gorumense Schltr.

Distribution: Indonesia, Papua New Guinea

Bulbophyllum savaiense Schltr. subsp. **subcubicum** (J. J. Sm.) J. J. Verm.
Bulbophyllum foveatum Schltr.
Bulbophyllum microtatanthum Schltr.
Bulbophyllum quadratum Schltr.
Bulbophyllum subcubicum J. J. Sm.

Bulbophyllum subcubicum J. J. Sm. var. *coccineum* J. J. Sm.

Distribution: Indonesia, Papua New Guinea, Philippines

Bulbophyllum sawiense J. J. Sm.

Distribution: Indonesia

Bulbophyllum scaberulum (Rolfe) Bolus
Bulbophyllum bambiliense De Wild.
Bulbophyllum chevalieri De Wild.
Bulbophyllum clarkei (Rolfe) Schltr.
Bulbophyllum congolanum Schltr.
Bulbophyllum ealaense De Wild.
Bulbophyllum eburneum (Pfitzer.) De Wild.
Bulbophyllum jesperseni De Wild.
Bulbophyllum pobeguenii (Finet) De Wild
Bulbophyllum summerhayesii A. D. Hawkes
Bulbophyllum zobiaense De Wild.
Megaclinium bambiliense De Wild.
Megaclinium chevalieri De Wild.
Megaclinium clarkei Rolfe
Megaclinium ealaense De Wild.
Megaclinium eburneum Pfitzer
Megaclinium jesperseni De Wild.
Megaclinium pobeguinii Finet
Megaclinium scaberulum Rolfe
Megaclinium zobiaense De Wild.

Distribution: Angola, Cameroon, Central African Republic, Congo, Côte d'Ivoire, Democratic Republic of the Congo, Ethiopia, Gabon, Ghana, Guinea, India, Kenya, Malawi, Mozambique, Nigeria, Sierra Leone, South Africa, Sudan, United Republic of Tanzania, Zambia, Zimbabwe

Bulbophyllum scaberulum (Rolfe) Bolus var. **album** Pérez-Vera

Distribution: Côte d'Ivoire

Bulbophyllum scaberulum (Rolfe) Bolus var. **crotalicaudatum** J. J. Verm.
Distribution: United Republic of Tanzania

Bulbophyllum scaberulum (Rolfe) Bolus var. **fuerstenbergianum** (De Wild.) J. J. Verm.
Bulbophyllum fuerstenbergianum (De Wild.) De Wild.
Megaclinium fuerstenbergianum De Wild.

Distribution: Cameroon, Democratic Republic of the Congo, Equatorial Guinea, Nigeria

Bulbophyllum scabratum Rchb. f.
Bulbophyllum confertum Hook. f.
Cirrhopetalum caespitosum Wall. ex Lindl.
Phyllorkis conferta (Hook. f.) Kuntze

Distribution: Bangladesh, Bhutan, India

Bulbophyllum scabrum J. J. Verm. & A. L. Lamb

Distribution: Indonesia

Bulbophyllum scaphiforme J. J. Verm.

Distribution: China, Thailand, Viet Nam

Bulbophyllum scaphosepalum Ridl.
  Hapalochilus scaphosepalum (Ridl.) Garay & W. Kittredge

Distribution: Indonesia

Bulbophyllum scariosum Summerh.

Distribution: Côte d'Ivoire, Equatorial Guinea, Guinea, Liberia, Sierra Leone

Bulbophyllum sceliphron J. J. Verm.

Distribution: Papua New Guinea

Bulbophyllum schefferi (Kuntze) Schltr.
  Bulbophyllum bilobipetalum J. J. Sm.
  Bulbophyllum corticicola Schltr.
  Bulbophyllum corticicola Schltr. var. *minor* Schltr.
  Bulbophyllum gracile (Blume) Lindl.
  Bulbophyllum marcidum Ames
  Diphyes gracilis Blume
  Phyllorkis schefferi Kuntze

Distribution: Indonesia, Philippines

Bulbophyllum schillerianum Rchb. f.
  Bulbophyllum aurantiacum F. Muell.
  Bulbophyllum aurantiacum F. Muell. var. *wuttsii* F. M. Bailey
  Dendrobium aurantiacum (F. Muell.) F. Muell.
  Dendrobium shepherdii F. Muell. var. *platyphyllum* F. Muell.
  Oxysepala schilleriana (Rchb. f.) D. L. Jones & M. A. Clem.
  Phyllorkis aurantiaca (F. Muell.) Kuntze

Distribution: Australia

Bulbophyllum schimperianum Kraenzl.
  Bulbophyllum acutisepalum De Wild.
  Bulbophyllum xanthoglossum Schltr.

Distribution: Cameroon, Central African Republic, Democratic Republic of the Congo, Gabon, Nigeria, Uganda

Part II: Accepted Names / Noms Acceptés / Nombres Aceptado

Bulbophyllum schinzianum Kraenzl.
Bulbophyllum gentilii Rolfe

Distribution: Cameroon, Côte d'Ivoire, Democratic Republic of the Congo, Gabon, Liberia, Nigeria

Bulbophyllum schinzianum Kraenzl. var. **irigaleae** (P. J. Cribb & Perez-Vera) J. J. Verm.
Bulbophyllum irigaleae P. J. Cribb & Perez-Vera

Distribution: Côte d'Ivoire, Liberia

Bulbophyllum schinzianum Kraenzl. var. **phaeopogon** (Schltr.) J. J. Verm.
Bulbophyllum phaeopogon Schltr.

Distribution: Cameroon, Côte d'Ivoire, Democratic Republic of the Congo, Ghana, Nigeria

Bulbophyllum schistopetalum Schltr.

Distribution: Papua New Guinea

Bulbophyllum schizopetalum L. O. Williams
Hapalochilus schizopetalum (L. O. Williams) Garay & W. Kittr.

Distribution: Indonesia

Bulbophyllum schmidii Garay
Hyalosema schmidii (Garay) Rysy

Distribution: Indonesia

Bulbophyllum schmidtianum Rchb. f.
Phyllorkis schmidtiana (Rchb. f.) Kuntze

Distribution: India

Bulbophyllum schwarzii Sieder
Cirrhopetalum roseopunctatum Garay, Hamer & Siegerist

Distribution: Viet Nam

Bulbophyllum sciaphile Bosser

Distribution: Madagascar

Bulbophyllum scintilla Ridl.

Distribution: Malaysia

Part II: Accepted Names / Noms Acceptés / Nombres Aceptado

Bulbophyllum scitulum Ridl.
Hapalochilus scitulus (Ridl.) Garay & W. Kittredge

Distribution: Indonesia

Bulbophyllum scopa J. J. Verm.

Distribution: Indonesia, Papua New Guinea

Bulbophyllum scopula Schltr.

Distribution: Papua New Guinea

Bulbophyllum scotifolium J. J. Sm.

Distribution: Indonesia

Bulbophyllum scotinochiton J.J.Verm. & P.O'Byrne

Distribution: Indonesia

Bulbophyllum scrobiculilabre J. J. Sm.

Distribution: Indonesia

Bulbophyllum scutiferum J. J. Verm.
Peltopus scutifer (J. J. Verm.) Szlach. & Marg.

Distribution: Indonesia, Papua New Guinea

Bulbophyllum scyphochilus Schltr.
Hapalochilus scyphochilus (Schltr.) Garay & W. Kittr.

Distribution: Papua New Guinea

Bulbophyllum scyphochilus Schltr. var. **phaeanthum** Schltr.

Distribution: Papua New Guinea

Bulbophyllum secundum Hook. f.
Bulbophyllum subparviflorum Z. H. Tsi & S. C. Chen
Phyllorkis secunda (Hook.) Kuntze

Distribution: Bhutan, China, India, Lao People's Democratic Republic, Malaysia, Myanmar, Nepal, Thailand, Viet Nam

Bulbophyllum seidenfadenii A. D. Kerr

Distribution: Lao People's Democratic Republic

Bulbophyllum semiasperum J. J. Sm.

Distribution: Indonesia

Bulbophyllum semiteres Schltr.

Distribution: Papua New Guinea

Bulbophyllum semiteretifolium Gagnep.

Distribution: Viet Nam

Bulbophyllum sempiternum Ames

Distribution: Philippines

Bulbophyllum sensile Ames

Distribution: Philippines

Bulbophyllum sepikense W. Kittr.
Bulbophyllum cuspidipetalum Schltr.

Distribution: Papua New Guinea

Bulbophyllum septatum Schltr.
Bulbophyllum ambreae H. Perrier
Bulbophyllum serratum H. Perrier

Distribution: Madagascar

Bulbophyllum septemtrionale (J. J. Sm.) J. J. Sm.
Bulbophyllum digoelense J. J. Sm. var. *septemtrionale* J. J. Sm.

Distribution: Indonesia

Bulbophyllum serra Schltr.

Distribution: Indonesia, Papua New Guinea

Bulbophyllum serratotruncatum Seidenf.
Bulbophyllum ochraceum (Ridl.) Ridl.
Cirrhopetalum ochraceum Ridl.

Distribution: Malaysia, Myanmar

Bulbophyllum serripetalum Schltr.

Distribution: Papua New Guinea

Bulbophyllum serrulatifolium J. J. Sm.

Distribution: Indonesia

Bulbophyllum serrulatum Schltr.

Distribution: Papua New Guinea

Bulbophyllum setaceum T. P. Lin

 Distribution: Taiwan, Province of China

Bulbophyllum setigerum Lindl.
  Phyllorkis setigera (Lindl.) Kuntze

Distribution: Guyana

Bulbophyllum setuliferum Verm. J.J. & Saw L.G.

Distribution: Malaysia

Bulbophyllum shanicum King & Pantl.

Distribution: Thailand

Bulbophyllum shepherdi (F. Muell.) Rchb. f.
  Bulbophyllum crassulifolium (A. Cunn.) Rupp
  Dendrobium shepherdii F. Muell.
  Oxysepala shepherdii (F. Muell.) D. L. Jones & M. A. Clem.
  Phyllorkis shepherdii (F. Muell.) Kuntze

Distribution: Australia

Bulbophyllum shweliense W. W. Sm.
  Bulbophyllum craibianum Kerr

Distribution: Bhutan, China, Thailand, Viet Nam

Part II: Accepted Names / Noms Acceptés / Nombres Aceptado

Bulbophyllum siamense Rchb. f.
Bulbophyllum lobbii Lindl. var. *siamense* (Rchb. f.) Rchb.f.

Distribution: Cambodia, India, Lao People's Democratic Republic, Malaysia, Myanmar, Thailand

Bulbophyllum sibuyanense Ames
Cirrhopetalum sibuyanense (Ames) Garay, Hamer & Siegerist

Distribution: Philippines

Bulbophyllum sicyobulbon C. S. P. Parish & Rchb. f.
Phyllorkis sicyobulbon (C. S. P. Parish & Rchb. f.) Kuntze

Distribution: Malaysia, Myanmar, Thailand

Bulbophyllum siederi Garay

Distribution: Indonesia

Bulbophyllum sigaldiae Guillaumin

Distribution: Viet Nam

Bulbophyllum sigmoideum Ames & C. Schweinf.

Distribution: Indonesia, Philippines

Bulbophyllum signatum J. J. Verm.

Distribution: Malaysia

Bulbophyllum sikapingense J. J. Sm.

Distribution: Indonesia

Bulbophyllum sikkimense (King & Pantl.) J. J. Sm.
Cirrhopetalum sikkimense King & Pantl.

Distribution: India

Bulbophyllum silentvalliensis M. P. Sharma & S. K. Srivastrava

Distribution: India

Bulbophyllum sillemianum Rchb. f.
  Phyllorkis silleniana (Rchb. f.) Kuntze

Distribution: Myanmar, Thailand

Bulbophyllum similare Garay, Hamer & Siegerist

Distribution: Indonesia

Bulbophyllum simile Schltr.
  Bulbophyllum erythrochilum Schltr.

Distribution: Indonesia, Papua New Guinea

Bulbophyllum similissimum J. J. Verm.

Distribution: Indonesia

Bulbophyllum simmondsii Kores

 Distribution: Fiji

Bulbophyllum simondii Gagnep.

Distribution: Viet Nam

Bulbophyllum simplex J. J. Verm. & P. O'Byrne

Distribution: Indonesia

Bulbophyllum simplicilabellum Seidenf.

Distribution: Thailand

Bulbophyllum simulacrum Schltr.

Distribution: Madagascar

Bulbophyllum sinapis J. J. Verm. & P. O'Byrne

Distribution: Papua New Guinea

Bulbophyllum singaporeanum Schltr.
  Bulbophyllum densiflorum Ridl.

Distribution: Indonesia, Malaysia, Singapore

Bulbophyllum singulare Schltr.
Hyalosema singulare (Schltr.) Rolfe

Distribution: Papua New Guinea

Bulbophyllum singuliflorum W. Kittr.
Bulbophyllum nemorosum Schltr.

Distribution: Papua New Guinea

Bulbophyllum skeatianum Ridl.
Cirrhopetalum skeatianum (Ridl.) Garay, Hamer & Siegerist

Distribution: Malaysia

Bulbophyllum smithianum Schltr.
Bulbophyllum angustifolium (Blume) Lindl. var. *nanum* J. J. Sm.
Bulbophyllum angustifolium (Blume) Lindl. var. *pavum* J. J. Sm.
Bulbophyllum demissum Ridl.

Distribution: Indonesia

Bulbophyllum smitinandii Seidenf. & Thorut

Distribution: Thailand, Viet Nam

Bulbophyllum sociale Rolfe

Distribution: Indonesia

Bulbophyllum solteroi R. G. Tamayo

Distribution: Mexico

Bulbophyllum sopoetanense Schltr.
Bulbophyllum conspectum J. J. Sm.

Distribution: Indonesia

Bulbophyllum sordidum Lindl.
Bolbophyllaria sordida (Lindl.) Rchb. f.
Bulbophyllum ecuadorense Schltr.
Bulbophyllum ecuadoriensis (Schltr.) Gilli
Phyllorkis sordida (Lindl.) Kuntze

Distribution: Ecuador, Panama

Bulbophyllum sororculum J. J. Verm.

Distribution: Indonesia

276

Part II: Accepted Names / Noms Acceptés / Nombres Aceptado

Bulbophyllum spadiciflorum Tixier
Osyricera spadiciflora (Tixier) Garay, Hamer & Siegerist

Distribution: Viet Nam

Bulbophyllum spaerobulbum H. Perrier

Distribution: Madagascar

Bulbophyllum spathaceum Rolfe

Distribution: Myanmar

Bulbophyllum spathilingue J. J. Sm.

Distribution: Indonesia

Bulbophyllum spathipetalum J. J. Sm.

Distribution: Indonesia

Bulbophyllum spathulatum (Rolfe ex E. Cooper) Seidenf.
Cirrhopetalum spathulatum Rolfe ex Cooper
Rhytionanthos spathulatum (Rolfe ex E. Cooper) Garay, Hamer & Siegerist

Distribution: Bhutan, India, Lao People's Democratic Republic, Myanmar, Thailand, Viet Nam

Bulbophyllum speciosum Schltr.
Hapalochilus speciosus (Schltr.) Garay & W. Kittr.

Distribution: Papua New Guinea

Bulbophyllum sphaeracron Schltr.

Distribution: Papua New Guinea

Bulbophyllum sphaericum Z. H. Tsi & H. Li
Rhytionanthos sphaericum (Z. H. Tsi & H. Li) Garay, Hamer & Siegerist

Distribution: China

Bulbophyllum sphaerobulbum H. Perrier

Distribution: Madagascar

Bulbophyllum spiesii Garay, Hamer & Siegerist

Distribution: Papua New Guinea

Bulbophyllum spissum J. J. Verm.

Distribution: Indonesia, Malaysia

Bulbophyllum spongiola J. J. Verm.

Distribution: Papua New Guinea

Bulbophyllum stabile J. J. Sm.
  Hapalochilus stabilis (J. J. Sm.) Garay & W. Kittr.

Distribution: Indonesia

Bulbophyllum steffensii Schltr.

Distribution: Indonesia

Bulbophyllum stelis J. J. Sm.

Distribution: Indonesia

Bulbophyllum stelis J. J. Sm. var. **humile** J. J. Sm.

Distribution: Indonesia

Bulbophyllum stellatum Ames

Distribution: Philippines

Bulbophyllum stellula Ridl.
  Hapalochilus stellula (Ridl.) Garay & W. Kittredge

Distribution: Indonesia

Bulbophyllum stenobulbon C. S. P. Parish & Rchb. f.
  Bulbophyllum clarkeanum King & Pantl.
  Bulbophyllum hornense Govaerts
  Bulbophyllum youngsayeanum S. Y. Hu & G. Barretto
  Phyllorkis stenobulbon (C. S. P. Parish & Rchb. f.) Kuntze

Distribution: Bhutan, Lao People's Democratic Republic, Myanmar, Thailand, Viet Nam

Bulbophyllum stenochilum Schltr.

Distribution: Papua New Guinea, Solomon Islands

Bulbophyllum stenophyllum Schltr.
Hapalochilus stenophyton Garay & W. Kittr.

Distribution: Papua New Guinea, Solomon Islands, Vanuatu

Bulbophyllum stenorhopalos Schltr.
Epicranthes stenorhopalon (Schltr.) Garay & W. Kittr.

Distribution: Papua New Guinea

Bulbophyllum stenurum J. J. Verm. & P. O'Byrne

Distribution: Indonesia

Bulbophyllum sterile (Lam.) Suresh
Bulbophyllum caudatum Lindl.
Cirrhopetalum caudatum (Lindl.) King & Pantl.
Epidendrum sterile Lam.
Phyllorkis caudata (Lindl.) Kuntze

Distribution: Bhutan

Bulbophyllum steyermarkii Foldats
Bulbophyllum bracteosum C. Schweinf.

Distribution: Ecuador, Venezuela

Bulbophyllum stictanthum Schltr.
Hapalochilus stictanthus (Schltr.) Garay & W. Kittr.

Distribution: Papua New Guinea

Bulbophyllum stictosepalum Schltr.

Distribution: Papua New Guinea

Bulbophyllum stipitatibulbum J. J. Sm.

Distribution: Indonesia

Bulbophyllum stipulaceum Schltr.
Bulbophyllum absconditum J. J. Sm. var. *gautierense* J. J. Sm.
Bulbophyllum absconditum J. J. Sm. var. *neoguineese* J. J. Sm.
Bulbophyllum pelma J. J. Sm.
Bulbophyllum pelma J. J. Sm. var. *gautierense* J. J. Sm.

Distribution: Indonesia, Papua New Guinea

Bulbophyllum stolleanum Schltr.

Distribution: Papua New Guinea

279

Bulbophyllum stolzii Schltr.

Distribution: Malawi, United Republic of Tanzania

Bulbophyllum stormii J. J. Sm.
Bulbophyllum araniferum Ridl.
Bulbophyllum longistelidium Ridl.
Bulbophyllum stormii J. J. Sm. var. *pengadangaense* J. J. Sm.
Bulbophyllum tapirus J. J. Sm.
Bulbophyllum tristriatum Carr

Distribution: Indonesia, Malaysia

Bulbophyllum streptotriche J. J. Verm.

Distribution: Indonesia

Bulbophyllum striatellum Ridl.
Phyllorkis striatella (Ridl.) Kuntze

Distribution: Indonesia, Malaysia, Singapore

Bulbophyllum striatum (Griff.) Rchb. f.
Bulbophyllum striatitepalum Seidenf.
Dendrobium striatum Griff.
Phyllorkis striata (Griff.) Kuntze
Sarcopodium striatum (Griff.) Lindl.

Distribution: Bhutan, India, Thailand, Viet Nam

Bulbophyllum strigosum (Garay) Sieder & Kiehn †
Rhytionanthos strigosum Garay

Distribution: Viet Nam

Bulbophyllum sturmhoefelii Hoehne

Distribution: Brazil

Bulbophyllum suavissimum Rolfe
Bulbophyllum parryae Summerh. ex Parry
Phyllorkis suavissima (Rolfe) Kuntze

Distribution: Myanmar, Thailand

Bulbophyllum subaequale Ames

Distribution: Philippines

Part II: Accepted Names / Noms Acceptés / Nombres Aceptado

Bulbophyllum subapetalum J. J. Sm.
Peltopus subapetalus (J. J. Sm.) Szlach. & Marg.

Distribution: Indonesia

Bulbophyllum subapproximatum H. Perrier

Distribution: Madagascar

Bulbophyllum subbullatum J. J. Verm.

Distribution: Malaysia

Bulbophyllum subclausum J. J. Sm.

Distribution: Indonesia

Bulbophyllum subclavatum Schltr.

Distribution: Madagascar

Bulbophyllum subcrenulatum Schltr.

Distribution: Madagascar

Bulbophyllum subebulbum Gagnep.

Distribution: Viet Nam

Bulbophyllum subligaculiferum J. J. Verm.

Distribution: Gabon

Bulbophyllum submarmoratum J. J. Sm.

Distribution: Indonesia

Bulbophyllum subpatulum J. J. Verm.
Bulbophyllum muscicola Schltr. (1913)

Distribution: Papua New Guinea

Bulbophyllum subsecundum Schltr.

Distribution: Madagascar

Bulbophyllum subsessile Schltr.

Distribution: Madagascar

Bulbophyllum subtenellum Seidenf.

Distribution: Thailand

Bulbophyllum subtrilobatum Schltr.

Distribution: Papua New Guinea

Bulbophyllum subuliferum Schltr.

Distribution: Indonesia

Bulbophyllum subulifolium Schltr.

Distribution: Papua New Guinea

Bulbophyllum subumbellatum Ridl.

Distribution: Indonesia, Malaysia

Bulbophyllum subverticillatum Ridl.

Distribution: Indonesia

Bulbophyllum succedaneum J. J. Sm.

Distribution: Indonesia

Bulbophyllum sukhakulii Seidenf.

Distribution: Thailand

Bulbophyllum sulawesii Garay, Hamer & Siegerist

Distribution: Indonesia

Bulbophyllum sulcatum (Blume) Lindl.
  Bulbophyllum modestum Hook. f.
  Bulbophyllum ochranthum Ridl.
  Bulbophyllum paullum Ridl.
  Diphyes sulcata Blume
  Phyllorkis modesta (Hook. f.) Kuntze

Part II: Accepted Names / Noms Acceptés / Nombres Aceptado

Phyllorkis sulcata (Blume) Kuntze

Distribution: Indonesia

Bulbophyllum sulfureum Schltr.

Distribution: Madagascar

Bulbophyllum sumatranum Garay, Hamer & Siegerist
Bulbophyllum lobbii Lindl. var. *breviflorum* J. J. Sm.

Distribution: Indonesia

Bulbophyllum superfluum Kraenzl.

Distribution: Philippines

Bulbophyllum superpositum Schltr.

Distribution: Papua New Guinea

Bulbophyllum supervacaenum Kraenzl.

Distribution: Indonesia

Bulbophyllum surigaense Ames & Quisumb.

Distribution: Philippines

Bulbophyllum sutepense (Rolfe ex Downie) Seidenf. & Smitinand
Cirrhopetalum sutepense Rolfe ex Downie

Distribution: China, Lao People's Democratic Republic, Thailand

Bulbophyllum systenochilum J. J. Verm.
Peltopus systenochilus (J. J. Verm.) Szlach. & Marg.

Distribution: Indonesia

Bulbophyllum taeniophyllum C. S. P. Parish & Rchb. f.
Bulbophyllum fenestratum J. J. Sm.
Bulbophyllum mundulum (W. Bull) J. J. Sm.
Bulbophyllum punctatissimum Ridl.
Bulbophyllum rupicolum Ridl.
Bulbophyllum simillinum C. S. P. Parish & Rchb. f.
Cirrhopetalum fenestratum (J. J. Sm.) Garay, Hamer & Siegerist
Cirrhopetalum mundulum W. Bull
Cirrhopetalum punctatissimum Rolfe ex Ridl.
Cirrhopetalum simillimum Rchb. f.
Cirrhopetalum simullinum (C. S. P. Parish & Rchb. f.) Hook. f.

Part II: Accepted Names / Noms Acceptés / Nombres Aceptado

Cirrhopetalum stragularium Rchb. f.
Cirrhopetalum taeniophyllum (C. S. P. Parish & Rchb. f.) Hook. f.
Phyllorkis simillima (Lindl.) Kuntze
Phyllorkis taeniophylla (C. S. P. Parish & Rchb. f.) Kuntze

Distribution: China, Indonesia, Lao People's Democratic Republic, Malaysia, Myanmar, Thailand, Viet Nam

Bulbophyllum taeter J. J. Verm.

Distribution: Brunei Darussalam, Malaysia

Bulbophyllum tahanense Carr

Distribution: Malaysia

Bulbophyllum tahitense Nadeaud

Distribuition: French Polynesia

Bulbophyllum taiwanense (Fukuyama) Nakajima
Bulbophyllum taiwanense (Fukuyama) S. S. Ying
Bulbophyllum taiwanense (Fukuyama) Seidenf.
Cirrhopetalum taiwanense Fukuyama

Distribution: Taiwan, Province of China

Bulbophyllum talauense (J. J. Sm.) Carr
Bulbophyllum laxiflorum (Blume) Lindl. var. *taluense* J. J. Sm.

Distribution: Malaysia

Bulbophyllum tampoketsense H. Perrier

Distribution: Madagascar

Bulbophyllum tanystiche J. J. Verm.

Distribution: Papua New Guinea

Bulbophyllum tarantula Schuit. & De Vogel

Distribution: Papua New Guinea

Bulbophyllum tardeflorens Ridl.

Distribution: Indonesia

Part II: Accepted Names / Noms Acceptés / Nombres Aceptado

Bulbophyllum tectipes J. J. Verm. & P. O'Byrne

Distribution: Indonesia

Bulbophyllum tectipetalum J. J. Sm.

Distribution: Indonesia

Bulbophyllum tectipetalum J. J. Sm. var. **longisepalum** J. J. Sm.

Distribution: Indonesia

Bulbophyllum tectipetalum J. J. Sm. var. **maximum** J. J. Sm.

Distribution: Indonesia

Bulbophyllum tekuense Carr

Distribution: Malaysia

Bulbophyllum tenellum (Blume) Lindl.
  Diphyes tenella Blume
  Phyllorkis tenella (Blume) Kuntze

Distribution: Indonesia

Bulbophyllum tengchongense Z. H. Tsi

Distribution: China

Bulbophyllum tenompokense J. J. Sm.

Distribution: Indonesia, Malaysia

Bulbophyllum tentaculatum Schltr.

Distribution: Papua New Guinea

Bulbophyllum tentaculiferum Schltr.
  Bulbophyllum tentaculatum Schltr.

Distribution: Papua New Guinea

Bulbophyllum tenue Schltr.

Distribution: Papua New Guinea

Bulbophyllum tenuifolium (Blume) Lindl.
Bulbophyllum microstele Schltr.
Bulbophyllum nigromaculatum Holttum
Diphyes tenuifolia Blume
Phyllorkis tenuifolia (Blume) Kuntze

Distribution: Indonesia, Malaysia, Thailand

Bulbophyllum tenuipes Schltr.

Distribution: Indonesia, Papua New Guinea

Bulbophyllum teres Carr

Distribution: Indonesia

Bulbophyllum teresense Ruschi

Distribution: Brazil

Bulbophyllum teretibulbum H. Perrier

Distribution: Madagascar

Bulbophyllum teretifolium Schltr.

Distribution: Cameroon

Bulbophyllum teretilabre J. J. Sm.

Distribution: Indonesia

Bulbophyllum ternatense J. J. Sm.

Distribution: Indonesia

Bulbophyllum tetragonum Lindl.
Bulbophyllum wrightii Summerh.

Distribution: Cameroon, Côte d'Ivoire, Democratic Republic of the Congo, Ghana, Liberia, Sierra Leone

Bulbophyllum teysmannii J. J. Sm.

Distribution: Indonesia

Bulbophyllum thaiorum J. J. Sm.
Bulbophyllum papillosum (Rolfe) Seidenf. & Smitinand
Bulbophyllum thailandicum Seidenf. & Smitinand

286

Part II: Accepted Names / Noms Acceptés / Nombres Aceptado

Cirrhopetalum papillosum Rolfe

Distribution: Thailand, Viet Nam

Bulbophyllum theioglossum Schltr.

Distribution: Papua New Guinea

Bulbophyllum thelantyx J. J. Verm.
Peltopus thelantyx (J. J. Verm.) Szlach. & Marg.

Distribution: Papua New Guinea

Bulbophyllum therezienii Bosser

Distribution: Madagascar

Bulbophyllum thersites J. J. Verm.

Distribution: Papua New Guinea

Bulbophyllum theunissenii J. J. Sm.

Distribution: Indonesia

Bulbophyllum thiurum J.J.Verm. & P.O'Byrne

Distribution: Malaysia

Bulbophyllum thompsonii Ridl.
Phyllorkis thompsonii (Ridl.) Kuntze

Distribution: Madagascar

Bulbophyllum thrixspermiflorum J. J. Sm.

Distribution: Indonesia

Bulbophyllum thrixspermoides J. J. Sm.

Distribution: Indonesia

Bulbophyllum thwaitesii Rchb. f.
Cirrhopetalum thwaitesii (Rchb. f.) Hook. f.
Phyllorkis thwaitesii (Rchb. f.) Kuntze

Distribution: Sri Lanka

Part II: Accepted Names / Noms Acceptés / Nombres Aceptado

Bulbophyllum thymophorum J. J. Verm. & A. L. Lamb

Distribution: Indonesia

Bulbophyllum tigridium Hance
  Cirrhopetalum tigridum (Hance) Rolfe

Distribution: China

Bulbophyllum tinekeae Schuit. & De Vogel

Distribution: Papua New Guinea

Bulbophyllum titanea Ridl.

Distribution: Malaysia

Bulbophyllum tixieri Seidenf.

Distribution: Viet Nam

Bulbophyllum tjadasmalangense J. J. Sm.

Distribution: Indonesia

Bulbophyllum toilliezae Bosser

Distribution: Madagascar

Bulbophyllum tokioi Fukuyama
  Bulbophyllum derchianum S. S. Ying

Distribution: Taiwan, Province of China

Bulbophyllum toppingii Ames

Distribution: Philippines

Bulbophyllum toranum J. J. Sm.
  Bulbophyllum barbellatum Schltr.

Distribution: Indonesia, Papua New Guinea

Bulbophyllum torquatum J. J. Sm.

Distribution: Indonesia

Bulbophyllum torricellense Schltr.
Hapalochilus torricellensis (Schltr.) Garay & W. Kittr.

Distribution: Papua New Guinea

Bulbophyllum tortum Schltr.

Distribution: Papua New Guinea

Bulbophyllum tortuosum (Blume) Lindl.
Bulbophyllum indragirense Schltr.
Bulbophyllum listeri King & Pantl.
Diphyes tortuosa Blume
Phyllorkis tortuosa (Blume) Kuntze

Distribution: Bhutan, Indonesia, Lao People's Democratic Republic, Malaysia, Philippines, Thailand, Viet Nam

Bulbophyllum trachyanthum Kraenzl.
Hyalosema trachyanthum (Kraenzl.) Rolfe

Distribution: Fiji, Indonesia, Papua New Guinea, Samoa, Solomon Islands

Bulbophyllum trachybracteum Schltr.

Distribution: Papua New Guinea

Bulbophyllum trachyglossum Schltr.
Hapalochilus trachyglossus (Schltr.) Garay & W. Kittr.

Distribution: Papua New Guinea, Solomon Islands

Bulbophyllum trachypus Schltr.

Distribution: Indonesia, Papua New Guinea

Bulbophyllum tremulum Wight
Phyllorkis tremula (Wight) Kuntze

Distribution: India

Bulbophyllum triadenium (Lindl.) Rchb. f.
Epigeneium triadenium (Lindl.) Summerh.
Phyllorkis triadenia (Lindl.) Kuntze
Sarcopodium triadenium Lindl.

Distribution: Indonesia

Part II: Accepted Names / Noms Acceptés / Nombres Aceptado

Bulbophyllum triandrum Schltr.

Distribution: New Caledonia, Papua New Guinea

Bulbophyllum triaristella Schltr.

Distribution: Papua New Guinea

Bulbophyllum tricanaliferum J. J. Sm.
  Hyalosema tricanaliferum (J. J. Sm.) Rolfe

Distribution: Indonesia

Bulbophyllum tricarinatum Petch

Distribution: Sri Lanka

Bulbophyllum trichaete Schltr.

Distribution: Papua New Guinea

Bulbophyllum trichambon Schltr.

Distribution: Papua New Guinea

Bulbophyllum trichocephalum (Schltr.) T. Tang & F. T. Wang
  Cirrhopetalum trichocephalum Schltr.

Distribution: India, Thailand

Bulbophyllum trichocephalum (Schltr.) T. Tang & F. T. Wang var.
wallongense Agarwal Sabapathy & Chowdhery

Distribution: India

Bulbophyllum trichochlamys H. Perrier

Distribution: Madagascar

Bulbophyllum trichorhachis J. J. Verm. & P. O'Byrne

Distribution: Indonesia

Bulbophyllum trichromum Schltr.
  Hapalochilus trichromus (Schltr.) Garay & Sieder

Distribution: Papua New Guinea

Part II: Accepted Names / Noms Acceptés / Nombres Aceptado

Bulbophyllum triclavigerum J. J. Sm.

Distribution: Indonesia

Bulbophyllum tricolor L. B. Sm. & S. K. Harris

 Distribution: Bolivia

Bulbophyllum tricorne Seidenf. & Smitinand

 Distribution: Cambodia, India, Thailand

Bulbophyllum tricornoides Seidenf.

Distribution: Thailand

Bulbophyllum tridentatum Kraenzl.

Distribution: Thailand

Bulbophyllum trifarium Rolfe

Distribution: Madagascar

Bulbophyllum trifilum J. J. Sm.
 Bulbophyllum cavistigma J. J. Sm.
 Bulbophyllum fatuum J. J. Sm.
 Bulbophyllum recurvimarginatum J. J. Sm.

Distribution: Indonesia, Papua New Guinea

Bulbophyllum trifilum J. J. Sm. subsp. **filisepalum** (J. J. Sm.) J. J. Verm.
 Bulbophyllum filisepalum J. J. Sm.

Distribution: Indonesia, Papua New Guinea

Bulbophyllum triflorum (Breda) Blume
 Odontostyles triflora Breda
 Phyllorkis triflora (Breda) Kuntze

Distribution: Indonesia

Bulbophyllum trifolium Ridl.
 Bulbophyllum tacitum Carr

Distribution: Indonesia, Malaysia, Singapore

Part II: Accepted Names / Noms Acceptés / Nombres Aceptado

Bulbophyllum trigonidioides J. J. Sm.

Distribution: Indonesia

Bulbophyllum trigonobulbum Schltr. & J. J. Sm.

Distribution: Indonesia

Bulbophyllum trigonocarpum Schltr.
Hapalochilus trigonocarpus (Schltr.) Garay & W. Kittr.

Distribution: Papua New Guinea

Bulbophyllum trigonosepalum Kraenzl.

Distribution: Philippines

Bulbophyllum trilineatum H. Perrier

Distribution: Madagascar

Bulbophyllum trimeni (Hook. f. ex Trim.) J. J. Sm.
Cirrhopetalum trimeni Hook. f.
Phyllorkis trimenii (Hook. f.) Kuntze

Distribution: Sri Lanka

Bulbophyllum trinervium J. J. Sm.

Distribution: Indonesia

Bulbophyllum tripaleum Seidenf.

Distribution: Thailand

Bulbophyllum tripetalum Lindl.
Didactyle tripetala (Lindl.) Linden
Phyllorkis tripetala (Lindl.) Kuntze

Distribution: Brazil

Bulbophyllum tripudians C. S. P. Parish & Rchb. f.
Cirrhopetalum tripudians (C. S. P. Parish & Rchb. f.) C. S. P. Parish & Rchb. f.

Distribution: Lao People's Democratic Republic, Myanmar, Thailand, Viet Nam

Part II: Accepted Names / Noms Acceptés / Nombres Aceptado

Bulbophyllum trirhopalon Schltr.
Epicranthes trirhopalon (Schltr.) Garay & W. Kittr.

Distribution: Papua New Guinea

Bulbophyllum triste Rchb. f.
Bulbophyllum alopecurum Rchb. f.
Bulbophyllum mackeeanum Guillaumin
Bulbophyllum micranthum Hook. f.
Phyllorkis alopecurus (Rchb. f.) Kuntze
Phyllorkis micrantha (Hook. f.) Kuntze
Phyllorkis tristis (Rchb. f.) Kuntze

Distribution: Bhutan

Bulbophyllum tristelidium W. Kittr.
Bulbophyllum tridentatum Rolfe

Distribution: Indonesia

Bulbophyllum triurum Kraenzl.

Distribution: Indonesia

Bulbophyllum triviale Seidenf.

Distribution: Thailand

Bulbophyllum trulliferum J. J. Verm. & A. L. Lamb

Distribution: Malaysia

Bulbophyllum truncatum J. J. Sm.

Distribution: Indonesia

Bulbophyllum truncicola Schltr.

Distribution: Papua New Guinea

Bulbophyllum tryssum J. J. Verm. & A. L. Lamb

Distribution: Indonesia, Malaysia

Bulbophyllum tseanum (S. Y. Hu & Barretto) Z. H. Tsi
Cirrhopetalum tseanum S. Y. Hu & G. Barretto

Distribution: China

Part II: Accepted Names / Noms Acceptés / Nombres Aceptado

Bulbophyllum tuberculatum Colenso
Adelopetalum tuberculatum (Colenso) D. L. Jones, M. A. Clem. & Molloy

Distribution: Australia, New Zealand

Bulbophyllum tubilabrum J. J. Verm. & P. O'Byrne

Distribution: Indonesia

Bulbophyllum tumidum J. J. Verm.

Distribution: Indonesia

Bulbophyllum tumoriferum Schltr.

Distribution: Papua New Guinea

Bulbophyllum turgidum J. J. Verm.

Distribution: Indonesia

Bulbophyllum turkii Bosser & P. J. Cribb

Distribution: Madagascar

Bulbophyllum turpis J. J. Verm. & P. O'Byrne

Distribution: Malaysia

Bulbophyllum tylophorum Schltr.

Distribution: Indonesia

Bulbophyllum ulcerosum J. J. Sm.

Distribution: Indonesia, Papua New Guinea, Solomon Islands

Bulbophyllum umbellatum Lindl.
Bulbophyllopsis maculosa (Lindl.) Rchb. f.
Bulbophyllopsis morphologorum Rchb. f.
Bulbophyllum annamicum (Finet & Gagnep.) Bân & D. H. Duong
Bulbophyllum maculosum (Lindl.) Rchb. f.
Bulbophyllum saruwatarii Hayata
Bulbophyllum tibeticum Rolfe
Bulbophyllum tortisepalum Guillaumin
Cirrhopetalum annamicum (Finet ex Gagnep.) T. Tang & F. T. Wang
Cirrhopetalum bootanensis Griff.
Cirrhopetalum maculosum Lindl.
Cirrhopetalum maculosum Lindl. var. *annamicum* Finet
Cirrhopetalum saruwatarii (Hayata) Hayata

Cirrhopetalum umbellatum (Lindl.) Linden
Hippoglossum umbellatum Breda
Phyllorkis bootanensis (Griff.) Kuntze
Phyllorkis maculosa Kuntze
Phyllorkis umbellata (G. Forst.) Kuntze

Distribution: Bhutan, China, India, Myanmar, Taiwan, Province of China , Thailand, Viet Nam

Bulbophyllum umbellatum Lindl. var. **fuscescens** (Hook. F.)P. K. Sarkar
Cirrhopetalum maculosum Lindl. var. *fuscescens* Hook. f.

Distribution: Australia, India

Bulbophyllum umbonatum Kraenzl

Distribution: Unknown

Bulbophyllum umbraticola Schltr.

Distribution: Papua New Guinea

Bulbophyllum uncinatum J. J. Verm. & P. O'Byrne

Distribution: Indonesia

Bulbophyllum unciniferum Seidenf.
Cirrhopetalum unciniferum (Seidenf.) Senghas
Rhytionanthos unciniferum (Seidenf.) Garay, Hamer & Siegerist

Distribution: Thailand

Bulbophyllum undatilabre J. J. Sm.

Distribution: Indonesia

Bulbophyllum undecifilum J. J. Sm.
Epicranthes undecifila (J. J. Sm.) Garay & W. Kittr.

Distribution: Indonesia

Bulbophyllum unguiculatum Rchb. f.
Phyllorkis unguiculata (Rchb. f.) Kuntze

Distribution: Indonesia, Philippines

Bulbophyllum unguilabium Schltr.

Distribution: Papua New Guinea

Part II: Accepted Names / Noms Acceptés / Nombres Aceptado

Bulbophyllum unicaudatum Schltr.

Distribution: Papua New Guinea

Bulbophyllum unicaudatum Schltr. var. **xanthospaerum** Schltr.

Distribution: Papua New Guinea

Bulbophyllum uniflorum (Blume) Hassk.
 Bulbophyllum galbinum Ridl.
 Bulbophyllum hewetii Ridl.
 Bulbophyllum reinwardtii (Lindl.) Rchb. f.
 Bulbophyllum uniflorum (Blume) Hassk. var. *pluriflorum* Carr
 Bulbophyllum uniflorum (Blume) Hassk. var. *rubrum* (Ridl.) Carr
 Bulbophyllum uniflorum (Blume) Hassk. var. *variabile* (Ridl.) Carr
 Bulbophyllum variabile Ridl. var. *rubrum* Ridl.
 Cirrhopetalum compressum Lindl.
 Dendrobium grandiflorum Reinw. ex Hook. F.
 Ephippium uniflorum Blume
 Phyllorkis reinwardtii (Lindl.) Kuntze
 Phyllorkis uniflora (Blume) Kuntze
 Sarcopodium reinwardtii Lindl.

Distribution: Indonesia, Malaysia, Philippines

Bulbophyllum unifoliatum De Wild.

Distribution: Angola, Democratic Republic of the Congo, Rwanda, United Republic of Tanzania, Zambia

Bulbophyllum unifoliatum De Wild. subsp. **flectens** (P. J. Cribb & Taylor) J. J. Verm.
 Bulbophyllum flectens P. J. Cribb & P. Taylor

Distribution: Malawi, United Republic of Tanzania

Bulbophyllum unifoliatum De Wild. subsp. **infracarinatum** (G. Will.) J. J. Verm.
 Bulbophyllum carinatum G. Will.
 Bulbophyllum infracarinatum G. Will.

Distribution: Malawi, Mozambique, Zimbabwe

Bulbophyllum unitubum J. J. Sm.

Distribution: Indonesia

Bulbophyllum univenum J. J. Verm.

Distribution: Indonesia

Part II: Accepted Names / Noms Acceptés / Nombres Aceptado

Bulbophyllum uroglossum Schltr.

Distribution: Papua New Guinea

Bulbophyllum urosepalum Schltr.

Distribution: Papua New Guinea

Bulbophyllum ustusfortiter J. J. Verm.

Distribution: Papua New Guinea

Bulbophyllum uviflorum O'Byrne

Distribution: Indonesia

Bulbophyllum vaccinioides Schltr.

Distribution: Papua New Guinea

Bulbophyllum vagans Ames & Rolfe
  Bulbophyllum vagans Ames & Rolfe var. *angustum* Ames
  Bulbophyllum vagans Ames & Rolfe var. *linearifolium* Ames

Distribution: Philippines

Bulbophyllum vaginatum (Lindl.) Rchb. f.
  Bulbophyllum whiteanum (Rolfe) J. J. Sm.
  Cirrhopetalum caudatum Wight
  Cirrhopetalum stramineum Teijsm. & Binn.
  Cirrhopetalum vaginatum Lindl.
  Cirrhopetalum whiteanum Rolfe
  Phyllorkis vaginata (Lindl.) Kuntze

Distribution: Indonesia, Malaysia, Philippines, Thailand

Bulbophyllum valeryi J. J. Verm. & P. O'Byrne

Distribution: Indonesia

Bulbophyllum validum Carr

Distribution: Indonesia

Bulbophyllum vanum J. J. Verm.

Distribution: Cameroon, Democratic Republic of the Congo, Gabon

Part II: Accepted Names / Noms Acceptés / Nombres Aceptado

Bulbophyllum vanvuurenii J. J. Sm.

Distribution: Indonesia

Bulbophyllum vareschii Foldats

Distribution: Venezuela

Bulbophyllum variabile Ridl.

Distribution: Malaysia

Bulbophyllum variegatum Thouars
Phyllorkis variegata Thouars

Distribution: Comoros, Madagascar, Mauritius, Réunion

Bulbophyllum vaughanii Brade

Distribution: Brazil

Bulbophyllum ventriosum H. Perrier

Distribution: Madagascar

Bulbophyllum vermiculare Hook. f.
Bulbophyllum brookeanum Kraenzl.
Phyllorkis vermicularis (Hook. f.) Kuntze

Distribution: Indonesia, Malaysia, Philippines

Bulbophyllum verruciferum Schltr.

Distribution: Papua New Guinea

Bulbophyllum verruciferum Schltr. var. **carinatisepalum** Schltr.

Distribution: Papua New Guinea

Bulbophyllum verruculatum Schltr.

Distribution: Papua New Guinea

Bulbophyllum verruculiferum H. Perrier

Distribution: Madagascar

Part II: Accepted Names / Noms Acceptés / Nombres Aceptado

Bulbophyllum versteegii J. J. Sm.

Distribution: Indonesia

Bulbophyllum vesiculosum J. J. Sm.
  Bulbophyllum corneri Carr
  Epicranthes corneri (Carr) Garay & W. Kittr.
  Epicranthes vesiculosa (J. J. Sm.) Garay & W. Kittr.

Distribution: Indonesia, Malaysia

Bulbophyllum vestitum Bosser

Distribution: Madagascar

Bulbophyllum vestitum Bosser var. **meridionale** Bosser

Distribution: Madagascar

Bulbophyllum vexillarium Ridl.

Distribution: Indonesia

Bulbophyllum vietnamense Seidenf.
  Bulbophyllum dalatensis Seidenf.
  Cirrhopetalum sigaldii Guillaumin

Distribution: Viet Nam

Bulbophyllum viguieri Schltr.

Distribution: Madagascar

Bulbophyllum vinaceum Ames & C. Schweinf.

Distribution: Indonesia, Philippines

Bulbophyllum violaceolabellum Seidenf.

Distribution: China, Lao People's Democratic Republic, Thailand

Bulbophyllum violaceum (Blume) Lindl.
  Diphyes violacea Blume
  Phyllorkis violacea (Blume) Kuntze

Distribution: Indonesia

Part II: Accepted Names / Noms Acceptés / Nombres Aceptado

Bulbophyllum virescens J. J. Sm.
Bulbophyllum maximum (Ridl.) Ridl.
Bulbophyllum ridleyanum Garay, Hamer & Siegerist
Cirrhopetalum maximum Ridl.

Distribution: Indonesia, Malaysia

Bulbophyllum viridescens Ridl.

Distribution: Indonesia, Malaysia

Bulbophyllum viridiflorum (Hook. f.) Schltr.
Bulbophyllum viridiflorum (Hook. f.) J. J. Sm.
Cirrhopetalum viridiflorum Hook. f.
Phyllorkis viridiflora (Hook. f.) Kuntze

Distribution: Bhutan, India

Bulbophyllum vitellinum Ridl.

Distribution: Indonesia, Malaysia

Bulbophyllum volkensii Schltr.

Distribution: Palau

Bulbophyllum vulcanicum Kraenzl.

Distribution: Burundi, Cameroon, Democratic Republic of the Congo, Kenya, Rwanda, Uganda

Bulbophyllum vulcanorum H. Perrier

Distribution: Madagascar

Bulbophyllum vutimenaense B. A. Lewis

Distribution: Vanuatu

Bulbophyllum wadsworthii Dockr.
Oxysepala wadsworthii (Dockr.) D. L. Jones & M. A. Clem.

Distribution: Australia

Bulbophyllum wagneri Schltr.

Distribution: Colombia, Ecuador, Panama

Bulbophyllum wakoi Howcroft

Distribution: Papua New Guinea

Bulbophyllum wallichi Rchb. f.
Bulbophyllum refractoides Seidenf.
Cirrhopetalum refractum Zoll. var. *laciniatum* Finet

Distribution: Bhutan, China, India, Thailand, Viet Nam

Bulbophyllum wangkaense Seidenf.

Distribution: Thailand

Bulbophyllum warianum Schltr.
Hapalochilus warianus (Schltr.) Garay & W. Kittr.

Distribution: Papua New Guinea

Bulbophyllum warmingianum Cogn.
Bulbophyllum vittatum Rchb. f. & Warm.
Phyllorkis vittata (Rchb. f. & Warm.) Kuntze

Distribution: Brazil

Bulbophyllum weberbauerianum Kraenzl.

Distribution: Bolivia, Peru

Bulbophyllum weberbauerianum Kraenzl. var. **angustius** C. Schweinf.

Distribution: Bolivia

Bulbophyllum weberi Ames
Bulbophyllum baucoense Ames
Cirrhopetalum baucoense (Ames) Garay, Hamer & Siegerist
Cirrhopetalum weberi (Ames) Senghas

Distribution: Philippines

Bulbophyllum wechsbergii Sieder & Kiehn †
Bulbophyllum rhynchoglossum Schltr. (1912)
Hapalochilus reflexus Garay & W. Kittr.

Distribution: Papua New Guinea

Bulbophyllum weddelii (Lindl.) Rchb. f.
Didactyle weddelii Lindl.
Phyllorkis weddellii (Lindl.) Kuntze
Xiphizusa weddelii (Lindl.) Rchb. f.

Distribution: Brazil

Bulbophyllum weinthalii R. S. Rogers
   Adelopetalum weinthalii (R. S. Rogers) D. L. Jones & M. A. Clem.
   Spilorchis weinthalii (R. S. Rogers) D. L. Jones & M. A. Clem.

Distribution: Australia

Bulbophyllum weinthalii R. S. Rogers subsp. **striatum** D. L. Jones & M. A. Clem.
   Adelopetalum weinthalii (R. S. Rogers) D. L. Jones & M. A. Clem. subsp. *striatum* (D. L. Jones) D. L. Jones & M. A. Clem.
   Spilorchis weinthalii (R. S. Rogers) D. L. Jones & M. A. Clem. subsp. *striatum* (D. L. Jones) D. L. Jones & M. A. Clem.

Distribution: Australia

Bulbophyllum wendlandianum (Kraenzl.) Dammer
   Bulbophyllum wendlandianum (Kraenzl.) J. J. Sm.
   Cirrhopetalum collettianum Hemsl. ex Collett & Hemsl.
   Cirrhopetalum collettii Hemsl.
   Cirrhopetalum fastuosum Rchb. f.
   Cirrhopetalum proliferum Hort.
   Cirrhopetalum wendlandianum Kraenzl.
   Phyllorkis collettii (Hemsl.) Kuntze.

Distribution: Myanmar, Thailand

Bulbophyllum wenzelii Ames

Distribution: Philippines

Bulbophyllum werneri Schltr.

Distribution: Papua New Guinea

Bulbophyllum whitei T. E. Hunt & Rupp
   Serpenticaulis whitei (T. E. Hunt & Rupp) M. A. Clem. & D. L. Jones

Distribution: Australia

Bulbophyllum whitfordii Rolfe ex Ames

Distribution: Philippines

Bulbophyllum wightii Rchb. f.
   Bulbophyllum pingtungense S. S. Ying & Chen
   Cirrhopetalum grandiflorum Wight
   Cirrhopetalum pingtungense (Ying & Chen) Garay, Hamer & Siegerist
   Phyllorkis wightii (Rchb. f.) Kuntze

Distribution: Myanmar, Sri Lanka

Bulbophyllum wilkianum T. E. Hunt
Adelopetalum wilkianum (T. E. Hunt) M. A. Clem.
Bulbophyllum exiguum F. Muell. var. *dallachyi* Benth.

Distribution: Australia

Bulbophyllum williamsii A. D. Hawkes
Bulbophyllum caudatum L. O. Williams

Distribution: Philippines

Bulbophyllum windsorense B. Gray & D. L. Jones
Oxysepala windsorensis (B. Gray & D. L. Jones) D. L. Jones & M. A. Clem.

Distribution: Australia

Bulbophyllum woelfliae Garay, Senghas & Lemcke

Distribution: Philippines

Bulbophyllum wolfei B. Gray & D. L. Jones
Serpenticaulis wolfei (B. Gray & D. L. Jones) M. A. Clem. & D. L. Jones

Distribution: Australia

Bulbophyllum wollastonii Ridl.

Distribution: Indonesia

Bulbophyllum wrayi Hook. f.
Bulbophyllum pachyphyllum J. J. Sm.
Phyllorkis wrayi (Hook. f.) Kuntze

Distribution: Indonesia, Malaysia

Bulbophyllum wuzhishanensis X. H. Jin

Distribution: China

Bulbophyllum xantanthum Schltr.

Distribution: Indonesia

Part II: Accepted Names / Noms Acceptés / Nombres Aceptado

Bulbophyllum xanthoacron J. J. Sm.
Hapalochilus xanthoacron (J. J. Sm.) Garay & W. Kittr.

Distribution: Indonesia

Bulbophyllum xanthobulbum Schltr.

Distribution: Madagascar

Bulbophyllum xanthochlamys Schltr.
Bulbophyllum lamprochlamys Schltr.
Bulbophyllum unigibbum J. J. Sm.

Distribution: Indonesia, Papua New Guinea

Bulbophyllum xanthophaeum Schltr.
Hapalochilus xanthophaeus (Schltr.) Garay & W. Kittr.

Distribution: Papua New Guinea

Bulbophyllum xanthornis Schuit. & De Vogel

Distribution: Papua New Guinea

Bulbophyllum xanthotes Schltr.

Distribution: Papua New Guinea

Bulbophyllum xanthum Ridl.

Distribution: Malaysia

Bulbophyllum xenosum J. J. Verm.

Distribution: Brunei Darussalam

Bulbophyllum xiphion J. J. Verm.

Distribution: Indonesia

Bulbophyllum xylocarpi J. J. Sm.

Distribution: Indonesia, Malaysia

Bulbophyllum xylophyllum C. S. P. Parish & Rchb. f.
Bulbophyllum agastymalayanum R. Gopalan & A. N. Henry

Part II: Accepted Names / Noms Acceptés / Nombres Aceptado

Phyllorkis xylophylla (C. S. P. Parish & Rchb. f.) Kuntze

Distribution: India, Myanmar, Thailand, Viet Nam

Bulbophyllum yoksunense J. J. Sm.
Bulbophyllum brachypodium A. S. Rao & Balakr. var. *parviflorum* A. S. Rao & Balakr.
Bulbophyllum brachypodum A. S. Rao & Balakr.
Bulbophyllum yoksunense J. J. Sm. var. *parviflorum* (Rao & Balak.) S. S. R. Bennet
Cirrhopetalum brevipes Hook. f.

Distribution: Bhutan, India, Malaysia

Bulbophyllum yuanyangense Z. H. Tsi

Distribution: China

Bulbophyllum yunnanense Rolfe

Distribution: China

Bulbophyllum zambalense Ames

Distribution: Philippines

Bulbophyllum zamboangense Ames
Cirrhopetalum zamboangense (Ames) Garay, Hamer & Siegerist

Distribution: Philippines

Bulbophyllum zaratananae Schltr.

Distribution: Madagascar

Bulbophyllum zaratananae Schltr. subsp. **disjunctum** H. Perrier

Distribution: Madagascar

Bulbophyllum zebrinum J. J. Sm.
Bulbophyllum ornithoglossum Schltr.

Distribution: Indonesia

Part II: Accepted Names / Noms Acceptés / Nombres Aceptado

CHASEELLA BINOMIALS IN CURRENT USE

CHASEELLA BINOMES ACTUELLEMENT EN USAGE

CHASEELLA BINOMIALES UTILIZADOS NORMALMENTE

Chaseella pseudohydra Summerh.

Distribution: Kenya, Zimbabwe

CODONOSIPHON BINOMIALS IN CURRENT USE

CODONOSIPHON BINOMES ACTUELLEMENT EN USAGE

CODONOSIPHON BINOMIALES UTILIZADOS NORMALMENTE

Codonosiphon campanulatum Schltr.

Distribution: Papua New Guinea

Codonosiphon codonanthum (Schltr.) Schltr.
Bulbophyllum codonanthum Schltr.

Distribution: Indonesia

Codonosiphon papuanum Schltr.

Distribution: Papua New Guinea

DRYMODA BINOMIALS IN CURRENT USE

DRYMODA BINOMES ACTUELLEMENT EN USAGE

DRYMODA BINOMIALES UTILIZADOS NORMALMENTE

Drymoda digitata (J. J. Sm.) Garay, Hamer & Siegerist
  Bulbophyllum digitatum J. J. Sm.
  Monomeria digitata (J. J. Sm.) W. Kittr.

Distribution: Indonesia

Drymoda picta Lindl.

Distribution: Lao People's Democratic Republic, Myanmar, Thailand

Drymoda siamensis Schltr.

Distribution: Lao People's Democratic Republic, Thailand

Part II: Accepted Names / Noms Acceptés / Nombres Aceptado

MONOMERIA BINOMIALS IN CURRENT USE

MONOMERIA BINOMES ACTUELLEMENT EN USAGE

MONOMERIA BINOMIALES UTILIZADOS NORMALMENTE

Monomeria barbata Lindl.
Epicranthes barbata (Lindl.) Rchb. f.
Monomeria crabro C. S. P. Parish & Rchb. f.

Distribution: Thailand, Viet Nam

Monomeria longipes (Rchb. f.) Aver.
Bulbophyllum longipes Rchb. f.
Henosis longipes (Rchb. f.) Hook. f.
Monomeria longipes (Rchb. f.) Garay, Hamer & Siegerist

Distribution: Myanmar, Thailand

MONOSEPALUM BINOMIALS IN CURRENT USE

MONOSEPALUM BINOMES ACTUELLEMENT EN USAGE

MONOSEPALUM BINOMIALES UTILIZADOS NORMALMENTE

Monosepalum dischorense Schltr.

Distribution: Papua New Guinea

Monosepalum muricatum (J. J. Sm.) Schltr.
  Bulbophyllum muricatum J. J. Sm.

Distribution: Papua New Guinea

Monosepalum torricellense Schltr.

Distribution: Papua New Guinea

Part II: Accepted Names / Noms Acceptés / Nombres Aceptado

Part II: Accepted Names / Noms Acceptés / Nombres Aceptado

PEDILOCHILUS BINOMIALS IN CURRENT USE

PEDILOCHILUS BINOMES ACTUELLEMENT EN USAGE

PEDILOCHILUS BINOMIALES UTILIZADOS NORMALMENTE

Pedilochilus alpinum P. Royen

Distribution: Papua New Guinea

Pedilochilus alpinum P. Royen var. **fasciculatum** P. Royen

Distribution: Papua New Guinea

Pedilochilus angustifolius Schltr.

Distribution: Papua New Guinea

Pedilochilus augustifolium Schltr.

Distribution: Papua New Guinea

Pedilochilus bantaengensis J. J. Sm.

Distribution: Indonesia

Pedilochilus brachiatus Schltr.

Distribution: Papua New Guinea

Pedilochilus brachypus Schltr.

Distribution: Papua New Guinea

Pedilochilus ciliolatum Schltr.

Distribution: Papua New Guinea, Solomon Islands

Pedilochilus clemensiae L. O. Williams

Distribution: Indonesia

Part II: Accepted Names / Noms Acceptés / Nombres Aceptado

Pedilochilus coiloglossum (Schltr.) Schltr.
Bulbophyllum coiloglossum Schltr.

Distribution: Indonesia, Papua New Guinea

Pedilochilus cyatheicola P. Royen

Distribution: Papua New Guinea

Pedilochilus dischorense Schltr.

Distribution: Papua New Guinea

Pedilochilus flavum Schltr.

Distribution: Papua New Guinea

Pedilochilus grandifolium P. Royen

Distribution: Papua New Guinea

Pedilochilus guttulatum Schltr.

Distribution: Papua New Guinea

Pedilochilus hermonii P. J. Cribb & B. Lewis

Distribution: Vanuatu

Pedilochilus humile J. J. Sm.

Distribution: Indonesia

Pedilochilus kermesinostriatum J. J. Sm.

Distribution: Indonesia

Pedilochilus longipes Schltr.

Distribution: Papua New Guinea

Pedilochilus macrorrhinum P. Royen

Distribution: Indonesia

Part II: Accepted Names / Noms Acceptés / Nombres Aceptado

Pedilochilus majus J. J. Sm.

Distribution: Indonesia

Pedilochilus montana Ridl.

Distribution: Indonesia

Pedilochilus obovatum J. J. Sm.
  Pedilochilus oreadum P. Royen

Distribution: Indonesia

Pedilochilus papuanum Schltr.

Distribution: Papua New Guinea

Pedilochilus parvulum Schltr.

Distribution: Papua New Guinea

Pedilochilus perpusillum P. Royen

Distribution: Indonesia

Pedilochilus petiolatum Schltr.

Distribution: Papua New Guinea

Pedilochilus petrophilum P. Royen

Distribution: Papua New Guinea

Pedilochilus piundaundense P. Royen

Distribution: Papua New Guinea

Pedilochilus psychrophilum (F. Muell.) Ormerod
  Dendrobium psychrophilum F. Muell.

Distribution: Papua New Guinea

Pedilochilus pumilio Ridl.

Distribution: Indonesia

Part II: Accepted Names / Noms Acceptés / Nombres Aceptado

Pedilochilus pusillum Schltr.

Distribution: Papua New Guinea

Pedilochilus sarawakatensis P. Royen

Distribution: Papua New Guinea

Pedilochilus stictanthum Schltr.

Distribution: Papua New Guinea

Pedilochilus subalpinum P. Royen

Distribution: Papua New Guinea

Pedilochilus sulphureum J. J. Sm.

Distribution: Indonesia

Pedilochilus terrestre J. J. Sm.

Distribution: Indonesia

Part II: Accepted Names / Noms Acceptés / Nombres Aceptado

SACCOGLOSSUM BINOMIALS IN CURRENT USE

SACCOGLOSSUM BINOMES ACTUELLEMENT EN USAGE

SACCOGLOSSUM BINOMIALES UTILIZADOS NORMALMENTE

Saccoglossum lanceolatum L. O. Williams

Distribution: Indonesia

Saccoglossum maculatum Schltr.

Distribution: Papua New Guinea

Saccoglossum papuanum Schltr.

Distribution: Papua New Guinea

Saccoglossum takeuchii Howcroft

Distribution: Papua New Guinea

Saccoglossum verrucosum L. O. Williams

Distribution: Indonesia

Part II: Accepted Names / Noms Acceptés / Nombres Aceptado

Part II: Accepted Names / Noms Acceptés / Nombres Aceptado

SUNIPIA BINOMIALS IN CURRENT USE

SUNIPIA BINOMES ACTUELLEMENT EN USAGE

SUNIPIA BINOMIALES UTILIZADOS NORMALMENTE

Sunipia andersonii (King & Pantl.) P. F. Hunt
Ione andersoni King & Pantl.
Ione andersoni King & Pantl. var. *flavescens* (Rolfe) Tang & Wang
Ione bifurcatoflorens Fukuyama
Ione flavescens Rolfe
Ione purpurata Braid
Ione sasakii Hayata
Sunipia bifurcatoflorens (Fukuy.) P. F. Hunt
Sunipia flavescens (Rolfe) P. F. Hunt
Sunipia purpurata (Braid) P. F. Hunt
Sunipia sasakii (Hayata) P. F. Hunt

Distribution: Bhutan, China, India, Myanmar, Taiwan, Province of China , Thailand, Viet Nam

Sunipia angustipetala Seidenf.
Ione angustipetala (Seidenf.) Seidenf.

Distribution: Thailand

Sunipia annamensis (Ridl.) P. F. Hunt
Ione annamensis Ridl.

Distribution: Thailand, Viet Nam

Sunipia australis (Seidenf) P. F. Hunt
Ione australis Seidenf.

Distribution: Thailand

Sunipia bicolor Lindl.
Bulbophyllum bicolor (Lindl.) Hook. f.
Dipodium khasyanum Griff.
Ione bicolor (Lindl.) Lindl.
Ione khasiana (Griff.) Lindl.

Distribution: Bhutan, China, India, Myanmar, Nepal, Thailand

Sunipia candida (Lindl.) P. F. Hunt
Ione candida Lindl.

Distribution: Bhutan

Sunipia cirrhata (Lindl.) P. F. Hunt
 Bulbophyllum mishmeense Hook. f.
 Bulbophyllum paleaceum Benth. & Hook. f.
 Ione cirrhata Lindl.
 Ione fusco-purpurea Lindl.
 Ione paleacea Lindl.
 Phyllorkis mischmeensis Kuntze
 Phyllorkis paleacea (Benth.) Kunze
 Sunipia fuscopurpurea (Lindl.) P. F. Hunt
 Sunipia paleacea (Lindl.) P. F. Hunt

Distribution: China, India, Nepal, Viet Nam

Sunipia cumberlegei (Seidenf) P. F. Hunt
 Ione cumberlegei Seidenf.

Distribution: Thailand

Sunipia dichroma (Rolfe) Bân & D. H. Duong
 Bulbophyllum dichromum Rolfe
 Bulbophyllum jacquetii Gagnep.
 Ione dichroma (Rolfe) Gagnep.
 Monomeria dichroma (Rolfe) Schltr.

Distribution: Viet Nam

Sunipia grandiflora (Rolfe) P. F. Hunt
 Ione grandiflora Rolfe

Distribution: Thailand

Sunipia hainanensis Z. H. Tsi

Distribution: China

Sunipia intermedia (King & Pantl.) P. F. Hunt
 Ione intermedia King & Pantl.

Distribution: Bhutan

Sunipia jainii T. M. Hynniewta & C. L. Malhotra
 Ione jainii (T. M. Hynniewta & C. L. Malhotra) Seidenf.

Distribution: India

Sunipia kachinensis Seidenf.
 Ione kachinensis (Seidenf.) Seidenf.

Distribution: Thailand

324

Part II: Accepted Names / Noms Acceptés / Nombres Aceptado

Sunipia minor (Seidenf) P. F. Hunt
Ione minor Seidenf.

Distribution: Thailand

Sunipia pallida (Aver.) Aver.
Ione pallida Aver.

Distribution: Viet Nam

Sunipia racemosa (Sm.) T. Tang & F. T. Wang
Ione scariosa (Lindl.) King & Pantl. var. *magnibracteatum* Kerr
Stelis racemosa J. E. Sm.

Distribution: China, India, Lao People's Democratic Republic, Myanmar, Nepal, Thailand, Viet Nam

Sunipia rimannii (Rchb. f.) Seidenf.
Acrochaene rimanni Rchb. f.
Ione rimannii (Rchb. f.) Seidenf.
Ione salweenensis Phillimore & W. W. Sm.
Monomeria rimannii (Rchb.) Schltr.
Sunipia salweenensis (Phil. & W. W. Sm.) P. F. Hunt

Distribution: Myanmar, Thailand

Sunipia scariosa Lindl.
Ione scariosa (Lindl.) King & Pantl.
Ione siamensis Rolfe

Distribution: Bhutan, Myanmar, Viet Nam

Sunipia soidaoensis (Seidenf) P. F. Hunt
Ione soidaoensis Seidenf.

Distribution: Thailand

Sunipia thailandica (Seidenf & Smit.) P. Hunt
Ione thailandica Seidenf. & Smitinand

Distribution: Thailand

Sunipia virens (Lindl.) P. F. Hunt
Bulbophyllum virens Hook. f.
Ione virens Lindl.
Phyllorkis virens (Lindl.) Kuntze

Distribution: Thailand

Sunipia viridis (Seidenf) P. F. Hunt
  Ione viridis Seidenf.

Distribution: Thailand

Part II: Accepted Names / Noms Acceptés / Nombres Aceptado

TRIAS BINOMIALS IN CURRENT USE

TRIAS BINOMES ACTUELLEMENT EN USAGE

TRIAS BINOMIALES UTILIZADOS NORMALMENTE

Trias antheae J. J. Verm. & A. L. Lamb

Distribution: Malaysia

Trias bonaccordensis C. Sathish Kumar

Distribution: India

Trias cambodiana Christenson

 Distribution: Cambodia

Trias disciflora (Rolfe) Rolfe
  Bulbophyllum disciflorum Rolfe

Distribution: Lao People's Democratic Republic

Trias intermedia Seidenf. & Smitinand

Distribution: Thailand

Trias mollis Seidenf.

Distribution: Thailand

Trias nana Seidenf.

Distribution: Thailand

Trias nasuta (Rchb. f.) Stapf
  Bulbophyllum nasutum Rchb. f.
  Phyllorkis nasuta (Rchb. f.) Kuntze
  Trias vitrina Rolfe

Distribution: Myanmar, Thailand, Viet Nam

Trias oblonga Lindl.
  Bulbophyllum moulmeinense Rchb. f.
  Bulbophyllum oblongum (Lindl.) Rchb. f.
  Trias ovata Lindl.

Distribution: Myanmar, Thailand

Part II: Accepted Names / Noms Acceptés / Nombres Aceptado

Trias picta (C. S. P. Parish & Rchb. f.) C. S. P. Parish ex Hemsl.
Bulbophyllum pictum C. S. P. Parish & Rchb. f.

Distribution: Myanmar, Thailand

Trias pusilla J. Deka & H. Deka
Jejosephia pusilla (J. Deka & H. Deka) A. N. Rao & Mani

Distribution: India

Trias rosea (Ridl.) Seidenf.
Bulbophyllum roseum Ridl.
Trias rolfei Stapf

Distribution: Thailand

Trias stocksii Benth. ex Hook. f.

Distribution: India

Trias tothastes (J. J. Verm.) J. J. Wood
Bulbophyllum tothastes J. J. Verm.

Distribution: Indonesia

Part III: Country Checklist / Liste par Pays / Lista por Paises

Part III: Country Checklist / Liste par Pays / Lista por Paises

PART III: COUNTRY CHECKLIST
For the genera:

Acrochaene, Bulbophyllum, Chaseella, Codonosiphon, Drymoda, Monomeria, Monosepalum, Pedilochilus, Saccoglossum, Sunipia and *Trias*

TROISIÈME PARTIE: LISTE PAR PAYS
Pour les genre:

Acrochaene, Bulbophyllum, Chaseella, Codonosiphon, Drymoda, Monomeria, Monosepalum, Pedilochilus, Saccoglossum, Sunipia et *Trias*

PARTE III: LISTA POR PAÍSES
Para el genero:

Acrochaene, Bulbophyllum, Chaseella, Codonosiphon, Drymoda, Monomeria, Monosepalum, Pedilochilus, Saccoglossum, Sunipia y *Trias*

PART III: COUNTRY CHECKLIST FOR THE GENERA:
Acrochaene, Bulbophyllum, Chaseella, Codonosiphon, Drymoda, Monomeria, Monosepalum, Pedilochilus, Saccoglossum, Sunipia and *Trias*

TROISIÈME PARTIE: LISTE PAR PAYS POUR LES GENRE:
Acrochaene, Bulbophyllum, Chaseella, Codonosiphon, Drymoda, Monomeria, Monosepalum, Pedilochilus, Saccoglossum, Sunipia et *Trias*

PARTE III: LISTA POR PAISES PARA EL GENERO:
Acrochaene, Bulbophyllum, Chaseella, Codonosiphon, Drymoda, Monomeria, Monosepalum, Pedilochilus, Saccoglossum, Sunipia y *Trias*

ANGOLA / ANGOLA (L') / ANGOLA

Bulbophyllum cocoinum Bateman ex Lindl.
Bulbophyllum colubrinum (Rchb. f.) Rchb. f.
Bulbophyllum fuscum Lindl.
Bulbophyllum intertextum Lindl.
Bulbophyllum maximum (Lindl.) Rchb. f.
Bulbophyllum scaberulum (Rolfe) Bolus
Bulbophyllum unifoliatum De Wild.

AUSTRALIA / AUSTRALIE (L.) / AUSTRALIA

Bulbophyllum argyropus (Endl.) Rchb. f.
Bulbophyllum baileyi F. Muell.
Bulbophyllum boonjee B. Gray & D. L. Jones
Bulbophyllum bowkettae F. M. Bailey
Bulbophyllum bracteatum F. M. Bailey
Bulbophyllum caldericola G. F. Walsh
Bulbophyllum elizae (F. Muell.) Benth.
Bulbophyllum evasum Hunt & Rupp
Bulbophyllum exiguum F. Muell.
Bulbophyllum gadgarrense Rupp
Bulbophyllum globuliforme Nicholls
Bulbophyllum gracillimum (Rolfe) Rolfe
Bulbophyllum grandimesense B. Gray & D. L. Jones
Bulbophyllum intermedium F. M. Bailey
Bulbophyllum johnsonii T. E. Hunt
Bulbophyllum kirkwoodae T. E. Hunt
Bulbophyllum lageniforme F. M. Bailey
Bulbophyllum lamingtonense D. L. Jones
Bulbophyllum lewisense B. Gray & D. L. Jones
Bulbophyllum lilianae Rendle
Bulbophyllum longiflorum Thouars
Bulbophyllum macphersoni Rupp
Bulbophyllum macphersoni Rupp var. **spathulatum** Dockrill & St. Cloud
Bulbophyllum minutissimum (F. Muell.) F. Muell.
Bulbophyllum nematopodum F. Muell.
Bulbophyllum newportii (F. M. Bailey) Rolfe
Bulbophyllum radicans F. M. Bailey
Bulbophyllum schillerianum Rchb. f.

Bulbophyllum shepherdi (F. Muell.) Rchb. f.
Bulbophyllum tuberculatum Colenso
Bulbophyllum umbellatum Lindl. var. **fuscescens** P. K. Sarkar
Bulbophyllum wadsworthii Dockr.
Bulbophyllum weinthalii R. S. Rogers
Bulbophyllum weinthalii R. S. Rogers subsp. **striatum** D. L. Jones & M. A. Clem.
Bulbophyllum whitei T. E. Hunt & Rupp
Bulbophyllum wilkianum T. E. Hunt
Bulbophyllum windsorense B. Gray & D. L. Jones
Bulbophyllum wolfei B. Gray & D. L. Jones

BANGLADESH / BANGLADESH (LE) / BANGLADESH

Bulbophyllum liliacinum Ridl.
Bulbophyllum neilgherrense Wight
Bulbophyllum reptans (Lindl.) Lindl.
Bulbophyllum scabratum Rchb. f.

BELIZE / BELIZE (LE) / BELICE

Bulbophyllum aristatum (Rchb. f.) Hemsl.
Bulbophyllum oerstedii (Rchb. f.) Hemsl.
Bulbophyllum pachyrachis (A. Rich.) Griseb.

BHUTAN / BHOUTAN (LE) / BHUTÁN

Acrochaene punctata Lindl.
Bulbophyllum affine Lindl.
Bulbophyllum amplifolium (Rolfe) Balakr. & Sud.Chowdhury
Bulbophyllum andersonii (Hook. f.) J. J. Sm.
Bulbophyllum apodum Hook. f.
Bulbophyllum appendiculatum (Rolfe) J. J. Sm.
Bulbophyllum bisetum Lindl.
Bulbophyllum careyanum (Hook. f.) Sprengel
Bulbophyllum cariniflorum Rchb. f.
Bulbophyllum cauliflorum Hook. f.
Bulbophyllum cauliflorum Hook. f. var. **sikkimense** N. Peace & P. J. Cribb
Bulbophyllum cornu-cervi King & Pantl.
Bulbophyllum crassipes Hook. f.
Bulbophyllum cylindraceum Lindl.
Bulbophyllum elatum (Hook. f.) J. J. Sm.
Bulbophyllum emarginatum (Finet) J. J. Sm.
Bulbophyllum euplepharum Rchb. f.
Bulbophyllum gracilipes King & Pantl.
Bulbophyllum griffithii (Lindl.) Rchb. f.
Bulbophyllum guttulatum (Hook. f.) Balakr.
Bulbophyllum gymnopus Hook. f.
Bulbophyllum helenae (Kuntze) J. J. Sm.
Bulbophyllum hirtum (J. E. Sm.) Lindl.
Bulbophyllum hymenanthum Hook. f.
Bulbophyllum khasyanum Griff.
Bulbophyllum leopardinum (Wall.) Lindl.
Bulbophyllum leopardinum (Wall.) Lindl. var. **tuberculatum** N. P. Balakrishnan &

S. Chowdhury

Bulbophyllum leptanthum Hook. f.
Bulbophyllum monanthum (Kuntze) J. J. Sm.
Bulbophyllum odoratissimum (J. E. Sm.) Lindl.
Bulbophyllum odoratissimum (J. E. Sm.) Lindl. var. **racemosum** N. P. Balakr.
Bulbophyllum otoglossum Tuyama
Bulbophyllum pantlingii S. Z. Lucksom
Bulbophyllum parviflorum C. S. P. Parish & Rchb. f.
Bulbophyllum penicillium C. S. P. Parish & Rchb. f.
Bulbophyllum piluliferum King & Pantl.
Bulbophyllum polyrhizum Lindl.
Bulbophyllum reptans (Lindl.) Lindl.
Bulbophyllum retusiusculum Rchb. f.
Bulbophyllum rigidum King & Pantl.
Bulbophyllum rolfei (Kuntze) Seidenf.
Bulbophyllum rothschildianum (O' Brien) J. J. Sm.
Bulbophyllum roxburghii (Lindl.) Rchb. f.
Bulbophyllum sarcophylloides Garay, Hamer & Siegerist
Bulbophyllum sarcophyllum (King & Pantl.) J. J. Sm.
Bulbophyllum scabratum Rchb. f.
Bulbophyllum secundum Hook. f.
Bulbophyllum shweliense W. W. Sm.
Bulbophyllum spathulatum (Rolfe ex E. Cooper) Seidenf.
Bulbophyllum stenobulbon C. S. P. Parish & Rchb. f.
Bulbophyllum sterile (Lam.) Suresh
Bulbophyllum striatum (Griff.) Rchb. f.
Bulbophyllum tortuosum (Blume) Lindl.
Bulbophyllum triste Rchb. f.
Bulbophyllum umbellatum Lindl.
Bulbophyllum viridiflorum (Hook. f.) Schltr.
Bulbophyllum wallichi Rchb. f.
Bulbophyllum yoksunense J. J. Sm.
Sunipia andersonii (King & Pantl.) P. F. Hunt
Sunipia bicolor Lindl.
Sunipia candida (Lindl.) P. F. Hunt
Sunipia intermedia (King & Pantl.) P. F. Hunt
Sunipia scariosa Lindl.

BOLIVIA/ BOLIVIE (LA) / BOLIVIA

Bulbophyllum amazonicum L. O. Williams
Bulbophyllum bolivianum Schltr.
Bulbophyllum bracteolatum Lindl.
Bulbophyllum exaltatum Lindl.
Bulbophyllum morenoi Dodson & Vasquez
Bulbophyllum pachyrachis (A. Rich.) Griseb.
Bulbophyllum tricolor L. B. Sm. & S. K. Harris
Bulbophyllum weberbauerianum Kraenzl.
Bulbophyllum weberbauerianum Kraenzl. var. **angustius** C. Schweinf.

BRAZIL / BRÉSIL (LE) / BRASIL (EL)

Bulbophyllum adiamantinum Brade
Bulbophyllum arianeae Fraga & E.C.Smidt
Bulbophyllum atropurpureum Barb. Rodr.
Bulbophyllum barbatum Barb. Rodr.
Bulbophyllum bidentatum (Barb. Rodr.) Cogn.
Bulbophyllum boudetiana Fraga
Bulbophyllum bracteolatum Lindl.
Bulbophyllum campos-portoi Brade
Bulbophyllum cantagallense (Barb. Rodr.) Cogn.
Bulbophyllum chloroglossum Rchb. f.
Bulbophyllum chloropterum Rchb. f.
Bulbophyllum ciluliae Bianch. & J.A.N.Bat.
Bulbophyllum clausseni Rchb. f.
Bulbophyllum cogniauxianum (Kraenzl.) J. J. Sm.
Bulbophyllum correae Pabst
Bulbophyllum cribbianum Toscano
Bulbophyllum dusenii Kraenzl.
Bulbophyllum epiphytum Barb. Rodr.
Bulbophyllum exaltatum Lindl.
Bulbophyllum filifolium Borba & E.C.Smidt
Bulbophyllum geraense Rchb. f.
Bulbophyllum gladiatum Lindl.
Bulbophyllum glutinosum (Barb. Rodr.) Cogn.
Bulbophyllum gomesii Fraga
Bulbophyllum granulosum Barb. Rodr.
Bulbophyllum insectiferum Barb. Rodr.
Bulbophyllum involutum L. E. Borba & J. Semir
Bulbophyllum ipanemense Hoehne
Bulbophyllum jaguariahyvae Kraenzl.
Bulbophyllum kautskyi Toscano
Bulbophyllum kegelii Hamer & Garay
Bulbophyllum laciniatum (Barb. Rodr.) Cogn.
Bulbophyllum longipetalum Pabst
Bulbophyllum longispicatum Cogn.
Bulbophyllum luederwaldtii Hoehne & Schltr.
Bulbophyllum lundianum Rchb. f. & Warm.
Bulbophyllum macroceras Barb. Rodr.
Bulbophyllum malachadenia Cogn.
Bulbophyllum melloi Pabst
Bulbophyllum mentosum Barb. Rodr.
Bulbophyllum micranthum Barb. Rodr.
Bulbophyllum micropetaliforme Leite
Bulbophyllum mirandaianum Hoehne
Bulbophyllum mucronifolium Rchb. f. & Warm.
Bulbophyllum napelli Lindl.
Bulbophyllum napelloides Kraenzl.
Bulbophyllum nemorosum (Barb. Rodr.) Cogn.
Bulbophyllum ochraceum (Barb. Rodr.) Cogn.
Bulbophyllum oerstedii (Rchb. f.) Hemsl.
Bulbophyllum pabstii Garay
Bulbophyllum paranaense Schltr.
Bulbophyllum paranaense Schltr. var. **pauloense** Hoehne & Schltr.
Bulbophyllum perii Schltr.

Bulbophyllum plumosum (Barb. Rodr.) Cogn.
Bulbophyllum proencai Leite
Bulbophyllum punctatum Barb. Rodr.
Bulbophyllum quadricolor (Barb. Rodr.) Cogn.
Bulbophyllum quadrisetum Lindl.
Bulbophyllum regnelli Rchb. f.
Bulbophyllum ricaldonei Leite
Bulbophyllum roraimense Rolfe
Bulbophyllum rupicola Barb. Rodr.
Bulbophyllum sanderianum Rolfe
Bulbophyllum sturmhoefelii Hoehne
Bulbophyllum teresense Ruschi
Bulbophyllum tripetalum Lindl.
Bulbophyllum vaughanii Brade
Bulbophyllum warmingianum Cogn.
Bulbophyllum weddelii (Lindl.) Rchb. f.

BRUNEI DARUSSALAM / BRUNÉI DARUSSALAM (LE) / BRUNEI DARUSSALAM

Bulbophyllum aithorhachis J. J. Verm.
Bulbophyllum camptochilum J. J. Verm.
Bulbophyllum elongatum (Blume) Hassk.
Bulbophyllum limbatum Lindl.
Bulbophyllum mirabile Hallier f.
Bulbophyllum praetervisum J. J. Verm.
Bulbophyllum taeter J. J. Verm.
Bulbophyllum xenosum J. J. Verm.

BURUNDI / BURUNDI (LE) / BURUNDI

Bulbophyllum cochleatum Lindl. var. **brachyanthum** (Summerh.) J. J. Verm.
Bulbophyllum encephalodes Summerh.
Bulbophyllum josephi (Kuntze) Summerh.
Bulbophyllum vulcanicum Kraenzl.

CAMBODIA / CAMBODGE (LE) / CAMBOYA

Bulbophyllum dayanum Rchb. f.
Bulbophyllum lobbii Lindl.
Bulbophyllum moniliforme C. S. P. Parish & Rchb. f.
Bulbophyllum pinicolum Gagnep.
Bulbophyllum rufinum Rchb. f.
Bulbophyllum siamense Rchb. f.
Bulbophyllum tricorne Seidenf. & Smitinand
Trias cambodiana Christenson

CAMEROON / CAMEROUN (LE) / CAMERÚN (EL)

Bulbophyllum acutebracteatum De Wild.
Bulbophyllum acutebracteatum De Wild. var. **rubrobrunneopapillosum** (De Wild.) J. J. Verm.
Bulbophyllum alinae Szlach. & Olszewski
Bulbophyllum barbigerum Lindl.
Bulbophyllum bidenticulatum J. J. Verm.
Bulbophyllum bifarium Hook. f.
Bulbophyllum blepharochilum Garay
Bulbophyllum calvum Summerh.
Bulbophyllum calyptratum Kraenzl.
Bulbophyllum capituliflorum Rolfe
Bulbophyllum carnosilabium Summerh.
Bulbophyllum carnosisepalum J. J. Verm.
Bulbophyllum cochleatum Lindl.
Bulbophyllum cochleatum Lindl. var. **bequaertii** (De Wild.) J. J. Verm.
Bulbophyllum cochleatum Lindl. var. **tenuicaule** (Lindl.) J. J. Verm.
Bulbophyllum cocoinum Bateman ex Lindl.
Bulbophyllum colubrinum (Rchb. f.) Rchb. f.
Bulbophyllum dolabriforme J. J. Verm.
Bulbophyllum encephalodes Summerh.
Bulbophyllum falcatum (Lindl.) Rchb. f.
Bulbophyllum falcatum (Lindl.) Rchb. f. var. **bufo** (Lindl.) J. J. Verm.
Bulbophyllum falcatum (Lindl.) Rchb. f. var. **velutinum** (Lindl.) J. J. Verm.
Bulbophyllum falcipetalum Lindl.
Bulbophyllum fayi J. J. Verm.
Bulbophyllum finetii Szlach. & Olszewski
Bulbophyllum fuscum Lindl.
Bulbophyllum fuscum Lindl. var. **melinostachyum** (Schltr) J. J. Verm.
Bulbophyllum hirsutissimum Kraenzl. (1912)
Bulbophyllum imbricatum Lindl.
Bulbophyllum intertextum Lindl.
Bulbophyllum ivorense P. J. Cribb & Perez-Vera
Bulbophyllum jaapii Szlach. & Olszewski
Bulbophyllum josephi (Kuntze) Summerh.
Bulbophyllum josephi (Kuntze) Summerh. var. **mahonii** (Rolfe) J. J. Verm.
Bulbophyllum kupense P. J. Cribb & B. J. Pollard
Bulbophyllum lupulinum Lindl.
Bulbophyllum magnibracteatum Summerh.
Bulbophyllum maximum (Lindl.) Rchb. f.
Bulbophyllum mediocre Summerh.
Bulbophyllum nigericum Summerh.
Bulbophyllum nigritianum Rendle
Bulbophyllum nummularia (Wendl. & Kraenzl.) Rolfe
Bulbophyllum oreonastes Rchb. f.
Bulbophyllum oxychilum Schltr.
Bulbophyllum pipio Rchb. f.
Bulbophyllum porphyrostachys Summerh.
Bulbophyllum pumilum (Sw.) Lindl.
Bulbophyllum purpureorhachis (De Wild.) Schltr.
Bulbophyllum renkinianum (Laurent) De Wild.
Bulbophyllum resupinatum Ridl.
Bulbophyllum resupinatum Ridl. var. **filiforme** (Kraenzl.) J. J. Verm.
Bulbophyllum saltatorium Lindl.

Bulbophyllum saltatorium Lindl. var. **albociliatum** (Finet) J. J. Verm.
Bulbophyllum saltatorium Lindl. var. **calamarium** (Lindl.) J. J. Verm.
Bulbophyllum sandersonii (Hook. f.) Rchb. f.
Bulbophyllum sandersonii (Hook. f.) Rchb. f. subsp. **stenopetalum** (Kraenzl.) J. J. Verm.
Bulbophyllum scaberulum (Rolfe) Bolus
Bulbophyllum scaberulum (Rolfe) Bolus var. **fuerstenbergianum** (De Wild.) J. J. Verm.
Bulbophyllum schimperianum Kraenzl.
Bulbophyllum schinzianum Kraenzl.
Bulbophyllum schinzianum Kraenzl. var. **phaeopogon** (Schltr.) J. J. Verm.
Bulbophyllum teretifolium Schltr.
Bulbophyllum tetragonum Lindl.
Bulbophyllum vanum J. J. Verm.
Bulbophyllum vulcanicum Kraenzl.

CENTRAL AFRICAN REPUBLIC / RÉPUBLIQUE CENTRAFRICAINE (LA) / REPÚBLICA CENTROAFRICANA (LA)

Bulbophyllum barbigerum Lindl.
Bulbophyllum falcatum (Lindl.) Rchb. f.
Bulbophyllum fuscum Lindl.
Bulbophyllum imbricatum Lindl.
Bulbophyllum magnibracteatum Summerh.
Bulbophyllum maximum (Lindl.) Rchb. f.
Bulbophyllum oreonastes Rchb. f.
Bulbophyllum oxychilum Schltr.
Bulbophyllum saltatorium Lindl. var. **albociliatum** (Finet) J. J. Verm.
Bulbophyllum scaberulum (Rolfe) Bolus
Bulbophyllum schimperianum Kraenzl.

CHINA / CHINE (LA) / CHINA

Bulbophyllum affine Lindl.
Bulbophyllum ambrosia (Hance) Schltr.
Bulbophyllum andersonii (Hook. f.) J. J. Sm.
Bulbophyllum bicolor Lindl.
Bulbophyllum bomiensis Z. H. Tsi
Bulbophyllum boninense (Schltr.) J. J. Sm.
Bulbophyllum brevispicatum Z. H. Tsi & H. C. Chen
Bulbophyllum chinense (Lindl.) Rchb. f.
Bulbophyllum chondriophorum (Gagnep.) Seidenf.
Bulbophyllum colomaculosum Z. H. Tsi & H. C. Chen
Bulbophyllum corallinum Tixier & Guillaumin
Bulbophyllum delitescens Hance
Bulbophyllum depressum King & Pantl.
Bulbophyllum drymoglossum Maxim.
Bulbophyllum electrinum Seidenf.
Bulbophyllum emarginatum (Finet) J. J. Sm.
Bulbophyllum forresti Seidenf.
Bulbophyllum funingense Z. H. Tsi & H. C. Chen
Bulbophyllum gongshanense Z. H. Tsi
Bulbophyllum guttulatum (Hook. f.) Balakr.

Bulbophyllum hainanense Z. H. Tsi
Bulbophyllum helenae (Kuntze) J. J. Sm.
Bulbophyllum henanense J. L. Lu
Bulbophyllum hirundinis (Gagnep.) Seidenf.
Bulbophyllum japonicum (Makino) Makino
Bulbophyllum khasyanum Griff.
Bulbophyllum kwangtungense Schltr.
Bulbophyllum ledungense T. Tang & F. T. Wang
Bulbophyllum levinei Schltr.
Bulbophyllum longibrachiatum Z. H. Tsi
Bulbophyllum longidens (Rolfe) Seidenf.
Bulbophyllum melanoglossum Hayata
Bulbophyllum menghaiense Z. H. Tsi
Bulbophyllum menglunense Z. H. Tsi & Y. Z. Ma
Bulbophyllum nigrescens Rolfe
Bulbophyllum nigripetalum Rolfe
Bulbophyllum obtusiangulum Z. H. Tsi
Bulbophyllum odoratissimum (J. E. Sm.) Lindl.
Bulbophyllum orientale Seidenf.
Bulbophyllum pectenveneris (Gagnep.) Seidenf.
Bulbophyllum pectinatum Finet
Bulbophyllum poilanei Gagnep.
Bulbophyllum psittacoglossum Rchb. f.
Bulbophyllum quadrangulum Z. H. Tsi
Bulbophyllum reptans (Lindl.) Lindl.
Bulbophyllum retusiusculum Rchb. f.
Bulbophyllum retusiusculum Rchb. f. var. **oreogenes** (W. W. Sm.) Z. H. Tsi
Bulbophyllum retusiusculum Rchb. f. var. **tigridum** (Hance) Z. H. Tsi
Bulbophyllum riyanum Fukuyama
Bulbophyllum scaphiforme J. J. Verm.
Bulbophyllum secundum Hook. f.
Bulbophyllum shweliense W. W. Sm.
Bulbophyllum sphaericum Z. H. Tsi & H. Li
Bulbophyllum sutepense (Rolfe ex Downie) Seidenf. & Smitinand
Bulbophyllum taeniophyllum C. S. P. Parish & Rchb. f.
Bulbophyllum tengchongense Z. H. Tsi
Bulbophyllum tigridium Hance
Bulbophyllum tseanum (S. Y. Hu & Barretto) Z. H. Tsi
Bulbophyllum umbellatum Lindl.
Bulbophyllum violaceolabellum Seidenf.
Bulbophyllum wallichi Rchb. f.
Bulbophyllum wuzhishanensis X. H. Jin
Bulbophyllum yuanyangense Z. H. Tsi
Bulbophyllum yunnanense Rolfe
Sunipia andersonii (King & Pantl.) P. F. Hunt
Sunipia bicolor Lindl.
Sunipia cirrhata (Lindl.) P. F. Hunt
Sunipia hainanensis Z. H. Tsi
Sunipia racemosa (Sm.) T. Tang & F. T. Wang

COLOMBIA / COLOMBIE (LA) / COLOMBIA

Bulbophyllum antioquiense Kraenzl.
Bulbophyllum lehmannianum Kraenzl.

Bulbophyllum meridense Rchb. f.
Bulbophyllum oerstedii (Rchb. f.) Hemsl.
Bulbophyllum popayanense Leme & Kraenzl.
Bulbophyllum wagneri Schltr.

COMOROS (THE) / COMORES (LES) / COMORAS (LAS)

Bulbophyllum clavatum Thouars
Bulbophyllum comorianum H. Perrier
Bulbophyllum coriophorum Ridl.
Bulbophyllum hyalinum Schltr.
Bulbophyllum leoni Kraenzl.
Bulbophyllum megalonyx Rchb. f.
Bulbophyllum variegatum Thouars

CONGO (THE) / CONGO (LE) / CONGO (EL)

Bulbophyllum barbigerum Lindl.
Bulbophyllum calyptratum Kraenzl.
Bulbophyllum capituliflorum Rolfe
Bulbophyllum colubrinum (Rchb. f.) Rchb. f.
Bulbophyllum imbricatum Lindl.
Bulbophyllum maximum (Lindl.) Rchb. f.
Bulbophyllum mayombeense Garay
Bulbophyllum pumilum (Sw.) Lindl.
Bulbophyllum purpureorhachis (De Wild.) Schltr.
Bulbophyllum saltatorium Lindl. var. **albociliatum** (Finet) J. J. Verm.
Bulbophyllum saltatorium Lindl. var. **calamarium** (Lindl.) J. J. Verm.
Bulbophyllum scaberulum (Rolfe) Bolus

DEMOCRATIC REPUBLIC OF THE CONGO / RÉPUBLIQUE DÉMOCRATIQUE DU CONGO (LA) / REPÚBLICA DEMOCRÁTICA DEL CONGO (LA)

Bulbophyllum acutebracteatum De Wild.
Bulbophyllum acutebracteatum De Wild. var. **rubrobrunneopapillosum** (De Wild.) J. J. Verm.
Bulbophyllum barbigerum Lindl.
Bulbophyllum burttii Summerh.
Bulbophyllum calyptratum Kraenzl.
Bulbophyllum calyptratum Kraenzl. var. **graminifolium** (Summerh.) J. J. Verm.
Bulbophyllum capituliflorum Rolfe
Bulbophyllum carnosilabium Summerh.
Bulbophyllum carnosisepalum J. J. Verm.
Bulbophyllum cochleatum Lindl. var. **bequaertii** (De Wild.) J. J. Verm.
Bulbophyllum cochleatum Lindl. var. **brachyanthum** (Summerh.) J. J. Verm.
Bulbophyllum cochleatum Lindl. var. **tenuicaule** (Lindl.) J. J. Verm.
Bulbophyllum cocoinum Bateman ex Lindl.
Bulbophyllum colubrinum (Rchb. f.) Rchb. f.
Bulbophyllum elliottii Rolfe
Bulbophyllum encephalodes Summerh.
Bulbophyllum expallidum J. J. Verm.

Bulbophyllum falcatum (Lindl.) Rchb. f. var. **bufo** (Lindl.) J. J. Verm.
Bulbophyllum falcatum (Lindl.) Rchb. f. var. **velutinum** (Lindl.) J. J. Verm.
Bulbophyllum fuscum Lindl.
Bulbophyllum fuscum Lindl. var. **melinostachyum** (Schltr) J. J. Verm.
Bulbophyllum gravidum Lindl.
Bulbophyllum horridulum J. J. Verm.
Bulbophyllum imbricatum Lindl.
Bulbophyllum injoloense De Wild.
Bulbophyllum injoloense De Wild. subsp. **pseudoxypterum** (J. J. Verm.) J. J. Verm.
Bulbophyllum intertextum Lindl.
Bulbophyllum ivorense P. J. Cribb & Perez-Vera
Bulbophyllum josephi (Kuntze) Summerh.
Bulbophyllum josephi (Kuntze) Summerh. var. **mahonii** (Rolfe) J. J. Verm.
Bulbophyllum longiflorum Thouars
Bulbophyllum lupulinum Lindl.
Bulbophyllum magnibracteatum Summerh.
Bulbophyllum maximum (Lindl.) Rchb. f.
Bulbophyllum nigritianum Rendle
Bulbophyllum oreonastes Rchb. f.
Bulbophyllum oxychilum Schltr.
Bulbophyllum prorepens Summerh.
Bulbophyllum pumilum (Sw.) Lindl.
Bulbophyllum purpureorhachis (De Wild.) Schltr.
Bulbophyllum renkinianum (Laurent) De Wild.
Bulbophyllum resupinatum Ridl.
Bulbophyllum resupinatum Ridl. var. **filiforme** (Kraenzl.) J. J. Verm.
Bulbophyllum saltatorium Lindl.
Bulbophyllum saltatorium Lindl. var. **albociliatum** (Finet) J. J. Verm.
Bulbophyllum saltatorium Lindl. var. **calamarium** (Lindl.) J. J. Verm.
Bulbophyllum sandersonii (Hook. f.) Rchb. f.
Bulbophyllum sandersonii (Hook. f.) Rchb. f. subsp. **stenopetalum** (Kraenzl.) J. J. Verm.
Bulbophyllum scaberulum (Rolfe) Bolus
Bulbophyllum scaberulum (Rolfe) Bolus var. **fuerstenbergianum** (De Wild.) J. J. Verm.
Bulbophyllum schimperianum Kraenzl.
Bulbophyllum schinzianum Kraenzl.
Bulbophyllum schinzianum Kraenzl. var. **phaeopogon** (Schltr.) J. J. Verm.
Bulbophyllum tetragonum Lindl.
Bulbophyllum unifoliatum De Wild.
Bulbophyllum vanum J. J. Verm.
Bulbophyllum vulcanicum Kraenzl.

COOK ISLANDS / ILES COOK (LES) / ISLAS COOK (LAS)

Bulbophyllum longiflorum Thouars

COSTA RICA / COSTA RICA (LE) / COSTA RICA

Bulbophyllum aristatum (Rchb. f.) Hemsl.
Bulbophyllum oerstedii (Rchb. f.) Hemsl.
Bulbophyllum pachyrachis (A. Rich.) Griseb.

CÔTE D'IVOIRE / CÔTE D'IVOIRE (LA) / CÔTE D'IVOIRE (LA)

Bulbophyllum barbigerum Lindl.
Bulbophyllum bidenticulatum J. J. Verm.
Bulbophyllum calyptratum Kraenzl.
Bulbophyllum calyptratum Kraenzl. var. **graminifolium** (Summerh.) J. J. Verm.
Bulbophyllum calyptratum Kraenzl. var. **lucifugum** (Summerh.) J. J. Verm.
Bulbophyllum carnosisepalum J. J. Verm.
Bulbophyllum cochleatum Lindl.
Bulbophyllum cocoinum Bateman ex Lindl.
Bulbophyllum colubrinum (Rchb. f.) Rchb. f.
Bulbophyllum comatum Lindl. var. **inflatum** (Rolfe) J. J. Verm.
Bulbophyllum danii Pérez-Vera
Bulbophyllum denticulatum Rolfe
Bulbophyllum falcatum (Lindl.) Rchb. f.
Bulbophyllum falcatum (Lindl.) Rchb. f. var. **bufo** (Lindl.) J. J. Verm.
Bulbophyllum falcatum (Lindl.) Rchb. f. var. **velutinum** (Lindl.) J. J. Verm.
Bulbophyllum falcipetalum Lindl.
Bulbophyllum fuscum Lindl.
Bulbophyllum fuscum Lindl. var. **melinostachyum** (Schltr) J. J. Verm.
Bulbophyllum imbricatum Lindl.
Bulbophyllum intertextum Lindl.
Bulbophyllum ivorense P. J. Cribb & Perez-Vera
Bulbophyllum josephi (Kuntze) Summerh.
Bulbophyllum josephi (Kuntze) Summerh. var. **mahonii** (Rolfe) J. J. Verm.
Bulbophyllum lupulinum Lindl.
Bulbophyllum magnibracteatum Summerh.
Bulbophyllum maximum (Lindl.) Rchb. f.
Bulbophyllum maximum (Lindl.) Rchb. f. var. **oxypterum** (Lindl.) Pérez-Vera
Bulbophyllum nigericum Summerh.
Bulbophyllum nigritianum Rendle
Bulbophyllum oreonastes Rchb. f.
Bulbophyllum oxychilum Schltr.
Bulbophyllum pipio Rchb. f.
Bulbophyllum pumilum (Sw.) Lindl.
Bulbophyllum purpureorhachis (De Wild.) Schltr.
Bulbophyllum resupinatum Ridl.
Bulbophyllum resupinatum Ridl. var. **filiforme** (Kraenzl.) J. J. Verm.
Bulbophyllum saltatorium Lindl.
Bulbophyllum saltatorium Lindl. var. **albociliatum** (Finet) J. J. Verm.
Bulbophyllum saltatorium Lindl. var. **calamarium** (Lindl.) J. J. Verm.
Bulbophyllum sandersonii (Hook. f.) Rchb. f. subsp. **stenopetalum** (Kraenzl.) J. J. Verm.
Bulbophyllum scaberulum (Rolfe) Bolus
Bulbophyllum scaberulum (Rolfe) Bolus var. **album** Pérez-Vera
Bulbophyllum scariosum Summerh.
Bulbophyllum schinzianum Kraenzl.
Bulbophyllum schinzianum Kraenzl. var. **irigaleae** (P. J. Cribb & Perez-Vera) J. J. Verm.
Bulbophyllum schinzianum Kraenzl. var. **phaeopogon** (Schltr.) J. J. Verm.
Bulbophyllum tetragonum Lindl.

Part III: Country Checklist / Liste par Pays / Lista por Paises

CUBA / CUBA / CUBA

Bulbophyllum pachyrachis (A. Rich.) Griseb.

ECUADOR / EQUATEUR (L.) / ECUADOR (EL) ECUADOR

Bulbophyllum meridense Rchb. f.
Bulbophyllum oerstedii (Rchb. f.) Hemsl.
Bulbophyllum pachyrachis (A. Rich.) Griseb.
Bulbophyllum sordidum Lindl.
Bulbophyllum steyermarkii Foldats
Bulbophyllum wagneri Schltr.

EL SALVADOR / EL SALVADOR (L.) / EL SALVADOR

Bulbophyllum pachyrachis (A. Rich.) Griseb.

EQUATORIAL GUINEA / GUINÉE ÉQUATORIALE (LA) / GUINEA ECUATORIAL (LA)

Bulbophyllum acutebracteatum De Wild. var. **rubrobrunneopapillosum** (De Wild.) J. J. Verm.
Bulbophyllum calyptratum Kraenzl.
Bulbophyllum cochleatum Lindl.
Bulbophyllum cochleatum Lindl. var. **tenuicaule** (Lindl.) J. J. Verm.
Bulbophyllum comatum Lindl.
Bulbophyllum curvimentatum J. J. Verm.
Bulbophyllum falcatum (Lindl.) Rchb. f.
Bulbophyllum falcatum (Lindl.) Rchb. f. var. **velutinum** (Lindl.) J. J. Verm.
Bulbophyllum fuscum Lindl. var. **melinostachyum** (Schltr) J. J. Verm.
Bulbophyllum imbricatum Lindl.
Bulbophyllum intertextum Lindl.
Bulbophyllum josephi (Kuntze) Summerh. var. **mahonii** (Rolfe) J. J. Verm.
Bulbophyllum magnibracteatum Summerh.
Bulbophyllum oreonastes Rchb. f.
Bulbophyllum pumilum (Sw.) Lindl.
Bulbophyllum saltatorium Lindl.
Bulbophyllum saltatorium Lindl. var. **albociliatum** (Finet) J. J. Verm.
Bulbophyllum saltatorium Lindl. var. **calamarium** (Lindl.) J. J. Verm.
Bulbophyllum scaberulum (Rolfe) Bolus var. **fuerstenbergianum** (De Wild.) J. J. Verm.
Bulbophyllum scariosum Summerh.

ETHIOPIA / ETHIOPIE (L.) / ETIOPA

Bulbophyllum intertextum Lindl.
Bulbophyllum josephi (Kuntze) Summerh.
Bulbophyllum lupulinum Lindl.
Bulbophyllum scaberulum (Rolfe) Bolus

FIJI / FIDJI (LES) / FIJI

Bulbophyllum amplistigmaticum Kores
Bulbophyllum aphanopetalum Schltr.
Bulbophyllum aristopetalum Kores
Bulbophyllum betchei F. Muell.
Bulbophyllum ebulbe Schltr.
Bulbophyllum gracillimum (Rolfe) Rolfe
Bulbophyllum hassalli Kores
Bulbophyllum hexarhopalos Schltr.
Bulbophyllum incommodum Kores
Bulbophyllum longiflorum Thouars
Bulbophyllum longiscapum Rolfe
Bulbophyllum membranaceum Teijsm. & Binn.
Bulbophyllum pachyanthum Schltr.
Bulbophyllum phillipsianum Kores
Bulbophyllum quadricarinum Kores
Bulbophyllum rostriceps Rchb. f.
Bulbophyllum samoanum Schltr.
Bulbophyllum savaiense Schltr.
Bulbophyllum simmondsii Kores
Bulbophyllum trachyanthum Kraenzl.

FRENCH POLYNESIA / POLYNESIE FRANCAISE / POLINESIA FRANCESCA

Bulbophyllum longiflorum Thouars
Bulbophyllum tahitense Nadeaud

GABON / GABON (LE) / GABÓN (EL)

Bulbophyllum acutebracteatum De Wild.
Bulbophyllum acutebracteatum De Wild. var. **rubrobrunneopapillosum** (De Wild.) J. J. Verm.
Bulbophyllum barbigerum Lindl.
Bulbophyllum calyptratum Kraenzl.
Bulbophyllum capituliflorum Rolfe
Bulbophyllum carnosilabium Summerh.
Bulbophyllum carnosisepalum J. J. Verm.
Bulbophyllum cochleatum Lindl.
Bulbophyllum cocoinum Bateman ex Lindl.
Bulbophyllum colubrinum (Rchb. f.) Rchb. f.
Bulbophyllum comatum Lindl.
Bulbophyllum comatum Lindl. var. **inflatum** (Rolfe) J. J. Verm.
Bulbophyllum coriscense Rchb. f.
Bulbophyllum falcatum (Lindl.) Rchb. f. var. **velutinum** (Lindl.) J. J. Verm.
Bulbophyllum falcipetalum Lindl.
Bulbophyllum finetii Szlach. & Olszewski
Bulbophyllum fuscum Lindl.
Bulbophyllum fuscum Lindl. var. **melinostachyum** (Schltr) J. J. Verm.
Bulbophyllum imbricatum Lindl.
Bulbophyllum intertextum Lindl.
Bulbophyllum ivorense P. J. Cribb & Perez-Vera

Bulbophyllum magnibracteatum Summerh.
Bulbophyllum maximum (Lindl.) Rchb. f.
Bulbophyllum nigritianum Rendle
Bulbophyllum oreonastes Rchb. f.
Bulbophyllum oxychilum Schltr.
Bulbophyllum pandanetorum Summerh.
Bulbophyllum pumilum (Sw.) Lindl.
Bulbophyllum purpureorhachis (De Wild.) Schltr.
Bulbophyllum renkinianum (Laurent) De Wild.
Bulbophyllum resupinatum Ridl.
Bulbophyllum resupinatum Ridl. var. **filiforme** (Kraenzl.) J. J. Verm.
Bulbophyllum saltatorium Lindl.
Bulbophyllum saltatorium Lindl. var. **albociliatum** (Finet) J. J. Verm.
Bulbophyllum saltatorium Lindl. var. **calamarium** (Lindl.) J. J. Verm.
Bulbophyllum sandersonii (Hook. f.) Rchb. f.
Bulbophyllum sandersonii (Hook. f.) Rchb. f. subsp. **stenopetalum** (Kraenzl.) J. J. Verm.
Bulbophyllum sangae Schltr.
Bulbophyllum scaberulum (Rolfe) Bolus
Bulbophyllum schimperianum Kraenzl.
Bulbophyllum schinzianum Kraenzl.
Bulbophyllum subligaculiferum J. J. Verm.
Bulbophyllum vanum J. J. Verm.

GHANA / GHANA (LE) / GHANA

Bulbophyllum calyptratum Kraenzl.
Bulbophyllum calyptratum Kraenzl. var. **graminifolium** (Summerh.) J. J. Verm.
Bulbophyllum cocoinum Bateman ex Lindl.
Bulbophyllum colubrinum (Rchb. f.) Rchb. f.
Bulbophyllum falcatum (Lindl.) Rchb. f.
Bulbophyllum falcatum (Lindl.) Rchb. f. var. **bufo** (Lindl.) J. J. Verm.
Bulbophyllum falcatum (Lindl.) Rchb. f. var. **velutinum** (Lindl.) J. J. Verm.
Bulbophyllum falcipetalum Lindl.
Bulbophyllum imbricatum Lindl.
Bulbophyllum magnibracteatum Summerh.
Bulbophyllum maximum (Lindl.) Rchb. f.
Bulbophyllum nigritianum Rendle
Bulbophyllum oreonastes Rchb. f.
Bulbophyllum oxychilum Schltr.
Bulbophyllum pipio Rchb. f.
Bulbophyllum pumilum (Sw.) Lindl.
Bulbophyllum resupinatum Ridl.
Bulbophyllum saltatorium Lindl.
Bulbophyllum saltatorium Lindl. var. **albociliatum** (Finet) J. J. Verm.
Bulbophyllum saltatorium Lindl. var. **calamarium** (Lindl.) J. J. Verm.
Bulbophyllum sandersonii (Hook. f.) Rchb. f. subsp. **stenopetalum** (Kraenzl.) J. J. Verm.
Bulbophyllum scaberulum (Rolfe) Bolus
Bulbophyllum schinzianum Kraenzl. var. **phaeopogon** (Schltr.) J. J. Verm.
Bulbophyllum tetragonum Lindl.

GUAM, USA DEPENDENT TERRITORY / GUAM, TERRITORIO DEPENDIENT DE LOS ESTADOS UNIDOS DE AMÉRICA / GUAM, TERRITOIRE DÉPENDANT DES ETATS-UNIS D'AMÉRIQUE

Bulbophyllum guamense Ames
Bulbophyllum longiflorum Thouars

GUATEMALA / GUATEMALA (LE) / GUATEMALA

Bulbophyllum aristatum (Rchb. f.) Hemsl.
Bulbophyllum oerstedii (Rchb. f.) Hemsl.
Bulbophyllum pachyrachis (A. Rich.) Griseb.

GUINEA / GUINÉE (LA) / GUINEA

Bulbophyllum bidenticulatum J. J. Verm.
Bulbophyllum calyptratum Kraenzl.
Bulbophyllum calyptratum Kraenzl. var. **graminifolium** (Summerh.) J. J. Verm.
Bulbophyllum cochleatum Lindl.
Bulbophyllum falcatum (Lindl.) Rchb. f.
Bulbophyllum falcatum (Lindl.) Rchb. f. var. **bufo** (Lindl.) J. J. Verm.
Bulbophyllum fuscum Lindl.
Bulbophyllum intertextum Lindl.
Bulbophyllum josephi (Kuntze) Summerh. var. **mahonii** (Rolfe) J. J. Verm.
Bulbophyllum lupulinum Lindl.
Bulbophyllum maximum (Lindl.) Rchb. f.
Bulbophyllum oreonastes Rchb. f.
Bulbophyllum pumilum (Sw.) Lindl.
Bulbophyllum scaberulum (Rolfe) Bolus
Bulbophyllum scariosum Summerh.

GUYANA / GUYANA (LE) / GUYANA

Bulbophyllum bracteolatum Lindl.
Bulbophyllum exaltatum Lindl.
Bulbophyllum geraense Rchb. f.
Bulbophyllum setigerum Lindl.

HONDURAS / HONDURAS (LE) / HONDURAS

Bulbophyllum aristatum (Rchb. f.) Hemsl.
Bulbophyllum oerstedii (Rchb. f.) Hemsl.

INDIA / INDE (L.) / INDIA (LA)

Acrochaene punctata Lindl.
Bulbophyllum acutiflorum A. Rich.
Bulbophyllum affine Lindl.
Bulbophyllum albidum (Wight) Hook. f.
Bulbophyllum amplifolium (Rolfe) Balakr. & Sud.Chowdhury

Bulbophyllum andersonii (Hook. f.) J. J. Sm.
Bulbophyllum auratum (Lindl.) Rchb. f.
Bulbophyllum aureum (Hook. f.) J. J. Sm.
Bulbophyllum bisetum Lindl.
Bulbophyllum blepharistes Rchb. f.
Bulbophyllum candidum Hook. f.
Bulbophyllum capillipes C. S. P. Parish & Rchb. f.
Bulbophyllum careyanum (Hook. f.) Sprengel
Bulbophyllum cariniflorum Rchb. f.
Bulbophyllum cauliflorum Hook. f.
Bulbophyllum cirrhatum Hook. f.
Bulbophyllum collettii King & Pantl.
Bulbophyllum confusum (Garay, Hamer & Siegerist) Sieder & Kiehn †
Bulbophyllum cornu-cervi King & Pantl.
Bulbophyllum crassipes Hook. f.
Bulbophyllum cupreum Lindl.
Bulbophyllum cylindraceum Lindl.
Bulbophyllum cylindricum King & Pantl.
Bulbophyllum delitescens Hance
Bulbophyllum depressum King & Pantl.
Bulbophyllum ebulbe Schltr.
Bulbophyllum elassonotum Summerh.
Bulbophyllum elatum (Hook. f.) J. J. Sm.
Bulbophyllum elegans Gardner ex Thwaites
Bulbophyllum elegantulum (Rolfe) J. J. Sm.
Bulbophyllum emarginatum (Finet) J. J. Sm.
Bulbophyllum euplepharum Rchb. f.
Bulbophyllum fallax Rolfe
Bulbophyllum fimbriatum (Lindl.) Rchb. f.
Bulbophyllum fischeri Seidenf.
Bulbophyllum forresti Seidenf.
Bulbophyllum fusco-purpureum Wight
Bulbophyllum gracilipes King & Pantl.
Bulbophyllum griffithii (Lindl.) Rchb. f.
Bulbophyllum guttulatum (Hook. f.) Balakr.
Bulbophyllum gymnopus Hook. f.
Bulbophyllum gyrochilum Seidenf.
Bulbophyllum helenae (Kuntze) J. J. Sm.
Bulbophyllum hirtulum Ridl.
Bulbophyllum hirtum (J. E. Sm.) Lindl.
Bulbophyllum hookeri (Duthie) J. J. Sm.
Bulbophyllum hymenanthum Hook. f.
Bulbophyllum iners Rchb. f.
Bulbophyllum kaitiense Rchb. f.
Bulbophyllum keralensis Muktesh & Stephen
Bulbophyllum khasyanum Griff.
Bulbophyllum leopardinum (Wall.) Lindl.
Bulbophyllum leopardinum (Wall.) Lindl. var. **tuberculatum** N. P. Balakrishnan &
S. Chowdhury
Bulbophyllum leptanthum Hook. f.
Bulbophyllum liliacinum Ridl.
Bulbophyllum lobbii Lindl.
Bulbophyllum macraei (Lindl.) Rchb. f.
Bulbophyllum macranthum Lindl.
Bulbophyllum maskeliyense Livera

Bulbophyllum moniliforme C. S. P. Parish & Rchb. f.
Bulbophyllum mysorense (Rolfe) J. J. Sm.
Bulbophyllum neilgherrense Wight
Bulbophyllum nodosum (Rolfe) J. J. Sm.
Bulbophyllum odoratissimum (J. E. Sm.) Lindl.
Bulbophyllum orectopetalum Garay, Hamer & Siegerist
Bulbophyllum ornatissimum (Rchb. f.) J. J. Sm.
Bulbophyllum orezii Sath. Kumar
Bulbophyllum parviflorum C. S. P. Parish & Rchb. f.
Bulbophyllum pectinatum Finet
Bulbophyllum penicillium C. S. P. Parish & Rchb. f.
Bulbophyllum picturatum (Lodd.) Rchb.f.
Bulbophyllum piluliferum King & Pantl.
Bulbophyllum polyrhizum Lindl.
Bulbophyllum protractum Hook. f.
Bulbophyllum proudlockii (King & Pantl.) J. J. Sm.
Bulbophyllum psychoon Rchb. f.
Bulbophyllum pulchrum (N. E. Br.) J. J. Sm.
Bulbophyllum putidum (Teijsm. & Binn.) J. J. Sm.
Bulbophyllum raui Arora
Bulbophyllum repens Griff.
Bulbophyllum reptans (Lindl.) Lindl.
Bulbophyllum retusiusculum Rchb. f.
Bulbophyllum rheedei K. S. Manilal & Kumar Sathish
Bulbophyllum rigidum King & Pantl.
Bulbophyllum rolfei (Kuntze) Seidenf.
Bulbophyllum rosemarianum C. S. Kumar, P. C. S. Kumar & Saleem
Bulbophyllum rothschildianum (O' Brien) J. J. Sm.
Bulbophyllum roxburghii (Lindl.) Rchb. f.
Bulbophyllum rufinum Rchb. f.
Bulbophyllum sarcophyllum (King & Pantl.) J. J. Sm.
Bulbophyllum scaberulum (Rolfe) Bolus
Bulbophyllum scabratum Rchb. f.
Bulbophyllum schmidtianum Rchb. f.
Bulbophyllum secundum Hook. f.
Bulbophyllum siamense Rchb. f.
Bulbophyllum sikkimense (King & Pantl.) J. J. Sm.
Bulbophyllum silentvalliensis M. P. Sharma & S. K. Srivastrava
Bulbophyllum spathulatum (Rolfe ex E. Cooper) Seidenf.
Bulbophyllum striatum (Griff.) Rchb. f.
Bulbophyllum tremulum Wight
Bulbophyllum trichocephalum (Schltr.) T. Tang & F. T. Wang
Bulbophyllum trichocephalum (Schltr.) T. Tang & F. T. Wang var. **wallongense** Agarwal, Sabapathy & Chowdhery
Bulbophyllum tricorne Seidenf. & Smitinand
Bulbophyllum umbellatum Lindl.
Bulbophyllum umbellatum Lindl. var. **fuscescens** P. K. Sarkar
Bulbophyllum viridiflorum (Hook. f.) Schltr.
Bulbophyllum wallichi Rchb. f.
Bulbophyllum xylophyllum C. S. P. Parish & Rchb. f.
Bulbophyllum yoksunense J. J. Sm.
Sunipia andersonii (King & Pantl.) P. F. Hunt
Sunipia bicolor Lindl.
Sunipia cirrhata (Lindl.) P. F. Hunt

Part III: Country Checklist / Liste par Pays / Lista por Paises

Sunipia jainii T. M. Hynniewta & C. L. Malhotra
Sunipia racemosa (Sm.) T. Tang & F. T. Wang
Trias bonaccordensis C. Sathish Kumar
Trias pusilla J. Deka & H. Deka
Trias stocksii Benth. ex Hook. f.

INDONESIA / INDONÉSIA (L.) / INDONESIA

Bulbophyllum aberrans Schltr.
Bulbophyllum absconditum J. J. Sm.
Bulbophyllum absconditum J. J. Sm. subsp. **hastula** J. J. Verm.
Bulbophyllum acuminatifolium J. J. Sm.
Bulbophyllum acuminatum (Ridl.) Ridl.
Bulbophyllum acutilingue J. J. Sm.
Bulbophyllum adelphidium J. J. Verm.
Bulbophyllum aechmophorum J. J. Verm.
Bulbophyllum aemulum Schltr.
Bulbophyllum agapethoides Schltr.
Bulbophyllum aithorhachis J. J. Verm.
Bulbophyllum alatum J. J. Verm.
Bulbophyllum algidum Ridl.
Bulbophyllum alkmaarense J. J. Sm.
Bulbophyllum alliifolium J. J. Sm.
Bulbophyllum alticaule Ridl.
Bulbophyllum amblyoglossum Schltr.
Bulbophyllum amplebracteatum Teijsm. & Binn.
Bulbophyllum anakbaruppui J. J. Verm. & P. O'Byrne
Bulbophyllum anceps Rolfe
Bulbophyllum anguipes Schltr.
Bulbophyllum angulatum J. J. Sm.
Bulbophyllum anguliferum Ames & C. Schweinf.
Bulbophyllum angustifolium (Blume) Lindl.
Bulbophyllum ankylochele J. J. Verm.
Bulbophyllum antennatum Schltr.
Bulbophyllum antenniferum (Lindl.) Rchb. f.
Bulbophyllum apertum Schltr.
Bulbophyllum apheles J. J. Verm.
Bulbophyllum apodum Hook. f.
Bulbophyllum apodum Hook. f. var. **lanceolatum** Ridl.
Bulbophyllum appressicaule Ridl.
Bulbophyllum arachnidium Ridl.
Bulbophyllum ardjunense J. J. Sm.
Bulbophyllum arfakense J. J. Sm.
Bulbophyllum arfakianum Kraenzl.
Bulbophyllum ariel Ridl.
Bulbophyllum aristilabre J. J. Sm.
Bulbophyllum armeniacum J. J. Sm.
Bulbophyllum arsoanum J. J. Sm.
Bulbophyllum ascochiloides J. J. Sm.
Bulbophyllum aspersum J. J. Sm.
Bulbophyllum asperulum J. J. Sm.
Bulbophyllum atratum J. J. Sm.
Bulbophyllum auratum (Lindl.) Rchb. f.
Bulbophyllum auricomum Lindl.

Bulbophyllum auriculatum J. J. Verm. & P. O'Byrne
Bulbophyllum auroreum J. J. Sm.
Bulbophyllum bacilliferum J. J. Sm.
Bulbophyllum baculiferum Ridl.
Bulbophyllum bakhuizenii van Steenis
Bulbophyllum balapiuense J. J. Sm.
Bulbophyllum batukauense J. J. Sm.
Bulbophyllum beccarii Rchb. f.
Bulbophyllum betchei F. Muell.
Bulbophyllum bidi Govaerts
Bulbophyllum biflorum Teijsm. & Binn.
Bulbophyllum bigibbosum J. J. Sm.
Bulbophyllum bigibbum Schltr.
Bulbophyllum binnendijkii J. J. Sm.
Bulbophyllum birugatum J. J. Sm.
Bulbophyllum bivalve J. J. Sm.
Bulbophyllum blumei (Lindl.) J. J. Sm.
Bulbophyllum botryophorum Ridl.
Bulbophyllum brassii J. J. Verm.
Bulbophyllum brastagiense Carr
Bulbophyllum brevibrachiatum (Schltr.) J. J. Sm.
Bulbophyllum brevicolumna J. J. Verm.
Bulbophyllum breviflorum Ridl. ex Stapf
Bulbophyllum brevilabium Schltr.
Bulbophyllum brienianum (Rolfe) Ames
Bulbophyllum bulliferum J. J. Sm.
Bulbophyllum caecilii J. J. Sm.
Bulbophyllum caecum J. J. Sm.
Bulbophyllum calceilabium J. J. Sm.
Bulbophyllum calceolus J. J. Verm.
Bulbophyllum callichroma Schltr.
Bulbophyllum callipes J. J. Sm.
Bulbophyllum capilligerum J. J. Sm.
Bulbophyllum capitatum (Blume) Lindl.
Bulbophyllum carinilabium J. J. Verm.
Bulbophyllum catenarium Ridl.
Bulbophyllum cateorum J. J. Verm.
Bulbophyllum caudatisepalum Ames & C. Schweinf.
Bulbophyllum caudipetalum J. J. Sm.
Bulbophyllum cavibulbum J. J. Sm.
Bulbophyllum cavipes J. J. Verm.
Bulbophyllum centrosemiflorum J. J. Sm.
Bulbophyllum cerambyx J. J. Sm.
Bulbophyllum ceratostylis J. J. Sm.
Bulbophyllum ceratostyloides Ridl.
Bulbophyllum cercanthum (Garay, Hamer & Siegerist) Sieder & Kiehn †
Bulbophyllum cernuum (Blume) Lindl.
Bulbophyllum cernuum Lindl. var. **vittata** (Teijsm. & Binn.) J. J. Sm.
Bulbophyllum chanii J. J. Verm. & A. L. Lamb
Bulbophyllum cheiri Lindl.
Bulbophyllum cheiropetalum Ridl.
Bulbophyllum chloranthum Schltr.
Bulbophyllum chlorascens J. J. Sm.
Bulbophyllum ciliatum (Blume) Lindl.

Bulbophyllum citrellum Ridl.
Bulbophyllum citricolor J. J. Sm.
Bulbophyllum citrinilabre J. J. Sm.
Bulbophyllum clandestinum Lindl.
Bulbophyllum cleistogamum Ridl.
Bulbophyllum cochlia Garay, Hamer & Siegerist
Bulbophyllum cochlioides J. J. Sm.
Bulbophyllum coelochilum J. J. Verm.
Bulbophyllum colliferum J. J. Sm.
Bulbophyllum coloratum J. J. Sm.
Bulbophyllum comberi J. J. Verm.
Bulbophyllum comberipictum J. J. Verm.
Bulbophyllum commissibulbum J. J. Sm.
Bulbophyllum compressum Teijsm. & Binn.
Bulbophyllum concavibasalis P. Royen
Bulbophyllum conchophyllum J. J. Sm.
Bulbophyllum concinnum Hook. f.
Bulbophyllum concolor J. J. Sm.
Bulbophyllum congestiflorum Ridl.
Bulbophyllum coniferum Ridl.
Bulbophyllum connatum Carr
Bulbophyllum conspersum J. J. Sm.
Bulbophyllum contortisepalum J. J. Sm.
Bulbophyllum coriaceum Ridl.
Bulbophyllum cornutum (Blume) Rchb. f.
Bulbophyllum corolliferum J. J. Sm.
Bulbophyllum crassinervium J. J. Sm.
Bulbophyllum crassissimum J. J. Sm.
Bulbophyllum crepidiferum J. J. Sm.
Bulbophyllum croceum (Blume) Lindl.
Bulbophyllum crocodilus J. J. Sm.
Bulbophyllum cruciatum J. J. Sm.
Bulbophyllum cruciferum J. J. Sm.
Bulbophyllum cruttwellii J. J. Verm.
Bulbophyllum cryptanthum Schltr.
Bulbophyllum culex Ridl.
Bulbophyllum cuniculiforme J. J. Sm.
Bulbophyllum cuspidipetalum J. J. Sm.
Bulbophyllum cycloglossum Schltr.
Bulbophyllum cyclopense J. J. Sm.
Bulbophyllum cyclophoroides J. J. Sm.
Bulbophyllum cylindraceum Lindl.
Bulbophyllum cylindrobulbum Schltr.
Bulbophyllum dawongense J. J. Sm.
Bulbophyllum dearei (Hort.) Rchb.f.
Bulbophyllum debruynii J. J. Sm.
Bulbophyllum decatriche J. J. Verm.
Bulbophyllum decurrentilobum J. J. Verm. & P. O'Byrne
Bulbophyllum decurviscapum J. J. Sm.
Bulbophyllum dekockii J. J. Sm.
Bulbophyllum delicatulum Schltr.
Bulbophyllum deltoideum Ames & C. Schweinf.
Bulbophyllum deminutum J. J. Sm.
Bulbophyllum dempoense J. J. Sm.
Bulbophyllum dendrobioides J. J. Sm.

Bulbophyllum dendrochiloides Schltr.
Bulbophyllum depressum King & Pantl.
Bulbophyllum desmotrichoides Schltr.
Bulbophyllum devium J. B. Comber
Bulbophyllum devogelii J. J. Verm.
Bulbophyllum dewildei J. J. Verm.
Bulbophyllum dianthum Schltr.
Bulbophyllum dichotomum J. J. Sm.
Bulbophyllum digoelense J. J. Sm.
Bulbophyllum diploncos Schltr.
Bulbophyllum dischidiifolium J. J. Sm.
Bulbophyllum discolor Schltr.
Bulbophyllum disjunctum Ames & C. Schweinf.
Bulbophyllum dolichoblepharon (Schltr.) J. J. Sm.
Bulbophyllum dransfieldii J. J. Verm.
Bulbophyllum dryas Ridl.
Bulbophyllum dubium J. J. Sm.
Bulbophyllum ebulbe Schltr.
Bulbophyllum echinolabium J. J. Sm.
Bulbophyllum ecornutum (J. J. Sm.) J. J. Sm.
Bulbophyllum elachanthe J. J. Verm.
Bulbophyllum elaphoglossum Schltr.
Bulbophyllum elbertii J. J. Sm.
Bulbophyllum elephantinum J. J. Sm.
Bulbophyllum elevatopunctatum J. J. Sm.
Bulbophyllum ellipticifolium J. J. Sm.
Bulbophyllum elodeiflorum J. J. Sm.
Bulbophyllum elongatum (Blume) Hassk.
Bulbophyllum ensiculiferum J. J. Sm.
Bulbophyllum epicrianthes Hook. f.
Bulbophyllum epicrianthes Hook. f. var. **sumatranum** (J. J. Sm.) J. J. Verm.
Bulbophyllum erinaceum Schltr.
Bulbophyllum exasperatum Schltr.
Bulbophyllum falcatocaudatum J. J. Sm.
Bulbophyllum falciferum J. J. Sm.
Bulbophyllum falculicorne J. J. Sm.
Bulbophyllum farinulentum J. J. Sm.
Bulbophyllum fasciculatum Schltr.
Bulbophyllum faunula Ridl.
Bulbophyllum fibrinum J. J. Sm.
Bulbophyllum filicaule J. J. Sm.
Bulbophyllum filovagans Carr
Bulbophyllum fissibrachium J. J. Sm.
Bulbophyllum flammuliferum Ridl.
Bulbophyllum flavescens (Blume) Lindl.
Bulbophyllum flavicolor J. J. Sm.
Bulbophyllum flavidiflorum Carr
Bulbophyllum flavofimbriatum J. J. Sm.
Bulbophyllum floribundum J. J. Sm.
Bulbophyllum foetidolens Carr
Bulbophyllum foetidum Schltr. var. **grandiflorum** J. J. Sm.
Bulbophyllum folliculiferum J. J. Sm.
Bulbophyllum fractiflexum J. J. Sm.
Bulbophyllum fraudulentum Garay, Hamer & Siegerist

Bulbophyllum fritillariiflorum J. J. Sm.
Bulbophyllum frustrans J. J. Sm.
Bulbophyllum fulgens J. J. Verm.
Bulbophyllum fulvibulbum J. J. Verm.
Bulbophyllum furcillatum J. J. Verm. & P. O'Byrne
Bulbophyllum futile J. J. Sm.
Bulbophyllum gajoense J. J. Sm.
Bulbophyllum galliaheneum P. Royen
Bulbophyllum gautierense J. J. Sm.
Bulbophyllum gemma-reginae J. J. Verm.
Bulbophyllum geniculiferum J. J. Sm.
Bulbophyllum gerlandianum Kraenzl.
Bulbophyllum gibbosum (Blume) Lindl.
Bulbophyllum gibbsiae Rolfe
Bulbophyllum giriwoensc J. J. Sm.
Bulbophyllum gjellerupii J. J. Sm.
Bulbophyllum glaucifolium J. J. Verm.
Bulbophyllum globiceps Schltr.
Bulbophyllum globulosum (Ridl.) Schuit. & de Vogel
Bulbophyllum goliathense J. J. Sm.
Bulbophyllum gomphreniflorum J. J. Sm.
Bulbophyllum gracilicaule W. Kittr.
Bulbophyllum gracillimum (Rolfe) Rolfe
Bulbophyllum gramineum Ridl.
Bulbophyllum grammopoma J. J. Verm.
Bulbophyllum grandiflorum Blume
Bulbophyllum grandilabre Carr
Bulbophyllum graveolens (F. M. Baill.) J. J. Sm.
Bulbophyllum groeneveldtii J. J. Sm.
Bulbophyllum grotianum J. J. Verm.
Bulbophyllum grudense J. J. Sm.
Bulbophyllum gusdorfii J. J. Sm.
Bulbophyllum habbemense P. Royen
Bulbophyllum habrotinum J. J. Verm. & A. L. Lamb
Bulbophyllum hahlianum Schltr.
Bulbophyllum hamatipes J. J. Sm.
Bulbophyllum hastiferum Schltr.
Bulbophyllum heldiorum J. J. Verm.
Bulbophyllum heliophilum J. J. Sm.
Bulbophyllum hirsutum (Blume) Lindl.
Bulbophyllum hirtulum Ridl.
Bulbophyllum hollandianum J. J. Sm.
Bulbophyllum holochilum J. J. Sm.
Bulbophyllum holochilum J. J. Sm. var. **aurantiacum** J. J. Sm.
Bulbophyllum holochilum J. J. Sm. var. **pubescens** J. J. Sm.
Bulbophyllum humiligibbum J. J. Sm.
Bulbophyllum hydrophilum J. J. Sm.
Bulbophyllum hymenanthum Hook. f.
Bulbophyllum hymenobracteum Schltr.
Bulbophyllum hymenobracteum Schltr. var. **giriwoense** J. J. Sm.
Bulbophyllum hymenochilum Kraenzl.
Bulbophyllum hystricinum Schltr.
Bulbophyllum idenburgense J. J. Sm.
Bulbophyllum igneum J. J. Sm.
Bulbophyllum ignobile J. J. Sm.

Bulbophyllum illecebrum J. J. Verm. & P. O'Byrne
Bulbophyllum illudens Ridl.
Bulbophyllum imbricans J. J. Sm.
Bulbophyllum impar Ridl.
Bulbophyllum inaequale (Blume) Lindl.
Bulbophyllum inaequale (Blume) Lindl. var. **angustifolium** J. J. Sm.
Bulbophyllum incisilabrum J. J. Verm. & P. O'Byrne
Bulbophyllum inclinatum J. J. Sm.
Bulbophyllum infundibuliforme J. J. Sm.
Bulbophyllum inquirendum J. J. Verm.
Bulbophyllum inunctum J. J. Sm.
Bulbophyllum ionophyllum J. J. Verm.
Bulbophyllum iterans J. J. Verm. & P. O'Byrne
Bulbophyllum janus J. J. Verm.
Bulbophyllum jensenii J. J. Sm.
Bulbophyllum jolandae J. J. Verm.
Bulbophyllum kemulense J. J. Sm.
Bulbophyllum kermesinum Ridl.
Bulbophyllum kestron J. J. Verm. & A. L. Lamb
Bulbophyllum kirroanthum Schltr.
Bulbophyllum kjellbergii J. J. Sm.
Bulbophyllum klabatense Schltr.
Bulbophyllum klossii Ridl.
Bulbophyllum korimense J. J. Sm.
Bulbophyllum korinchense Ridl.
Bulbophyllum korinchense Ridl. var. **grandflorum** Ridl.
Bulbophyllum korinchense Ridl. var. **parviflorum** Ridl.
Bulbophyllum korthalsii Schltr.
Bulbophyllum lacinulosum J. J. Sm.
Bulbophyllum lambii J. J. Verm.
Bulbophyllum lamelluliferum J. J. Sm.
Bulbophyllum lamii J. J. Sm.
Bulbophyllum languidum J. J. Sm.
Bulbophyllum lasianthum Lindl.
Bulbophyllum latibrachiatum J. J. Sm.
Bulbophyllum latibrachiatum J. J. Sm. var. **epilosum** J. J. Sm.
Bulbophyllum latipes J. J. Sm.
Bulbophyllum latisepalum Ames & C. Schweinf.
Bulbophyllum laxiflorum (Blume) Lindl.
Bulbophyllum laxiflorum (Blume) Lindl. var. **celebicum** Schltr.
Bulbophyllum lemniscatoides Rolfe
Bulbophyllum lemniscatoides Rolfe var. **exappendiculatum** J. J. Sm.
Bulbophyllum leniae J. J. Verm.
Bulbophyllum lepanthiflorum Schltr.
Bulbophyllum leproglossum J. J. Verm. & A. L. Lamb
Bulbophyllum leptoleucum Schltr.
Bulbophyllum leptophyllum W. Kittr.
Bulbophyllum levidense J. J. Sm.
Bulbophyllum leysenianum Burb.
Bulbophyllum ligulifolium J. J. Sm.
Bulbophyllum limbatum Lindl.
Bulbophyllum lineariflorum J. J. Sm.
Bulbophyllum linearilabium J. J. Sm.
Bulbophyllum lineatum (Teijsm. & Binn.) J. J. Sm.

Bulbophyllum linggense J. J. Sm.
Bulbophyllum lissoglossum J. J. Verm.
Bulbophyllum lobbii Lindl.
Bulbophyllum lokonense Schltr.
Bulbophyllum longerepens Ridl.
Bulbophyllum longhutense J. J. Sm.
Bulbophyllum longicaudatum (J. J. Sm.) J. J. Sm.
Bulbophyllum longiflorum Thouars
Bulbophyllum longimucronatum Ames & C. Schweinf.
Bulbophyllum longipedicellatum J. J. Sm.
Bulbophyllum longipedicellatum J. J. Sm. var. **gjellerupii** J. J. Sm.
Bulbophyllum longisepalum Rolfe
Bulbophyllum longivagans Carr
Bulbophyllum lorentzianum J. J. Sm.
Bulbophyllum lumbriciforme J. J. Sm.
Bulbophyllum luteopurpureum J. J. Sm.
Bulbophyllum lygeron J. J. Verm.
Bulbophyllum macranthoides Kraenzl.
Bulbophyllum macranthum Lindl.
Bulbophyllum macrobulbum J. J. Sm.
Bulbophyllum macrochilum Rolfe
Bulbophyllum mahakamense J. J. Sm.
Bulbophyllum major (Ridl.) van Royen
Bulbophyllum makoyanum (Rchb. f.) Ridl.
Bulbophyllum malleolabrum Carr
Bulbophyllum mamberamense J. J. Sm.
Bulbophyllum mandibulare Rchb. f.
Bulbophyllum marudiense Carr
Bulbophyllum masarangicum Schltr.
Bulbophyllum mastersianum (Rolfe) J. J. Sm.
Bulbophyllum mayrii J. J. Sm.
Bulbophyllum medusae (Lindl.) Rchb. f.
Bulbophyllum melilotus J. J. Sm.
Bulbophyllum meliphagirostrum P. Royen
Bulbophyllum membranaceum Teijsm. & Binn.
Bulbophyllum membranifolium Hook. f.
Bulbophyllum mentiferum J. J. Sm.
Bulbophyllum mesodon J. J. Verm.
Bulbophyllum micholitzianum Kraenzl.
Bulbophyllum micholitzii Rolfe
Bulbophyllum microglossum Ridl.
Bulbophyllum microlabium W. Kittr.
Bulbophyllum minahassae Schltr.
Bulbophyllum minutulum Ridl.
Bulbophyllum mirabile Hallier f.
Bulbophyllum mirum J. J. Sm.
Bulbophyllum montense Ridl.
Bulbophyllum moroides J. J. Sm.
Bulbophyllum morotaiense J. J. Sm.
Bulbophyllum mucronatum (Blume) Lindl.
Bulbophyllum mulderae J. J. Verm.
Bulbophyllum multiflexum J. J. Sm.
Bulbophyllum murkelense J. J. Sm.
Bulbophyllum mutabile (Blume) Lindl.
Bulbophyllum mutabile (Blume) Lindl. var. **obesum** J. J. Verm.

Bulbophyllum mutatum J. J. Sm.
Bulbophyllum myon J. J. Verm.
Bulbophyllum mystax Schuit. & De Vogel
Bulbophyllum nabawanense J. J. Wood & A. L. Lamb
Bulbophyllum nasica Schltr.
Bulbophyllum nematocaulon Ridl.
Bulbophyllum nemorale L. O. Williams
Bulbophyllum neoguinense J. J. Sm.
Bulbophyllum nieuwenhuisii J. J. Sm.
Bulbophyllum nigrilabium Schltr.
Bulbophyllum novaciae J. J. Verm. & P. O'Byrne
Bulbophyllum nubinatum J. J. Verm.
Bulbophyllum obliquum Schltr.
Bulbophyllum obovatifolium J. J. Sm.
Bulbophyllum obtusipetalum J. J. Sm.
Bulbophyllum obtusum (Blume) Lindl.
Bulbophyllum ochroleucum Schltr.
Bulbophyllum octarrhenipetalum J. J. Sm.
Bulbophyllum odoardi Pfitzer
Bulbophyllum odoratum (Blume) Lindl.
Bulbophyllum odoratum (Blume) Lindl. var. **grandiflorum** J. J. Sm. ex Bull
Bulbophyllum odoratum (Blume) Lindl. var. **obtusisepalum** J. J. Sm.
Bulbophyllum odoratum (Blume) Lindl. var. **polyarachne** J. J. Sm.
Bulbophyllum oligoblepharon Schltr.
Bulbophyllum olivinum J. J. Sm.
Bulbophyllum olorinum J. J. Sm.
Bulbophyllum orbiculare J. J. Sm.
Bulbophyllum orbiculare J. J. Sm. subsp. **cassideum** J. J. Verm.
Bulbophyllum orectopetalum Garay, Hamer & Siegerist
Bulbophyllum origami J. J. Verm.
Bulbophyllum ornithorhynchum (J. J. Sm.) Garay, Hamer & Siegerist
Bulbophyllum orohense J. J. Sm.
Bulbophyllum orsidice Ridl.
Bulbophyllum osyricera Schltr.
Bulbophyllum osyriceroides J. J. Sm.
Bulbophyllum otochilum J. J. Verm.
Bulbophyllum ovale Ridl.
Bulbophyllum ovalifolium (Blume) Lindl.
Bulbophyllum ovalitepalum J. J. Sm.
Bulbophyllum ovatolanceatum J. J. Sm.
Bulbophyllum oxysepaloides Ridl.
Bulbophyllum pachyacris J. J. Sm.
Bulbophyllum pachyneuron Schltr.
Bulbophyllum pachytelos Schltr.
Bulbophyllum pahudi (De Vriese) Rchb.f.
Bulbophyllum palilabre J. J. Sm.
Bulbophyllum paniscus Ridl.
Bulbophyllum papilio J. J. Sm.
Bulbophyllum papillatum J. J. Sm.
Bulbophyllum papuliferum Schltr.
Bulbophyllum pardalinum Ridl.
Bulbophyllum patens King ex Hook. f.
Bulbophyllum paucisetum J. J. Sm.
Bulbophyllum pelicanopsis J. J. Verm. & A. L. Lamb

Bulbophyllum peltopus Schltr.
Bulbophyllum penduliscapum J. J. Sm.
Bulbophyllum peperomiifolium J. J. Sm.
Bulbophyllum perductum J. J. Sm.
Bulbophyllum perductum J. J. Sm. var. **sebesiense** J. J. Sm.
Bulbophyllum perexiguum Ridl.
Bulbophyllum perforans J. J. Sm.
Bulbophyllum perparvulum Schltr.
Bulbophyllum perpendiculare Schltr.
Bulbophyllum petiolatum J. J. Sm.
Bulbophyllum peyerianum (Kranzlin) Seidenf.
Bulbophyllum phaeanthum Schltr.
Bulbophyllum phaeoneuron Schltr.
Bulbophyllum phalaenopsis J. J. Sm.
Bulbophyllum phormion J. J. Verm.
Bulbophyllum pidacanthum J. J. Verm.
Bulbophyllum pileatum Lindl.
Bulbophyllum piliferum J. J. Sm.
Bulbophyllum pisibulbum J. J. Sm.
Bulbophyllum placochilum J. J. Verm.
Bulbophyllum plagiatum Ridl.
Bulbophyllum planibulbe (Ridl.) Ridl.
Bulbophyllum planibulbe (Ridl.) Ridl. var. **sumatranum** J. J. Sm.
Bulbophyllum planitiae J. J. Sm.
Bulbophyllum pleurothallianthum Garay
Bulbophyllum plumatum Ames
Bulbophyllum pocillum J. J. Verm.
Bulbophyllum polycyclum J. J. Verm.
Bulbophyllum polygaliflorum J. J. Wood
Bulbophyllum porphyrotriche J. J. Verm.
Bulbophyllum posticum J. J. Sm.
Bulbophyllum praestans Kraenzl.
Bulbophyllum prianganense J. J. Sm.
Bulbophyllum pristis J. J. Sm.
Bulbophyllum pseudofilicaule J. J. Sm.
Bulbophyllum pseudopelma J. J. Verm. & P. O'Byrne
Bulbophyllum pseudoserrulatum J. J. Sm.
Bulbophyllum pubiflorum Schltr.
Bulbophyllum pugilanthum J. J. Wood
Bulbophyllum pugioniforme J. J. Sm.
Bulbophyllum puguahaanense Ames
Bulbophyllum pulchrum (N. E. Br.) J. J. Sm.
Bulbophyllum puntjakense J. J. Sm.
Bulbophyllum purpurascens Teijsm. & Binn.
Bulbophyllum purpurellum Ridl.
Bulbophyllum pustulatum Ridl.
Bulbophyllum putidum (Teijsm. & Binn.) J. J. Sm.
Bulbophyllum pyridion J. J. Verm.
Bulbophyllum quadrangulare J. J. Sm.
Bulbophyllum quadrangulare J. J. Sm. var. **latisepalum** J. J. Sm.
Bulbophyllum quadricaudatum J. J. Sm.
Bulbophyllum quadrifalciculatum J. J. Sm.
Bulbophyllum quadrisubulatum J. J. Sm.
Bulbophyllum quasimodo J. J. Verm.
Bulbophyllum racemosum Rolfe

Bulbophyllum rajanum J. J. Sm.
Bulbophyllum ramulicola Schuit. & De Vogel
Bulbophyllum rariflorum J. J. Sm.
Bulbophyllum rectilabre J. J. Sm.
Bulbophyllum recurviflorum J. J. Sm.
Bulbophyllum reductum J. J. Verm. & P. O'Byrne
Bulbophyllum reevei J. J. Verm.
Bulbophyllum reflexum Ames & C. Schweinf.
Bulbophyllum refractilingue J. J. Sm.
Bulbophyllum refractum (Zoll.) Rchb. f.
Bulbophyllum remiferum Carr
Bulbophyllum restrepia (Ridl.) Ridl.
Bulbophyllum reticulatum Bateman
Bulbophyllum retusiusculum Rchb. f.
Bulbophyllum rhizomatosum Ames & C. Schweinf.
Bulbophyllum rhodoleucum Schltr.
Bulbophyllum rhodosepalum Schltr.
Bulbophyllum rhynchoglossum Schltr.
Bulbophyllum rigidifilum J. J. Sm.
Bulbophyllum riparium J. J. Sm.
Bulbophyllum romburghii J. J. Sm.
Bulbophyllum roxburghii (Lindl.) Rchb. f.
Bulbophyllum rubiferum J. J. Sm.
Bulbophyllum ruficaudatum Ridl.
Bulbophyllum rugosum Ridl.
Bulbophyllum rugulosum J. J. Sm.
Bulbophyllum rupestre J. J. Sm.
Bulbophyllum saccolabioides J. J. Sm.
Bulbophyllum salaccense Rchb. f.
Bulbophyllum salebrosum J. J. Sm.
Bulbophyllum sarasinorum Schltr.
Bulbophyllum sarcanthiforme Ridl.
Bulbophyllum sarcodanthum Schltr.
Bulbophyllum sarcoscapum Teijsm. & Binn.
Bulbophyllum savaiense Schltr. subsp. **gorumense** (Schltr.) J. J. Verm.
Bulbophyllum savaiense Schltr. subsp. **subcubicum** (J. J. Sm.) J. J. Verm.
Bulbophyllum sawiense J. J. Sm.
Bulbophyllum scabrum J. J. Verm. & A. L. Lamb
Bulbophyllum scaphosepalum Ridl.
Bulbophyllum schefferi (Kuntze) Schltr.
Bulbophyllum schizopetalum L. O. Williams
Bulbophyllum schmidii Garay
Bulbophyllum scitulum Ridl.
Bulbophyllum scopa J. J. Verm.
Bulbophyllum scotifolium J. J. Sm.
Bulbophyllum scotinochiton J.J.Verm. & P.O'Byrne
Bulbophyllum scrobiculilabre J. J. Sm.
Bulbophyllum scutiferum J. J. Verm.
Bulbophyllum semiasperum J. J. Sm.
Bulbophyllum septemtrionale (J. J. Sm.) J. J. Sm.
Bulbophyllum serra Schltr.
Bulbophyllum serrulatifolium J. J. Sm.
Bulbophyllum siederi Garay
Bulbophyllum sigmoideum Ames & C. Schweinf.

Bulbophyllum sikapingense J. J. Sm.
Bulbophyllum similare Garay, Hamer & Siegerist
Bulbophyllum simile Schltr.
Bulbophyllum similissimum J. J. Verm.
Bulbophyllum simplex J. J. Verm. & P. O'Byrne
Bulbophyllum singaporeanum Schltr.
Bulbophyllum smithianum Schltr.
Bulbophyllum sociale Rolfe
Bulbophyllum sopoetanense Schltr.
Bulbophyllum sororculum J. J. Verm.
Bulbophyllum spathilingue J. J. Sm.
Bulbophyllum spathipetalum J. J. Sm.
Bulbophyllum spissum J. J. Verm.
Bulbophyllum stabile J. J. Sm.
Bulbophyllum steffensii Schltr.
Bulbophyllum stelis J. J. Sm.
Bulbophyllum stelis J. J. Sm. var. **humile** J. J. Sm.
Bulbophyllum stellula Ridl.
Bulbophyllum stenurum J. J. Verm. & P. O'Byrne
Bulbophyllum stipitatibulbum J. J. Sm.
Bulbophyllum stipulaceum Schltr.
Bulbophyllum stormii J. J. Sm.
Bulbophyllum streptotriche J. J. Verm.
Bulbophyllum striatellum Ridl.
Bulbophyllum subapetalum J. J. Sm.
Bulbophyllum subclausum J. J. Sm.
Bulbophyllum submarmoratum J. J. Sm.
Bulbophyllum subuliferum Schltr.
Bulbophyllum subumbellatum Ridl.
Bulbophyllum subverticillatum Ridl.
Bulbophyllum succedaneum J. J. Sm.
Bulbophyllum sulawesii Garay, Hamer & Siegerist
Bulbophyllum sulcatum (Blume) Lindl.
Bulbophyllum sumatranum Garay, Hamer & Siegerist
Bulbophyllum supervacaenum Kraenzl.
Bulbophyllum systenochilum J. J. Verm.
Bulbophyllum taeniophyllum C. S. P. Parish & Rchb. f.
Bulbophyllum tardeflorens Ridl.
Bulbophyllum tectipes J. J. Verm. & P. O'Byrne
Bulbophyllum tectipetalum J. J. Sm.
Bulbophyllum tectipetalum J. J. Sm. var. **longisepalum** J. J. Sm.
Bulbophyllum tectipetalum J. J. Sm. var. **maximum** J. J. Sm.
Bulbophyllum tenellum (Blume) Lindl.
Bulbophyllum tenompokense J. J. Sm.
Bulbophyllum tenuifolium (Blume) Lindl.
Bulbophyllum tenuipes Schltr.
Bulbophyllum teres Carr
Bulbophyllum teretilabre J. J. Sm.
Bulbophyllum ternatense J. J. Sm.
Bulbophyllum teysmannii J. J. Sm.
Bulbophyllum theunissenii J. J. Sm.
Bulbophyllum thrixspermiflorum J. J. Sm.
Bulbophyllum thrixspermoides J. J. Sm.
Bulbophyllum thymophorum J. J. Verm. & A. L. Lamb
Bulbophyllum tjadasmalangense J. J. Sm.

Bulbophyllum toranum J. J. Sm.
Bulbophyllum torquatum J. J. Sm.
Bulbophyllum tortuosum (Blume) Lindl.
Bulbophyllum trachyanthum Kraenzl.
Bulbophyllum trachypus Schltr.
Bulbophyllum triadenium (Lindl.) Rchb. f.
Bulbophyllum tricanaliferum J. J. Sm.
Bulbophyllum trichorhachis J. J. Verm. & P. O'Byrne
Bulbophyllum triclavigerum J. J. Sm.
Bulbophyllum trifilum J. J. Sm.
Bulbophyllum trifilum J. J. Sm. subsp. **filisepalum** (J. J. Sm.) J. J. Verm.
Bulbophyllum triflorum (Breda) Blume
Bulbophyllum trifolium Ridl.
Bulbophyllum trigonidioides J. J. Sm.
Bulbophyllum trigonobulbum Schltr. & J. J. Sm.
Bulbophyllum trinervium J. J. Sm.
Bulbophyllum tristelidium W. Kittr.
Bulbophyllum triurum Kraenzl.
Bulbophyllum truncatum J. J. Sm.
Bulbophyllum tryssum J. J. Verm. & A. L. Lamb
Bulbophyllum tubilabrum J. J. Verm. & P. O'Byrne
Bulbophyllum tumidum J. J. Verm.
Bulbophyllum turgidum J. J. Verm.
Bulbophyllum tylophorum Schltr.
Bulbophyllum ulcerosum J. J. Sm.
Bulbophyllum uncinatum J. J. Verm. & P. O'Byrne
Bulbophyllum undatilabre J. J. Sm.
Bulbophyllum undecifilum J. J. Sm.
Bulbophyllum unguiculatum Rchb. f.
Bulbophyllum uniflorum (Blume) Hassk.
Bulbophyllum unitubum J. J. Sm.
Bulbophyllum univenum J. J. Verm.
Bulbophyllum uviflorum O'Byrne
Bulbophyllum vaginatum (Lindl.) Rchb. f.
Bulbophyllum valeryi J. J. Verm. & P. O'Byrne
Bulbophyllum validum Carr
Bulbophyllum vanvuurenii J. J. Sm.
Bulbophyllum vermiculare Hook. f.
Bulbophyllum versteegii J. J. Sm.
Bulbophyllum vesiculosum J. J. Sm.
Bulbophyllum vexillarium Ridl.
Bulbophyllum vinaceum Ames & C. Schweinf.
Bulbophyllum violaceum (Blume) Lindl.
Bulbophyllum virescens J. J. Sm.
Bulbophyllum viridescens Ridl.
Bulbophyllum vitellinum Ridl.
Bulbophyllum wollastonii Ridl.
Bulbophyllum wrayi Hook. f.
Bulbophyllum xantanthum Schltr.
Bulbophyllum xanthoacron J. J. Sm.
Bulbophyllum xanthochlamys Schltr.
Bulbophyllum xiphion J. J. Verm.
Bulbophyllum xylocarpi J. J. Sm.
Bulbophyllum zebrinum J. J. Sm.

Codonosiphon codonanthum (Schltr.) Schltr.
Drymoda digitata (J. J. Sm.) Garay, Hamer & Siegerist
Pedilochilus bantaengensis J. J. Sm.
Pedilochilus clemensiae L. O. Williams
Pedilochilus coiloglossum (Schltr.) Schltr.
Pedilochilus humile J. J. Sm.
Pedilochilus kermesinostriatum J. J. Sm.
Pedilochilus macrorrhinum P. Royen
Pedilochilus majus J. J. Sm.
Pedilochilus montana Ridl.
Pedilochilus obovatum J. J. Sm.
Pedilochilus perpusillum P. Royen
Pedilochilus pumilio Ridl.
Pedilochilus sulphureum J. J. Sm.
Pedilochilus terrestre J. J. Sm.
Sarccoglossum lanceolatum L. O. Williams
Sarccoglossum verrucosum L. L. Williams
Trias tothastes (J. J. Verm.) J. J. Wood

JAMAICA / JAMAÏQUE (LA) / JAMAICA

Bulbophyllum jamaicense Cogn.
Bulbophyllum pachyrachis (A. Rich.) Griseb.

JAPAN / JAPON (LE) / JAPÓN (EL)

Bulbophyllum affine Lindl.
Bulbophyllum drymoglossum Maxim.
Bulbophyllum inconspicuum Maxim.
Bulbophyllum japonicum (Makino) Makino
Bulbophyllum macraei (Lindl.) Rchb. f.

KENYA / KENYA (LE) / KENYA

Bulbophyllum bidenticulatum J. J. Verm. var. **joyceae** J. J. Verm.
Bulbophyllum cochleatum Lindl.
Bulbophyllum cochleatum Lindl. var. **brachyanthum** (Summerh.) J. J. Verm.
Bulbophyllum cochleatum Lindl. var. **tenuicaule** (Lindl.) J. J. Verm.
Bulbophyllum encephalodes Summerh.
Bulbophyllum intertextum Lindl.
Bulbophyllum josephi (Kuntze) Summerh.
Bulbophyllum maximum (Lindl.) Rchb. f.
Bulbophyllum sandersonii (Hook. f.) Rchb. f.
Bulbophyllum scaberulum (Rolfe) Bolus
Bulbophyllum vulcanicum Kraenzl.
Chaseella pseudohydra Summerh.

KIRIBATI (REBUBLIC OF) / RÉPUBLIQUE DE KIRIBATA (LA) / REPÚBLICA DE KIRIBATA (LA)

Bulbophyllum micronesiacum Schltr.

KOREA (THE REPUBLIC OF) / RÉPUBLIQUE DE CORÉE (LA) / REPÚBLICA DE COREA (LA)

Bulbophyllum drymoglossum Maxim.

LAO PEOPLE'S DEMOCRATIC REPUBLIC (THE) / RÉPUBLIQUE DÉMOCRATIQUE POPULAIRE LAO (LA) / REPÚBLICA DEMOCRÁTICA POPULAR LAO (LA)

Bulbophyllum affine Lindl.
Bulbophyllum allenkerrii Seidenf.
Bulbophyllum blepharistes Rchb. f.
Bulbophyllum coweniorum J. J. Verm. & P. O'Byrne
Bulbophyllum dissitiflorum Seidenf.
Bulbophyllum haniffii Carr
Bulbophyllum laoticum Gagnep.
Bulbophyllum laxiflorum (Blume) Lindl.
Bulbophyllum longebracteatum Seidenf.
Bulbophyllum longibracteatum Seidenf.
Bulbophyllum moniliforme C. S. P. Parish & Rchb. f.
Bulbophyllum odoratissimum (J. E. Sm.) Lindl.
Bulbophyllum pectenveneris (Gagnep.) Seidenf.
Bulbophyllum proboscideum (Gagnep.) Seidenf. & Smitinand
Bulbophyllum putidum (Teijsm. & Binn.) J. J. Sm.
Bulbophyllum retusiusculum Rchb. f.
Bulbophyllum rufinum Rchb. f.
Bulbophyllum sanguineopunctatum Seidenf. & A. D. Kerr
Bulbophyllum secundum Hook. f.
Bulbophyllum seidenfadenii A. D. Kerr
Bulbophyllum siamense Rchb. f.
Bulbophyllum spathulatum (Rolfe ex E. Cooper) Seidenf.
Bulbophyllum stenobulbon C. S. P. Parish & Rchb. f.
Bulbophyllum sutepense (Rolfe ex Downie) Seidenf. & Smitinand
Bulbophyllum taeniophyllum C. S. P. Parish & Rchb. f.
Bulbophyllum tortuosum (Blume) Lindl.
Bulbophyllum tripudians C. S. P. Parish & Rchb. f.
Bulbophyllum violaceolabellum Seidenf.
Drymoda picta Lindl.
Drymoda siamensis Schltr.
Sunipia racemosa (Sm.) T. Tang & F. T. Wang
Trias disciflora (Rolfe) Rolfe

LIBERIA / LIBÉRIA (LE) / LIBERIA

Bulbophyllum acutebracteatum De Wild.
Bulbophyllum barbigerum Lindl.
Bulbophyllum bidenticulatum J. J. Verm.
Bulbophyllum calyptratum Kraenzl.
Bulbophyllum calyptratum Kraenzl. var. **graminifolium** (Summerh.) J. J. Verm.
Bulbophyllum calyptratum Kraenzl. var. **lucifugum** (Summerh.) J. J. Verm.
Bulbophyllum cochleatum Lindl.
Bulbophyllum cocoinum Bateman ex Lindl.

Bulbophyllum comatum Lindl. var. **inflatum** (Rolfe) J. J. Verm.
Bulbophyllum denticulatum Rolfe
Bulbophyllum falcatum (Lindl.) Rchb. f.
Bulbophyllum falcatum (Lindl.) Rchb. f. var. **bufo** (Lindl.) J. J. Verm.
Bulbophyllum falcatum (Lindl.) Rchb. f. var. **velutinum** (Lindl.) J. J. Verm.
Bulbophyllum fuscum Lindl.
Bulbophyllum fuscum Lindl. var. **melinostachyum** (Schltr) J. J. Verm.
Bulbophyllum imbricatum Lindl.
Bulbophyllum intertextum Lindl.
Bulbophyllum ivorense P. J. Cribb & Perez-Vera
Bulbophyllum josephi (Kuntze) Summerh. var. **mahonii** (Rolfe) J. J. Verm.
Bulbophyllum magnibracteatum Summerh.
Bulbophyllum maximum (Lindl.) Rchb. f.
Bulbophyllum nigritianum Rendle
Bulbophyllum oreonastes Rchb. f.
Bulbophyllum oxychilum Schltr.
Bulbophyllum pumilum (Sw.) Lindl.
Bulbophyllum resupinatum Ridl. var. **filiforme** (Kraenzl.) J. J. Verm.
Bulbophyllum saltatorium Lindl.
Bulbophyllum saltatorium Lindl. var. **albociliatum** (Finet) J. J. Verm.
Bulbophyllum saltatorium Lindl. var. **calamarium** (Lindl.) J. J. Verm.
Bulbophyllum sandersonii (Hook. f.) Rchb. f. subsp. **stenopetalum** (Kraenzl.) J. J. Verm.
Bulbophyllum scariosum Summerh.
Bulbophyllum schinzianum Kraenzl.
Bulbophyllum schinzianum Kraenzl. var. **irigaleae** (P. J. Cribb & Perez-Vera) J. J. Verm.
Bulbophyllum tetragonum Lindl.

MADAGASCAR / MADAGASCAR / MADAGASCAR

Bulbophyllum acutispicatum H. Perrier
Bulbophyllum afzelii Schltr.
Bulbophyllum afzelii Schltr. var. **microdoron** (Schltr.) Bosser
Bulbophyllum aggregatum Bosser
Bulbophyllum alexandrae Schltr.
Bulbophyllum alleizettei Schltr.
Bulbophyllum ambatoavense Bosser
Bulbophyllum ambrense H. Perrier
Bulbophyllum amoenum Bosser
Bulbophyllum amphorimorphum H. Perrier
Bulbophyllum analamazoatrae Schltr.
Bulbophyllum andohahelense H. Perrier
Bulbophyllum anjozorobeense Bosser
Bulbophyllum ankaizinense (Jum. & H. Perrier) Schltr.
Bulbophyllum ankaratranum Schltr.
Bulbophyllum antongilense Schltr.
Bulbophyllum approximatum Ridl.
Bulbophyllum aubrevillei Bosser
Bulbophyllum auriflorum H. Perrier
Bulbophyllum baronii Ridl.
Bulbophyllum bathieanum Schltr.
Bulbophyllum bicoloratum Schltr.
Bulbophyllum boiteaui H. Perrier

Bulbophyllum brachyphyton Schltr.
Bulbophyllum brachystachyum Schltr.
Bulbophyllum brevipetalum H. Perrier
Bulbophyllum callosum Bosser
Bulbophyllum calyptropus Schltr.
Bulbophyllum capuronii Bosser
Bulbophyllum cardiobulbum Bosser
Bulbophyllum cataractarum Schltr.
Bulbophyllum ceriodorum Boiteau
Bulbophyllum ciliatilabrum H. Perrier
Bulbophyllum cirrhoglossum H. Perrier
Bulbophyllum coccinatum H. Perrier
Bulbophyllum comorianum H. Perrier
Bulbophyllum complanatum H. Perrier
Bulbophyllum conchidioides Ridl.
Bulbophyllum coriophorum Ridl.
Bulbophyllum crassipetalum H. Perrier
Bulbophyllum cryptostachyum Schltr.
Bulbophyllum curvifolium Schltr.
Bulbophyllum cyclanthum Schltr.
Bulbophyllum cylindrocarpum Frapp ex Cordem.
Bulbophyllum cylindrocarpum Frapp. ex Cordem. var. **andringitrense** Bosser
Bulbophyllum debile Bosser
Bulbophyllum decaryanum H. Perrier
Bulbophyllum discilabium H. Perrier
Bulbophyllum divaricatum H. Perrier
Bulbophyllum edentatum H. Perrier
Bulbophyllum elliottii Rolfe
Bulbophyllum erectum Thouars
Bulbophyllum erythroglossum Bosser
Bulbophyllum erythrostachyum Rolfe
Bulbophyllum ferkoanum Schltr.
Bulbophyllum florulentum Schltr.
Bulbophyllum forsythianum Kraenzl.
Bulbophyllum francoisii H. Perrier
Bulbophyllum francoisii H. Perrier var. **andrangense** (H. Perrier) Bosser
Bulbophyllum gracile Thouars
Bulbophyllum hamelini W. Watson
Bulbophyllum hapalanthos Garay
Bulbophyllum henrici Schltr.
Bulbophyllum henrici Schltr. var. **rectangulare** H. Perrier
Bulbophyllum hildebrandtii Rchb. f.
Bulbophyllum hirsutiusculum H. Perrier
Bulbophyllum horizontale Bosser
Bulbophyllum hovarum Schltr.
Bulbophyllum humbertii Schltr.
Bulbophyllum humblottii Rolfe
Bulbophyllum hyalinum Schltr.
Bulbophyllum ikongoense H. Perrier
Bulbophyllum imerinense Schltr.
Bulbophyllum insolitum Bosser
Bulbophyllum johannis Kraenzl.
Bulbophyllum jumellanum Schltr.
Bulbophyllum kainochiloides H. Perrier

Bulbophyllum kieneri Bosser
Bulbophyllum labatii Bosser
Bulbophyllum lakatoense Bosser
Bulbophyllum lancisepalum H. Perrier
Bulbophyllum latipetalum H. Perrier
Bulbophyllum leandrianum H. Perrier
Bulbophyllum lecouflei Bosser
Bulbophyllum lemuraeoides H. Perrier
Bulbophyllum lemurense Bosser & P. J. Cribb
Bulbophyllum leoni Kraenzl.
Bulbophyllum leptochlamys Schltr.
Bulbophyllum leptostachyum Schltr.
Bulbophyllum lichenophylax Schltr.
Bulbophyllum lineariligulatum Schltr.
Bulbophyllum liparidioides Schltr.
Bulbophyllum longiflorum Thouars
Bulbophyllum longivaginans H. Perrier
Bulbophyllum lucidum Schltr.
Bulbophyllum luteobracteatum Jum. & H. Perrier
Bulbophyllum lyperocephalum Schltr.
Bulbophyllum lyperostachyum Schltr.
Bulbophyllum maleolens Kraenzl.
Bulbophyllum mananjarense Poiss.
Bulbophyllum mangenotii Bosser
Bulbophyllum marojejiense H. Perrier
Bulbophyllum maromanganum Schltr.
Bulbophyllum marovoense H. Perrier
Bulbophyllum masoalanum Schltr.
Bulbophyllum matitanense H. Perrier
Bulbophyllum matitanense H. Perrier subsp. **rostratum** H. Perrier
Bulbophyllum maudeae A. D. Hawkes
Bulbophyllum megalonyx Rchb. f.
Bulbophyllum melleum H. Perrier
Bulbophyllum metonymon Summerh.
Bulbophyllum minax Schltr.
Bulbophyllum minutilabrum H. Perrier
Bulbophyllum minutum Thouars
Bulbophyllum mirificum Schltr.
Bulbophyllum moldenkeanum A. D. Hawkes
Bulbophyllum molossus Rchb. f.
Bulbophyllum moratii Bosser
Bulbophyllum multiflorum Ridl.
Bulbophyllum multiligulatum H. Perrier
Bulbophyllum multivaginatum Jum. & H. Perrier
Bulbophyllum muscicola Schltr. (1913)
Bulbophyllum myrmecochilum Schltr.
Bulbophyllum namoronae Bosser
Bulbophyllum neglectum Bosser
Bulbophyllum nigriflorum H. Perrier
Bulbophyllum nitens Jum. & H. Perrier
Bulbophyllum nitens Jum. & H. Perrier var. **intermedium** H. Perrier
Bulbophyllum nitens Jum. & H. Perrier var. **majus** H. Perrier
Bulbophyllum nitens Jum. & H. Perrier var. **minus** H. Perrier
Bulbophyllum nitens Jum. & H. Perrier var. **pulverulentum** H. Perrier
Bulbophyllum nutans (Thouars) Thouars

Bulbophyllum nutans (Thouars) Thouars var. **variifolium** (Schltr.) Bosser
Bulbophyllum obscuriflorum H. Perrier
Bulbophyllum obtusatum (Jum. & H. Perrier) Schltr.
Bulbophyllum obtusilabium W. Kittr.
Bulbophyllum occlusum Ridl.
Bulbophyllum occultum Thouars
Bulbophyllum ochrochlamys Schltr. (1924)
Bulbophyllum onivense H. Perrier
Bulbophyllum ophiuchus Ridl.
Bulbophyllum ophiuchus Ridl. var. **baronianum** H. Perrier
Bulbophyllum oreodorum Schltr.
Bulbophyllum oxycalyx Schltr.
Bulbophyllum oxycalyx Schltr. var. **rubescens** (Schltr.) Bosser
Bulbophyllum pachypus Schltr.
Bulbophyllum paleiferum Schltr.
Bulbophyllum pallens (Jum. & H. Perrier) Schltr.
Bulbophyllum pandurella Schltr.
Bulbophyllum pantoblepharon Schltr.
Bulbophyllum pantoblepharon Schltr. var. **vestitum** H. Perrier
Bulbophyllum papangense H. Perrier
Bulbophyllum pentasticha (Pfitzer ex Kraenzl.) H. Perrier
Bulbophyllum percorniculatum H. Perrier
Bulbophyllum perpusillum Wendl. & Kraenzl.
Bulbophyllum perreflexum Bosser & P. J. Cribb
Bulbophyllum perrieri Schltr.
Bulbophyllum pervillei Rolfe ex Elliot
Bulbophyllum peyrotii Bosser
Bulbophyllum platypodum H. Perrier
Bulbophyllum pleiopterum Schltr.
Bulbophyllum pleurothallopsis Schltr.
Bulbophyllum protectum H. Perrier
Bulbophyllum ptiloglossum Wendl. & Kraenzl.
Bulbophyllum quadrialatum H. Perrier
Bulbophyllum quadrifarium Rolfe
Bulbophyllum ranomafanae Bosser & P. J. Cribb
Bulbophyllum rauhii Toill.-Gen. & Bosser
Bulbophyllum rauhii Toill.-Gen. & Bosser var. **andranobeense** Bosser
Bulbophyllum reflexiflorum H. Perrier
Bulbophyllum reflexiflorum H. Perrier subsp. **pogonochilum** Bosser
Bulbophyllum rhodostachys Schltr.
Bulbophyllum rienanense H. Perrier
Bulbophyllum rubiginosum Schltr.
Bulbophyllum rubrolabium Schltr.
Bulbophyllum rubrum Jum. & H. Perrier
Bulbophyllum ruginosum H. Perrier
Bulbophyllum rutenbergianum Schltr.
Bulbophyllum sambiranense Jum. & H. Perrier
Bulbophyllum sambiranense Jum. & H. Perrier var. **ankiranense** H. Perrier
Bulbophyllum sambiranense Jum. & H. Perrier var. **latibracteatum** H. Perrier
Bulbophyllum sandrangatense Bosser
Bulbophyllum sanguineum H. Perrier
Bulbophyllum sarcorhachis Schltr.
Bulbophyllum sarcorhachis Schltr. var. **beforonense** (Schltr.) H. Perrier
Bulbophyllum sarcorhachis Schltr. var. **flavemarginatum** H. Perrier

Bulbophyllum sciaphile Bosser
Bulbophyllum septatum Schltr.
Bulbophyllum simulacrum Schltr.
Bulbophyllum spaerobulbum H. Perrier
Bulbophyllum sphaerobulbum H. Perrier
Bulbophyllum subapproximatum H. Perrier
Bulbophyllum subclavatum Schltr.
Bulbophyllum subcrenulatum Schltr.
Bulbophyllum subsecundum Schltr.
Bulbophyllum subsessile Schltr.
Bulbophyllum sulfureum Schltr.
Bulbophyllum tampoketsense H. Perrier
Bulbophyllum teretibulbum H. Perrier
Bulbophyllum therezienii Bosser
Bulbophyllum thompsonii Ridl.
Bulbophyllum toilliezae Bosser
Bulbophyllum trichochlamys H. Perrier
Bulbophyllum trifarium Rolfe
Bulbophyllum trilineatum H. Perrier
Bulbophyllum turkii Bosser & P. J. Cribb
Bulbophyllum variegatum Thouars
Bulbophyllum ventriosum H. Perrier
Bulbophyllum verruculiferum H. Perrier
Bulbophyllum vestitum Bosser
Bulbophyllum vestitum Bosser var. **meridionale** Bosser
Bulbophyllum viguieri Schltr.
Bulbophyllum vulcanorum H. Perrier
Bulbophyllum xanthobulbum Schltr.
Bulbophyllum zaratananae Schltr.
Bulbophyllum zaratananae Schltr. subsp. **disjunctum** H. Perrier

MALAWI / MALAWI (LE) / MALAWI

Bulbophyllum bavonis J. J. Verm.
Bulbophyllum cochleatum Lindl.
Bulbophyllum elliottii Rolfe
Bulbophyllum encephalodes Summerh.
Bulbophyllum expallidum J. J. Verm.
Bulbophyllum fuscum Lindl. var. **melinostachyum** (Schltr) J. J. Verm.
Bulbophyllum humblottii Rolfe
Bulbophyllum intertextum Lindl.
Bulbophyllum josephi (Kuntze) Summerh.
Bulbophyllum josephi (Kuntze) Summerh. var. **mahonii** (Rolfe) J. J. Verm.
Bulbophyllum longiflorum Thouars
Bulbophyllum maximum (Lindl.) Rchb. f.
Bulbophyllum sandersonii (Hook. f.) Rchb. f.
Bulbophyllum scaberulum (Rolfe) Bolus
Bulbophyllum stolzii Schltr.
Bulbophyllum unifoliatum De Wild. subsp. **flectens** (P. J. Cribb & Taylor) J. J. Verm.
Bulbophyllum unifoliatum De Wild. subsp. **infracarinatum** (G. Will.) J. J. Verm.

MALAYSIA / MALAISIE (LA) / MALASIA

Bulbophyllum abbrevilabium Carr
Bulbophyllum acuminatum (Ridl.) Ridl.
Bulbophyllum alcicorne C. S. P. Parish & Rchb. f.
Bulbophyllum anaclastum J. J. Verm.
Bulbophyllum angustifolium (Blume) Lindl.
Bulbophyllum anisopterum J. J. Verm. & P. O'Byrne
Bulbophyllum annandalei Ridl.
Bulbophyllum apiferum Carr
Bulbophyllum apodum Hook. f.
Bulbophyllum armeniacum J. J. Sm.
Bulbophyllum atratum J. J. Sm.
Bulbophyllum auratum (Lindl.) Rchb. f.
Bulbophyllum bakhuizenii van Steenis
Bulbophyllum biflorum Teijsm. & Binn.
Bulbophyllum biseriale Carr
Bulbophyllum blepharistes Rchb. f.
Bulbophyllum blumei (Lindl.) J. J. Sm.
Bulbophyllum botryophorum Ridl.
Bulbophyllum brevipes Ridl.
Bulbophyllum brienianum (Rolfe) Ames
Bulbophyllum cameronense Garay, Hamer & Siegerist
Bulbophyllum capitatum (Blume) Lindl.
Bulbophyllum carinilabium J. J. Verm.
Bulbophyllum carrianum J. J. Verm.
Bulbophyllum catenarium Ridl.
Bulbophyllum caudatisepalum Ames & C. Schweinf.
Bulbophyllum cerebellum J. J. Verm.
Bulbophyllum cheiri Lindl.
Bulbophyllum cheiropetalum Ridl.
Bulbophyllum ciliatum (Blume) Lindl.
Bulbophyllum clandestinum Lindl.
Bulbophyllum cleistogamum Ridl.
Bulbophyllum comberi J. J. Verm.
Bulbophyllum comberipictum J. J. Verm.
Bulbophyllum concinnum Hook. f.
Bulbophyllum coniferum Ridl.
Bulbophyllum coriaceum Ridl.
Bulbophyllum corolliferum J. J. Sm.
Bulbophyllum crassipes Hook. f.
Bulbophyllum cupreum Lindl.
Bulbophyllum cuspidipetalum J. J. Sm.
Bulbophyllum cyanotriche J. J. Verm.
Bulbophyllum cyclosepalon Carr
Bulbophyllum dayanum Rchb. f.
Bulbophyllum dearei (Hort.) Rchb.f.
Bulbophyllum dentiferum Ridl.
Bulbophyllum dibothron J. J. Verm. & A. L. Lamb
Bulbophyllum diplantherum Carr
Bulbophyllum disjunctum Ames & C. Schweinf.
Bulbophyllum dracunculus J. J. Verm.
Bulbophyllum dryas Ridl.
Bulbophyllum elachanthe J. J. Verm.

Bulbophyllum elongatum (Blume) Hassk.
Bulbophyllum epicrianthes Hook. f.
Bulbophyllum evansii M. R. Henderson
Bulbophyllum farinulentum J. J. Sm.
Bulbophyllum farinulentum J. J. Sm. subsp. **densissimum** (Carr) J. J. Verm.
Bulbophyllum flammuliferum Ridl.
Bulbophyllum flavescens (Blume) Lindl.
Bulbophyllum flavorubellum J. J. Verm. & P. O'Byrne
Bulbophyllum foraminiferum J. J. Verm.
Bulbophyllum fulvibulbum J. J. Verm.
Bulbophyllum gemma-reginae J. J. Verm.
Bulbophyllum gibbosum (Blume) Lindl.
Bulbophyllum gilvum J. J. Verm. & A. L. Lamb
Bulbophyllum globulus Hook. f.
Bulbophyllum gracillimum (Rolfe) Rolfe
Bulbophyllum groeneveldtii J. J. Sm.
Bulbophyllum grudense J. J. Sm.
Bulbophyllum gusdorfii J. J. Sm.
Bulbophyllum habrotinum J. J. Verm. & A. L. Lamb
Bulbophyllum haniffii Carr
Bulbophyllum hemiprionotum J. J. Verm. & A. L. Lamb
Bulbophyllum hirtulum Ridl.
Bulbophyllum hodgsoni M. R. Henderson
Bulbophyllum ignevenosum Carr
Bulbophyllum inunctum J. J. Sm.
Bulbophyllum jolandae J. J. Verm.
Bulbophyllum khasyanum Griff.
Bulbophyllum korthalsii Schltr.
Bulbophyllum laetum J. J. Verm.
Bulbophyllum lasianthum Lindl.
Bulbophyllum lasiochilum C. S. P. Parish & Rchb. f.
Bulbophyllum laxiflorum (Blume) Lindl.
Bulbophyllum leptosepalum Hook. f.
Bulbophyllum liliacinum Ridl.
Bulbophyllum limbatum Lindl.
Bulbophyllum linearifolium King & Pantl.
Bulbophyllum lissoglossum J. J. Verm.
Bulbophyllum lobbii Lindl.
Bulbophyllum lohokii J. J. Verm. & A. L. Lamb
Bulbophyllum lordoglossum J. J. Verm. & A. L. Lamb
Bulbophyllum lumbriciforme J. J. Sm.
Bulbophyllum macranthum Lindl.
Bulbophyllum macrochilum Rolfe
Bulbophyllum makoyanum (Rchb. f.) Ridl.
Bulbophyllum malleolabrum Carr
Bulbophyllum medusae (Lindl.) Rchb. f.
Bulbophyllum membranaceum Teijsm. & Binn.
Bulbophyllum membranifolium Hook. f.
Bulbophyllum microglossum Ridl.
Bulbophyllum minutulum Ridl.
Bulbophyllum mirabile Hallier f.
Bulbophyllum mobilifilum Carr
Bulbophyllum monanthos Ridl.
Bulbophyllum moniliforme C. S. P. Parish & Rchb. f.
Bulbophyllum muscohaerens J. J. Verm. & A. L. Lamb

Bulbophyllum mutabile (Blume) Lindl.
Bulbophyllum nematocaulon Ridl.
Bulbophyllum nigropurpureum Carr
Bulbophyllum oblanceolatum King & Pantl.
Bulbophyllum obtusipetalum J. J. Sm.
Bulbophyllum obtusum (Blume) Lindl.
Bulbophyllum ochthodes J. J. Verm.
Bulbophyllum octorhopalon Seidenf.
Bulbophyllum odoratum (Blume) Lindl.
Bulbophyllum ovalifolium (Blume) Lindl.
Bulbophyllum pan Ridl.
Bulbophyllum papillosefilum Carr
Bulbophyllum patens King ex Hook. f.
Bulbophyllum penduliscapum J. J. Sm.
Bulbophyllum pileatum Lindl.
Bulbophyllum pilosum J. J. Verm.
Bulbophyllum placochilum J. J. Verm.
Bulbophyllum planibulbe (Ridl.) Ridl.
Bulbophyllum plumatum Ames
Bulbophyllum poekilon Carr
Bulbophyllum polycyclum J. J. Verm.
Bulbophyllum polygaliflorum J. J. Wood
Bulbophyllum praetervisum J. J. Verm.
Bulbophyllum proculcastris J. J. Verm.
Bulbophyllum puguahaanense Ames
Bulbophyllum pulchellum Ridl.
Bulbophyllum pulchrum Ridl. var. **brachysepalum** Ridl.
Bulbophyllum purpurascens Teijsm. & Binn.
Bulbophyllum pustulatum Ridl.
Bulbophyllum putidum (Teijsm. & Binn.) J. J. Sm.
Bulbophyllum rariflorum J. J. Sm.
Bulbophyllum retusiusculum Rchb. f.
Bulbophyllum rhizomatosum Ames & C. Schweinf.
Bulbophyllum ruficaudatum Ridl.
Bulbophyllum rugosum Ridl.
Bulbophyllum salaccense Rchb. f.
Bulbophyllum scintilla Ridl.
Bulbophyllum secundum Hook. f.
Bulbophyllum serratotruncatum Seidenf.
Bulbophyllum setuliferum J. J. Verm. & L. G. Saw
Bulbophyllum siamense Rchb. f.
Bulbophyllum sicyobulbon C. S. P. Parish & Rchb. f.
Bulbophyllum signatum J. J. Verm.
Bulbophyllum singaporeanum Schltr.
Bulbophyllum skeatianum Ridl.
Bulbophyllum spissum J. J. Verm.
Bulbophyllum stormii J. J. Sm.
Bulbophyllum striatellum Ridl.
Bulbophyllum subbullatum J. J. Verm.
Bulbophyllum subumbellatum Ridl.
Bulbophyllum taeniophyllum C. S. P. Parish & Rchb. f.
Bulbophyllum taeter J. J. Verm.
Bulbophyllum tahanense Carr
Bulbophyllum talauense (J. J. Sm.) Carr

Bulbophyllum tekuense Carr
Bulbophyllum tenompokense J. J. Sm.
Bulbophyllum tenuifolium (Blume) Lindl.
Bulbophyllum thiurum J.J.Verm. & P.O'Byrne
Bulbophyllum titanea Ridl.
Bulbophyllum tortuosum (Blume) Lindl.
Bulbophyllum trifolium Ridl.
Bulbophyllum trulliferum J. J. Verm. & A. L. Lamb
Bulbophyllum tryssum J. J. Verm. & A. L. Lamb
Bulbophyllum turpis J. J. Verm. & P. O'Byrne
Bulbophyllum uniflorum (Blume) Hassk.
Bulbophyllum vaginatum (Lindl.) Rchb. f.
Bulbophyllum variabile Ridl.
Bulbophyllum vermiculare Hook. f.
Bulbophyllum vesiculosum J. J. Sm.
Bulbophyllum virescens J. J. Sm.
Bulbophyllum viridescens Ridl.
Bulbophyllum vitellinum Ridl.
Bulbophyllum wrayi Hook. f.
Bulbophyllum xanthum Ridl.
Bulbophyllum xylocarpi J. J. Sm.
Bulbophyllum yoksunense J. J. Sm.
Trias antheae J. J. Verm. & A. L. Lamb

MAURITIUS / MAURICE / MAURICIO

Bulbophyllum caespitosum Thouars
Bulbophyllum clavatum Thouars
Bulbophyllum commersonii Thouars
Bulbophyllum densum Thouars
Bulbophyllum erectum Thouars
Bulbophyllum gracile Thouars
Bulbophyllum incurvum Thouars
Bulbophyllum longiflorum Thouars
Bulbophyllum nutans (Thouars) Thouars
Bulbophyllum occultum Thouars
Bulbophyllum pendulum Thouars
Bulbophyllum prismaticum Thouars
Bulbophyllum pusillum Thouars
Bulbophyllum variegatum Thouars

MEXICO / MEXIQUE (LE) / MÉXICO

Bulbophyllum aristatum (Rchb. f.) Hemsl.
Bulbophyllum cirrhosum L. O. Williams
Bulbophyllum nagelii L. O. Williams
Bulbophyllum oerstedii (Rchb. f.) Hemsl.
Bulbophyllum pachyrachis (A. Rich.) Griseb.
Bulbophyllum solteroi R. G. Tamayo

MICRONESIA (FEDERATED STATES OF) / MICRONÉSIA (ETATS FÉDÉRÉS DE) / MICRONESIA (ESTADOS FEDERADOS DE)

Bulbophyllum betchei F. Muell.

MOZAMBIQUE / MOZAMBIQUE (LE) / MOZAMBIQUE

Bulbophyllum fuscum Lindl. var. **melinostachyum** (Schltr) J. J. Verm.
Bulbophyllum josephi (Kuntze) Summerh.
Bulbophyllum maximum (Lindl.) Rchb. f.
Bulbophyllum oreonastes Rchb. f.
Bulbophyllum sandersonii (Hook. f.) Rchb. f.
Bulbophyllum scaberulum (Rolfe) Bolus
Bulbophyllum unifoliatum De Wild. subsp. **infracarinatum** (G. Will.) J. J. Verm.

MYANMAR / MYANMAR (LE) / MYANMAR

Acrochaene punctata Lindl.
Bulbophyllum alcicorne C. S. P. Parish & Rchb. f.
Bulbophyllum andersonii (Hook. f.) J. J. Sm.
Bulbophyllum auricomum Lindl.
Bulbophyllum birmense Schltr.
Bulbophyllum blepharistes Rchb. f.
Bulbophyllum bootanense C. S. P. Parish & Rchb. f.
Bulbophyllum burkilli Gage
Bulbophyllum capillipes C. S. P. Parish & Rchb. f.
Bulbophyllum careyanum (Hook. f.) Sprengel
Bulbophyllum cauliflorum Hook. f.
Bulbophyllum clandestinum Lindl.
Bulbophyllum comosum Collett & Hemsl.
Bulbophyllum crassipes Hook. f.
Bulbophyllum cupreum Lindl.
Bulbophyllum dayanum Rchb. f.
Bulbophyllum dickasonii Seidenf.
Bulbophyllum emarginatum (Finet) J. J. Sm.
Bulbophyllum farreri (W. W. Sm.) Seidenf.
Bulbophyllum gracillimum (Rolfe) Rolfe
Bulbophyllum guttulatum (Hook. f.) Balakr.
Bulbophyllum haniffii Carr
Bulbophyllum hirtum (J. E. Sm.) Lindl.
Bulbophyllum kanburiense Seidenf.
Bulbophyllum khasyanum Griff.
Bulbophyllum lasiochilum C. S. P. Parish & Rchb. f.
Bulbophyllum laxiflorum (Blume) Lindl.
Bulbophyllum lemniscatum C. S. P. Parish ex Hook. f.
Bulbophyllum limbatum Lindl.
Bulbophyllum lindleyanum Griff.
Bulbophyllum lineatum (Teijsm. & Binn.) J. J. Sm.
Bulbophyllum lobbii Lindl.
Bulbophyllum macranthum Lindl.
Bulbophyllum micropetalum Rchb. f.
Bulbophyllum moniliforme C. S. P. Parish & Rchb. f.
Bulbophyllum neilgherrense Wight
Bulbophyllum nigrescens Rolfe
Bulbophyllum odoratissimum (J. E. Sm.) Lindl.
Bulbophyllum oligoglossum Rchb. f.

Bulbophyllum ornatissimum (Rchb. f.) J. J. Sm.
Bulbophyllum parviflorum C. S. P. Parish & Rchb. f.
Bulbophyllum pectinatum Finet
Bulbophyllum penicillium C. S. P. Parish & Rchb. f.
Bulbophyllum picturatum (Lodd.) Rchb.f.
Bulbophyllum polyrhizum Lindl.
Bulbophyllum protractum Hook. f.
Bulbophyllum psittacoglossum Rchb. f.
Bulbophyllum pumilio C. S. P. Parish & Rchb. f.
Bulbophyllum purpurascens Teijsm. & Binn.
Bulbophyllum refractum (Zoll.) Rchb. f.
Bulbophyllum reichenbachii (Kuntze) Schltr.
Bulbophyllum repens Griff.
Bulbophyllum reptans (Lindl.) Lindl.
Bulbophyllum retusiusculum Rchb. f.
Bulbophyllum rufilabrum C. S. P. Parish ex Hook. f.
Bulbophyllum rufinum Rchb. f.
Bulbophyllum secundum Hook. f.
Bulbophyllum serratotruncatum Seidenf.
Bulbophyllum siamense Rchb. f.
Bulbophyllum sicyobulbon C. S. P. Parish & Rchb. f.
Bulbophyllum sillemianum Rchb. f.
Bulbophyllum spathaceum Rolfe
Bulbophyllum spathulatum (Rolfe ex E. Cooper) Seidenf.
Bulbophyllum stenobulbon C. S. P. Parish & Rchb. f.
Bulbophyllum suavissimum Rolfe
Bulbophyllum taeniophyllum C. S. P. Parish & Rchb. f.
Bulbophyllum tripudians C. S. P. Parish & Rchb. f.
Bulbophyllum umbellatum Lindl.
Bulbophyllum wendlandianum (Kraenzl.) Dammer
Bulbophyllum wightii Rchb. f.
Bulbophyllum xylophyllum C. S. P. Parish & Rchb. f.
Drymoda picta Lindl.
Monomeria longipes (Rchb. f.) Aver.
Sunipia andersonii (King & Pantl.) P. F. Hunt
Sunipia bicolor Lindl.
Sunipia racemosa (Sm.) T. Tang & F. T. Wang
Sunipia rimannii (Rchb. f.) Seidenf.
Sunipia scariosa Lindl.
Trias nasuta (Rchb. f.) Stapf
Trias oblonga Lindl.
Trias picta (C. S. P. Parish & Rchb. f.) C. S. P. Parish ex Hemsl.

NEPAL / NÉPAL / NEPAL

Bulbophyllum ambrosia (Hance) Schltr. subsp. **nepalensis** J. J. Wood
Bulbophyllum elatum (Hook. f.) J. J. Sm.
Bulbophyllum hirtulum Ridl.
Bulbophyllum odoratissimum (J. E. Sm.) Lindl.
Bulbophyllum retusiusculum Rchb. f.
Bulbophyllum secundum Hook. f.
Sunipia bicolor Lindl.
Sunipia cirrhata (Lindl.) P. F. Hunt
Sunipia racemosa (Sm.) T. Tang & F. T. Wang

NEW CALEDONIA (FRENCH) / NOUVELLE-CALEDONIE /NUEVA CALEDONIA

Bulbophyllum absconditum J. J. Sm.
Bulbophyllum aphanopetalum Schltr.
Bulbophyllum argyropus (Endl.) Rchb. f.
Bulbophyllum atrorubens Schltr.
Bulbophyllum baladeanum J. J. Sm.
Bulbophyllum betchei F. Muell.
Bulbophyllum cheiropetalum Ridl.
Bulbophyllum comptonii Rendle
Bulbophyllum ebulbe Schltr.
Bulbophyllum gracillimum (Rolfe) Rolfe
Bulbophyllum hexarhopalos Schltr.
Bulbophyllum keekee N. Halle
Bulbophyllum lingulatum Rendle
Bulbophyllum longiflorum Thouars
Bulbophyllum lophoglottis (Guillaumin) N. Halle
Bulbophyllum neo-caledonicum Schltr.
Bulbophyllum ngoyense Schltr.
Bulbophyllum pachyanthum Schltr.
Bulbophyllum pallidiflorum Schltr.
Bulbophyllum samoanum Schltr.
Bulbophyllum triandrum Schltr.

NEW ZEALAND / NOUVELLE-ZÉLANDE (LA) / NUEVA ZELANDIA

Bulbophyllum argyropus (Endl.) Rchb. f.
Bulbophyllum pygmaeum (Sm.) Lindl.
Bulbophyllum tuberculatum Colenso

NICARAGUA / NICARAGUIA (LE) / NICARAGUA

Bulbophyllum aristatum (Rchb. f.) Hemsl.
Bulbophyllum oerstedii (Rchb. f.) Hemsl.

NIGERIA / NIGÉRIA (LE) / NIGERIA

Bulbophyllum barbigerum Lindl.
Bulbophyllum calvum Summerh.
Bulbophyllum calyptratum Kraenzl.
Bulbophyllum cochleatum Lindl.
Bulbophyllum cochleatum Lindl. var. **tenuicaule** (Lindl.) J. J. Verm.
Bulbophyllum colubrinum (Rchb. f.) Rchb. f.
Bulbophyllum comatum Lindl.
Bulbophyllum dolabriforme J. J. Verm.
Bulbophyllum falcatum (Lindl.) Rchb. f.
Bulbophyllum falcatum (Lindl.) Rchb. f. var. **bufo** (Lindl.) J. J. Verm.
Bulbophyllum falcatum (Lindl.) Rchb. f. var. **velutinum** (Lindl.) J. J. Verm.
Bulbophyllum falcipetalum Lindl.

Bulbophyllum fuscum Lindl.
Bulbophyllum fuscum Lindl. var. **melinostachyum** (Schltr) J. J. Verm.
Bulbophyllum imbricatum Lindl.
Bulbophyllum intertextum Lindl.
Bulbophyllum ivorense P. J. Cribb & Perez-Vera
Bulbophyllum josephi (Kuntze) Summerh. var. **mahonii** (Rolfe) J. J. Verm.
Bulbophyllum lupulinum Lindl.
Bulbophyllum magnibracteatum Summerh.
Bulbophyllum maximum (Lindl.) Rchb. f.
Bulbophyllum nigericum Summerh.
Bulbophyllum nigritianum Rendle
Bulbophyllum oreonastes Rchb. f.
Bulbophyllum oxychilum Schltr.
Bulbophyllum pipio Rchb. f.
Bulbophyllum porphyrostachys Summerh.
Bulbophyllum pumilum (Sw.) Lindl.
Bulbophyllum resupinatum Ridl.
Bulbophyllum saltatorium Lindl.
Bulbophyllum saltatorium Lindl. var. **albociliatum** (Finet) J. J. Verm.
Bulbophyllum sandersonii (Hook. f.) Rchb. f. subsp. **stenopetalum** (Kraenzl.) J. J. Verm.
Bulbophyllum scaberulum (Rolfe) Bolus
Bulbophyllum scaberulum (Rolfe) Bolus var. **fuerstenbergianum** (De Wild.) J. J. Verm.
Bulbophyllum schimperianum Kraenzl.
Bulbophyllum schinzianum Kraenzl.
Bulbophyllum schinzianum Kraenzl. var. **phaeopogon** (Schltr.) J. J. Verm.

NIUE / NIOUEÉ (FEM) / NIUE

Bulbophyllum longiscapum Rolfe

PALAU / PALAOS / PALAU

Bulbophyllum betchei F. Muell.
Bulbophyllum desmanthum Tuyama
Bulbophyllum fukuyamae Tuyama
Bulbophyllum hatusimanum Tuyama
Bulbophyllum kusaiense Tuyama
Bulbophyllum membranaceum Teijsm. & Binn.
Bulbophyllum micronesiacum Schltr.
Bulbophyllum volkensii Schltr.

PANAMA / PANAMA (LE) / PANAMÁ

Bulbophyllum aristatum (Rchb. f.) Hemsl.
Bulbophyllum oerstedii (Rchb. f.) Hemsl.
Bulbophyllum pachyrachis (A. Rich.) Griseb.
Bulbophyllum sordidum Lindl.
Bulbophyllum wagneri Schltr.

PAPUA NEW GUINEA / PAPOUASIE-NOUVELLE-GUINÉE (LA) / PAPUA NUEVA GUINEA

Bulbophyllum ablepharon Schltr.
Bulbophyllum absconditum J. J. Sm.
Bulbophyllum absconditum J. J. Sm. subsp. **hastula** J. J. Verm.
Bulbophyllum acanthoglossum Schltr.
Bulbophyllum acropogon Schltr.
Bulbophyllum adenoblepharon Schltr.
Bulbophyllum adolphi Schltr.
Bulbophyllum aechmophorum J. J. Verm.
Bulbophyllum aemulum Schltr.
Bulbophyllum agastor Garay, Hamer & Siegerist
Bulbophyllum alabastraceus P. Royen
Bulbophyllum alticola Schltr.
Bulbophyllum alveatum J. J. Verm.
Bulbophyllum amblyacron Schltr.
Bulbophyllum amblyanthum Schltr.
Bulbophyllum andreeae A. D. Hawkes
Bulbophyllum ankylochele J. J. Verm.
Bulbophyllum ankylorhinon J. J. Verm.
Bulbophyllum antennatum Schltr.
Bulbophyllum antenniferum (Lindl.) Rchb. f.
Bulbophyllum aphanopetalum Schltr.
Bulbophyllum apiculatum Schltr.
Bulbophyllum appressum Schltr.
Bulbophyllum arachnoideum Schltr.
Bulbophyllum arfakianum Kraenzl.
Bulbophyllum arminii Sieder & Kiehn †
Bulbophyllum artostigma J. J. Verm.
Bulbophyllum asperilingue Schltr.
Bulbophyllum atrolabium Schltr.
Bulbophyllum aundense Ormerod
Bulbophyllum aureoapex Schltr.
Bulbophyllum aureobrunneum Schltr.
Bulbophyllum baculiferum Ridl.
Bulbophyllum bandischii Garay, Hamer & Siegerist
Bulbophyllum barbatum Barb. Rodr.
Bulbophyllum biantennatum Schltr.
Bulbophyllum bicaudatum Schltr.
Bulbophyllum bigibbum Schltr.
Bulbophyllum bisepalum Schltr.
Bulbophyllum bismarckense Schltr.
Bulbophyllum blepharicardium Schltr.
Bulbophyllum blephariglossum Schltr.
Bulbophyllum blepharopetalum Schltr.
Bulbophyllum bliteum J. J. Verm.
Bulbophyllum blumei (Lindl.) J. J. Sm.
Bulbophyllum brachychilum Schltr.
Bulbophyllum brachypetalum Schltr.
Bulbophyllum brassii J. J. Verm.
Bulbophyllum breve Schltr.
Bulbophyllum brevilabium Schltr.
Bulbophyllum bulhartii Sieder & Kiehn †

Bulbophyllum bulliferum J. J. Sm.
Bulbophyllum burfordiense Hort. ex Garay, Hamer & Siegerist
Bulbophyllum cadetioides Schltr.
Bulbophyllum callichroma Schltr.
Bulbophyllum caloglossum Schltr.
Bulbophyllum calviventer J. J. Verm.
Bulbophyllum caputgnomonis J. J. Verm.
Bulbophyllum cardiophyllum J. J. Verm.
Bulbophyllum cateorum J. J. Verm.
Bulbophyllum catillus J. J. Verm. & P. O'Byrne
Bulbophyllum cerinum Schltr.
Bulbophyllum chaetostroma Schltr.
Bulbophyllum chaunobulbon Schltr.
Bulbophyllum chaunobulbon Schltr. var. **ctenopetalum** Schltr.
Bulbophyllum chimaera Schltr.
Bulbophyllum chloranthum Schltr.
Bulbophyllum chlororhopalon Schltr.
Bulbophyllum chrysochilum Schltr.
Bulbophyllum chrysoglossum Schltr.
Bulbophyllum chrysotes Schltr.
Bulbophyllum ciliipetalum Schltr.
Bulbophyllum ciliolatum Schltr.
Bulbophyllum cimicinum J. J. Verm.
Bulbophyllum citrinilabre J. J. Sm.
Bulbophyllum colliferum J. J. Sm.
Bulbophyllum collinum Schltr.
Bulbophyllum cominsii Rolfe
Bulbophyllum compressilabellatum P. Royen
Bulbophyllum crenilabium W. Kittr.
Bulbophyllum crispatisepalum P. Royen
Bulbophyllum cruciatum J. J. Sm.
Bulbophyllum cruentum Garay, Hamer & Siegerist
Bulbophyllum cruttwellii J. J. Verm.
Bulbophyllum cryptanthoides J. J. Sm.
Bulbophyllum cryptanthum Schltr.
Bulbophyllum cucullatum Schltr.
Bulbophyllum curvicaule Schltr.
Bulbophyllum cuspidipetalum J. J. Sm.
Bulbophyllum cycloglossum Schltr.
Bulbophyllum cyclophyllum Schltr.
Bulbophyllum cylindrobulbum Schltr.
Bulbophyllum dasyphyllum Schltr.
Bulbophyllum decarhopalon Schltr.
Bulbophyllum decumbens Schltr.
Bulbophyllum decurvulum Schltr.
Bulbophyllum dekockii J. J. Sm.
Bulbophyllum dendrochiloides Schltr.
Bulbophyllum densibulbum W. Kittr.
Bulbophyllum densifolium Schltr.
Bulbophyllum dependens Schltr.
Bulbophyllum desmotrichoides Schltr.
Bulbophyllum dichaeoides Schltr.
Bulbophyllum dichilus Schltr.
Bulbophyllum dichotomum J. J. Sm.
Bulbophyllum dictyoneuron Schltr.

Bulbophyllum digoelense J. J. Sm.
Bulbophyllum dischorense Schltr.
Bulbophyllum discolor Schltr.
Bulbophyllum discolor Schltr. var. cubitale J. J. Verm.
Bulbophyllum distichum Schltr.
Bulbophyllum djamuense Schltr.
Bulbophyllum dolichoglottis Schltr.
Bulbophyllum drepanosepalum J. J. Verm.
Bulbophyllum dryadum Schltr.
Bulbophyllum dschischungarense Schltr.
Bulbophyllum ebulbe Schltr.
Bulbophyllum eciliatum Schltr.
Bulbophyllum elasmatopus Schltr.
Bulbophyllum elegantius Schltr.
Bulbophyllum ellipticum Schltr.
Bulbophyllum elongatum (Blume) Hassk.
Bulbophyllum endotrachys Schltr.
Bulbophyllum entomonopsis J. J. Verm.
Bulbophyllum epapillosum Schltr.
Bulbophyllum epibulbon Schltr.
Bulbophyllum erinaceum Schltr.
Bulbophyllum erioides Schltr.
Bulbophyllum erythrostictum Ormerod
Bulbophyllum euplepharum Rchb. f.
Bulbophyllum exasperatum Schltr.
Bulbophyllum exiguiflorum Schltr.
Bulbophyllum exilipes Schltr.
Bulbophyllum falcibracteum Schltr.
Bulbophyllum falcifolium Schltr.
Bulbophyllum fasciatum Schltr.
Bulbophyllum fasciculatum Schltr.
Bulbophyllum fasciculiferum Schltr.
Bulbophyllum filamentosum Schltr.
Bulbophyllum finisterrae Schltr.
Bulbophyllum fissipetalum Schltr.
Bulbophyllum flagellare Schltr.
Bulbophyllum flavum Schltr.
Bulbophyllum fletcherianum Rolfe
Bulbophyllum flexuosum Schltr.
Bulbophyllum foetidilabrum Ormerod
Bulbophyllum foetidum Schltr.
Bulbophyllum fonsflorum J. J. Verm.
Bulbophyllum forbesii Schltr.
Bulbophyllum fordii (Rolfe) J. J. Sm.
Bulbophyllum formosum Schltr.
Bulbophyllum fractiflexum J. J. Sm.
Bulbophyllum fractiflexum J. J. Sm. subsp. salomonense J. J. Verm. & A. L. Lamb
Bulbophyllum fruticicola Schltr.
Bulbophyllum fuscatum Schltr.
Bulbophyllum fusciflorum Schltr.
Bulbophyllum galactanthum Schltr.
Bulbophyllum glabrum Schltr.
Bulbophyllum glanduliferum Schltr.
Bulbophyllum glaucum Schltr.

Bulbophyllum globiceps Schltr.
Bulbophyllum globiceps Schltr. var. **boloboense** Schltr.
Bulbophyllum gobiense Schltr.
Bulbophyllum graciliscapum Schltr.
Bulbophyllum gracillimum (Rolfe) Rolfe
Bulbophyllum grammopoma J. J. Verm.
Bulbophyllum grandiflorum Blume
Bulbophyllum grandifolium Schltr.
Bulbophyllum guttatum Schltr.
Bulbophyllum gyaloglossum J. J. Verm.
Bulbophyllum hahlianum Schltr.
Bulbophyllum hamadryas Schltr.
Bulbophyllum hamadryas Schltr. var. **orientale** Schltr.
Bulbophyllum hans-meyeri J. J. Wood
Bulbophyllum harposepalum Schltr.
Bulbophyllum helix Schltr.
Bulbophyllum hellwigianum Kraenzl. ex Warb.
Bulbophyllum heteroblepharon Schltr.
Bulbophyllum heterorhopalon Schltr.
Bulbophyllum heterosepalum Schltr.
Bulbophyllum hexarhopalos Schltr.
Bulbophyllum hexurum Schltr.
Bulbophyllum hians Schltr.
Bulbophyllum hians Schltr. var. **alticola** Schltr.
Bulbophyllum hiljeae J. J. Verm.
Bulbophyllum hirudiniferum J. J. Verm.
Bulbophyllum howcroftii Garay, Hamer & Siegerist
Bulbophyllum hoyifolium J. J. Verm.
Bulbophyllum humile Schltr.
Bulbophyllum hydrophilum J. J. Sm.
Bulbophyllum hymenobracteum Schltr.
Bulbophyllum hystricinum Schltr.
Bulbophyllum ialibuense Ormerod
Bulbophyllum iboense Schltr.
Bulbophyllum icteranthum Schltr.
Bulbophyllum imitator J. J. Verm.
Bulbophyllum inaequisepalum Schltr.
Bulbophyllum inauditum Schltr. (1913)
Bulbophyllum inciferum J. J. Verm.
Bulbophyllum incumbens Schltr.
Bulbophyllum inquirendum J. J. Verm.
Bulbophyllum intersitum J. J. Verm.
Bulbophyllum inversum Schltr.
Bulbophyllum ischnopus Schltr.
Bulbophyllum ischnopus Schltr. var. **rhodoneuron** Schltr.
Bulbophyllum jadunae Schltr.
Bulbophyllum johannulii J. J. Verm.
Bulbophyllum kaniense Schltr.
Bulbophyllum kauloense Schltr.
Bulbophyllum kelelense Schltr.
Bulbophyllum kempfii Schltr.
Bulbophyllum kenae J. J. Verm.
Bulbophyllum kenejianum Schltr.
Bulbophyllum kenejiense W. Kittr.
Bulbophyllum latipes J. J. Sm.

Bulbophyllum laxum Schltr.
Bulbophyllum lemnifolium Schltr.
Bulbophyllum leontoglossum Schltr.
Bulbophyllum lepanthiflorum Schltr.
Bulbophyllum leptobulbon J. J. Verm.
Bulbophyllum leptoleucum Schltr.
Bulbophyllum leptophyllum W. Kittr.
Bulbophyllum leptopus Schltr.
Bulbophyllum leucorhodum Schltr.
Bulbophyllum leucothyrsus Schltr.
Bulbophyllum levatii Kraenzl. var. **mischanthum** J. J. Verm.
Bulbophyllum leve Schltr.
Bulbophyllum levyae Garay, Hamer & Siegerist
Bulbophyllum lichenoides Schltr.
Bulbophyllum ligulatum W. Kittr.
Bulbophyllum lineolatum Schltr.
Bulbophyllum lonchophyllum Schltr. (1913)
Bulbophyllum longiflorum Thouars
Bulbophyllum longilabre Schltr.
Bulbophyllum longirostre Schltr.
Bulbophyllum lophoton J. J. Verm.
Bulbophyllum loroglossum J. J. Verm.
Bulbophyllum louisiadum Schltr.
Bulbophyllum loxophyllum Schltr.
Bulbophyllum lyriforme J. J. Verm. & P. O'Byrne
Bulbophyllum maboroense Schltr.
Bulbophyllum macilentum J. J. Verm.
Bulbophyllum macneiceae Schuit. & De Vogel
Bulbophyllum macranthum Lindl.
Bulbophyllum macrobulbum J. J. Sm.
Bulbophyllum macrorhopalon Schltr.
Bulbophyllum macrourum Schltr.
Bulbophyllum maijenense Schltr.
Bulbophyllum manobulbum Schltr.
Bulbophyllum marginatum Schltr.
Bulbophyllum masonii (Senghas) J. J. Wood
Bulbophyllum mattesii Sieder & Kiehn †
Bulbophyllum maxillarioides Schltr.
Bulbophyllum melanoxanthum J. J. Verm. & B. A. Lewis
Bulbophyllum melinanthum Schltr.
Bulbophyllum melinoglossum Schltr.
Bulbophyllum membranaceum Teijsm. & Binn.
Bulbophyllum microblepharon Schltr.
Bulbophyllum microbulbon Schltr.
Bulbophyllum microcala P. F. Hunt
Bulbophyllum microdendron Schltr.
Bulbophyllum microrhombos Schltr.
Bulbophyllum microsphaerum Schltr.
Bulbophyllum microtes Schltr.
Bulbophyllum microthamnus Schltr.
Bulbophyllum mimiense Schltr.
Bulbophyllum minutibulbum W. Kittr.
Bulbophyllum minutipetalum Schltr.
Bulbophyllum mischobulbon Schltr.

Bulbophyllum monosema Schltr.
Bulbophyllum montanum Schltr.
Bulbophyllum mulderae J. J. Verm.
Bulbophyllum muriceum Schltr.
Bulbophyllum myolaense Garay, Hamer & Siegerist
Bulbophyllum myon J. J. Verm.
Bulbophyllum myrtillus Schltr.
Bulbophyllum mystrochilum Schltr.
Bulbophyllum mystrophyllum Schltr.
Bulbophyllum nannodes Schltr.
Bulbophyllum nasica Schltr.
Bulbophyllum nasilabium Schltr.
Bulbophyllum navicula Schltr.
Bulbophyllum nebularum Schltr.
Bulbophyllum nematorhizis Schltr.
Bulbophyllum neo-pommeranicum Schltr.
Bulbophyllum nephropetalum Schltr.
Bulbophyllum nigrilabium Schltr.
Bulbophyllum nitidum Schltr.
Bulbophyllum novae-hiberniae Schltr.
Bulbophyllum nubigenum Schltr.
Bulbophyllum nummularioides Schltr.
Bulbophyllum obyrnei Garay, Hamer & Siegerist
Bulbophyllum ochroleucum Schltr.
Bulbophyllum ochthochilum J. J. Verm.
Bulbophyllum odontoglossum Schltr.
Bulbophyllum odontopetalum Schltr.
Bulbophyllum oliganthum Schltr.
Bulbophyllum oligochaete Schltr.
Bulbophyllum olivinum J. J. Sm.
Bulbophyllum olivinum J. J. Sm. subsp. **linguiferum** J. J. Verm.
Bulbophyllum oobulbum Schltr.
Bulbophyllum orbiculare J. J. Sm.
Bulbophyllum orbiculare J. J. Sm. subsp. **cassideum** J. J. Verm.
Bulbophyllum oreocharis Schltr.
Bulbophyllum oreodoxa Schltr.
Bulbophyllum oreogenum Schltr.
Bulbophyllum origami J. J. Verm.
Bulbophyllum ornatum Schltr.
Bulbophyllum ortalis J. J. Verm.
Bulbophyllum orthosepalum J. J. Verm.
Bulbophyllum oxyanthum Schltr.
Bulbophyllum pachyglossum Schltr.
Bulbophyllum pachytelos Schltr.
Bulbophyllum papuliglossum Schltr.
Bulbophyllum papulipetalum Schltr.
Bulbophyllum parabates J. J. Verm.
Bulbophyllum patella J. J. Verm.
Bulbophyllum paululum Schltr.
Bulbophyllum peltopus Schltr.
Bulbophyllum pemae Schltr.
Bulbophyllum phaeoglossum Schltr.
Bulbophyllum phaeorhabdos Schltr.
Bulbophyllum phormion J. J. Verm.
Bulbophyllum phreatiopse J. J. Verm.

Bulbophyllum phymatum J. J. Verm.
Bulbophyllum pidacanthum J. J. Verm.
Bulbophyllum piestobulbon Schltr.
Bulbophyllum piliferum J. J. Sm.
Bulbophyllum plagiopetalum Schltr.
Bulbophyllum plicatum J. J. Verm.
Bulbophyllum plumula Schltr.
Bulbophyllum polyblepharon Schltr.
Bulbophyllum polyphyllum Schltr.
Bulbophyllum posticum J. J. Sm.
Bulbophyllum potamophila Schltr.
Bulbophyllum procerum Schltr.
Bulbophyllum pseudotrias J. J. Verm.
Bulbophyllum psilorhopalon Schltr.
Bulbophyllum ptilotes Schltr.
Bulbophyllum ptychantyx J. J. Verm.
Bulbophyllum pulvinatum Schltr.
Bulbophyllum punamense Schltr.
Bulbophyllum pungens Schltr.
Bulbophyllum pungens Schltr. var. **pachyphyllum** Schltr.
Bulbophyllum pyroglossum Schuit. & De Vogel
Bulbophyllum quadrichaete Schltr.
Bulbophyllum quasimodo J. J. Verm.
Bulbophyllum quinquelobum Schltr.
Bulbophyllum quinquelobum Schltr. var. **lancilabium** Schltr.
Bulbophyllum rarum Schltr.
Bulbophyllum reevei J. J. Verm.
Bulbophyllum reifii Sieder & Kiehn †
Bulbophyllum renipetalum Schltr.
Bulbophyllum rhodoglossum Schltr.
Bulbophyllum rhodoleucum Schltr.
Bulbophyllum rhodoneuron Schltr.
Bulbophyllum rhodostictum Schltr.
Bulbophyllum rhomboglossum Schltr.
Bulbophyllum rhopaloblepharon Schltr.
Bulbophyllum rhopalophorum Schltr.
Bulbophyllum rigidipes Schltr.
Bulbophyllum rivulare Schltr.
Bulbophyllum roseopunctatum Schltr.
Bulbophyllum rubipetalum P. Royen
Bulbophyllum rubrolineatum Schltr.
Bulbophyllum rubromaculatum W. Kittr.
Bulbophyllum sarcodanthum Schltr.
Bulbophyllum saronae Garay
Bulbophyllum sauguetiense Schltr.
Bulbophyllum savaiense Schltr. subsp. **gorumense** (Schltr.) J. J. Verm.
Bulbophyllum savaiense Schltr. subsp. **subcubicum** (J. J. Sm.) J. J. Verm.
Bulbophyllum sceliphron J. J. Verm.
Bulbophyllum schistopetalum Schltr.
Bulbophyllum scopa J. J. Verm.
Bulbophyllum scopula Schltr.
Bulbophyllum scutiferum J. J. Verm.
Bulbophyllum scyphochilus Schltr.
Bulbophyllum scyphochilus Schltr. var. **phaeanthum** Schltr.

Bulbophyllum semiteres Schltr.
Bulbophyllum sepikense W. Kittr.
Bulbophyllum serra Schltr.
Bulbophyllum serripetalum Schltr.
Bulbophyllum serrulatum Schltr.
Bulbophyllum simile Schltr.
Bulbophyllum sinapis J. J. Verm. & P. O'Byrne
Bulbophyllum singulare Schltr.
Bulbophyllum singuliflorum W. Kittr.
Bulbophyllum speciosum Schltr.
Bulbophyllum sphaeracron Schltr.
Bulbophyllum spiesii Garay, Hamer & Siegerist
Bulbophyllum spongiola J. J. Verm.
Bulbophyllum stenochilum Schltr.
Bulbophyllum stenophyllum Schltr.
Bulbophyllum stenorhopalos Schltr.
Bulbophyllum stictanthum Schltr.
Bulbophyllum stictosepalum Schltr.
Bulbophyllum stipulaceum Schltr.
Bulbophyllum stolleanum Schltr.
Bulbophyllum subpatulum J. J. Verm.
Bulbophyllum subtrilobatum Schltr.
Bulbophyllum subulifolium Schltr.
Bulbophyllum superpositum Schltr.
Bulbophyllum tanystiche J. J. Verm.
Bulbophyllum tarantula Schuit. & De Vogel
Bulbophyllum tentaculatum Schltr.
Bulbophyllum tentaculiferum Schltr.
Bulbophyllum tenue Schltr.
Bulbophyllum tenuipes Schltr.
Bulbophyllum theioglossum Schltr.
Bulbophyllum thelantyx J. J. Verm.
Bulbophyllum thersites J. J. Verm.
Bulbophyllum tinekeae Schuit. & De Vogel
Bulbophyllum toranum J. J. Sm.
Bulbophyllum torricellense Schltr.
Bulbophyllum tortum Schltr.
Bulbophyllum trachyanthum Kraenzl.
Bulbophyllum trachybracteum Schltr.
Bulbophyllum trachyglossum Schltr.
Bulbophyllum trachypus Schltr.
Bulbophyllum triandrum Schltr.
Bulbophyllum triaristella Schltr.
Bulbophyllum trichaete Schltr.
Bulbophyllum trichambon Schltr.
Bulbophyllum trichromum Schltr.
Bulbophyllum trifilum J. J. Sm.
Bulbophyllum trifilum J. J. Sm. subsp. **filisepalum** (J. J. Sm.) J. J. Verm.
Bulbophyllum trigonocarpum Schltr.
Bulbophyllum trirhopalon Schltr.
Bulbophyllum truncicola Schltr.
Bulbophyllum tumoriferum Schltr.
Bulbophyllum ulcerosum J. J. Sm.
Bulbophyllum umbraticola Schltr.
Bulbophyllum unguilabium Schltr.

Bulbophyllum unicaudatum Schltr.
Bulbophyllum unicaudatum Schltr. var. **xanthospaerum** Schltr.
Bulbophyllum uroglossum Schltr.
Bulbophyllum urosepalum Schltr.
Bulbophyllum ustusfortiter J. J. Verm.
Bulbophyllum vaccinioides Schltr.
Bulbophyllum verruciferum Schltr.
Bulbophyllum verruciferum Schltr. var. **carinatisepalum** Schltr.
Bulbophyllum verruculatum Schltr.
Bulbophyllum wakoi Howcroft
Bulbophyllum warianum Schltr.
Bulbophyllum wechsbergii Sieder & Kiehn †
Bulbophyllum werneri Schltr.
Bulbophyllum xanthochlamys Schltr.
Bulbophyllum xanthophaeum Schltr.
Bulbophyllum xanthornis Schuit. & De Vogel
Bulbophyllum xanthotes Schltr.
Codonosiphon campanulatum Schltr.
Codonosiphon papuanum Schltr.
Monosepalum dischorense Schltr.
Monosepalum muricatum (J. J. Sm.) Schltr.
Monosepalum torricellense Schltr.
Pedilochilus alpinum P. Royen
Pedilochilus alpinum P. Royen var. **fasciculatum** P. Royen
Pedilochilus angustifolius Schltr.
Pedilochilus augustifolium Schltr.
Pedilochilus brachiatus Schltr.
Pedilochilus brachypus Schltr.
Pedilochilus ciliolatum Schltr.
Pedilochilus coiloglossum (Schltr.) Schltr.
Pedilochilus cyatheicola P. Royen
Pedilochilus dischorense Schltr.
Pedilochilus flavum Schltr.
Pedilochilus grandifolium P. Royen
Pedilochilus guttulatum Schltr.
Pedilochilus longipes Schltr.
Pedilochilus papuanum Schltr.
Pedilochilus parvulum Schltr.
Pedilochilus petiolatum Schltr.
Pedilochilus petrophilum P. Royen
Pedilochilus piundaundense P. Royen
Pedilochilus psychrophilum (F. Muell.) Ormerod
Pedilochilus pusillum Schltr.
Pedilochilus sarawakatensis P. Royen
Pedilochilus stictanthum Schltr.
Pedilochilus subalpinum P. Royen
Saccoglossum maculatum Schltr.
Saccoglossum papuanum Schltr.
Saccoglossum takeuchii Howcroft

PARAGUAY / PARAGUAY (LE) / PARAGUAY (EL)

Bulbophyllum rojasii L. O. Williams

PERU / PÉROU / PERÚ

Bulbophyllum incarum Kraenzl.
Bulbophyllum machupicchuense D. E. Benn. & Christenson
Bulbophyllum meridense Rchb. f.
Bulbophyllum weberbauerianum Kraenzl.

PHILIPPINES (THE) / PHILIPPINES (LES) / FILIPINAS

Bulbophyllum absconditum J. J. Sm.
Bulbophyllum aeolium Ames
Bulbophyllum aestivale Ames
Bulbophyllum alagense Ames
Bulbophyllum albo-roseum Ames
Bulbophyllum alsiosum Ames
Bulbophyllum anguliferum Ames & C. Schweinf.
Bulbophyllum antenniferum (Lindl.) Rchb. f.
Bulbophyllum apodum Hook. f.
Bulbophyllum apoense Schuit. & De Vogel
Bulbophyllum arrectum Kraenzl.
Bulbophyllum auratum (Lindl.) Rchb. f.
Bulbophyllum basisetum J. J. Sm.
Bulbophyllum bataanense Ames
Bulbophyllum biflorum Teijsm. & Binn.
Bulbophyllum blumei (Lindl.) J. J. Sm.
Bulbophyllum bolsteri Ames
Bulbophyllum bontocense Ames
Bulbophyllum breviflorum Ridl. ex Stapf
Bulbophyllum calophyllum L. O. Williams
Bulbophyllum canlaonense Ames
Bulbophyllum careyanum (Hook. f.) Sprengel
Bulbophyllum carinatum (Teijsm. & Binn.) Naves
Bulbophyllum carunculatum Garay, Hamer & Siegerist
Bulbophyllum catenarium Ridl.
Bulbophyllum catenulatum Kraenzl.
Bulbophyllum caudatisepalum Ames & C. Schweinf.
Bulbophyllum cephalophorum Garay, Hamer & Siegerist
Bulbophyllum cheiri Lindl.
Bulbophyllum chrysendetum Ames
Bulbophyllum chryseum (Kraenzl.) Ames
Bulbophyllum clandestinum Lindl.
Bulbophyllum cleistogamum Ridl.
Bulbophyllum clemensiae Ames
Bulbophyllum colubrimodum Ames
Bulbophyllum coniferum Ridl.
Bulbophyllum cootesii Clements
Bulbophyllum copelandii Ames
Bulbophyllum coriaceum Ridl.
Bulbophyllum cornutum (Blume) Rchb. f.
Bulbophyllum costatum Ames
Bulbophyllum cryptophoranthus Garay
Bulbophyllum cubicum Ames

Bulbophyllum cumingii (Lindl.) Rchb. f.
Bulbophyllum cuneatum Rolfe
Bulbophyllum cupreum Lindl.
Bulbophyllum curranii Ames
Bulbophyllum dagamense Ames
Bulbophyllum dasypetalum Rolfe ex Ames
Bulbophyllum dearei (Hort.) Rchb.f.
Bulbophyllum debrincatiae J. J. Verm.
Bulbophyllum deceptum Ames
Bulbophyllum deltoideum Ames & C. Schweinf.
Bulbophyllum disjunctum Ames & C. Schweinf.
Bulbophyllum dissolutum Ames
Bulbophyllum doryphoroide Ames
Bulbophyllum ebracteolatum Kraenzl.
Bulbophyllum echinochilum Kraenzl.
Bulbophyllum elassoglossum Siegerist
Bulbophyllum elmeri Ames
Bulbophyllum elongatum (Blume) Hassk.
Bulbophyllum emiliorum Ames & Quisumb.
Bulbophyllum erosipetalum C. Schweinf.
Bulbophyllum erratum Ames
Bulbophyllum escritorii Ames
Bulbophyllum exile Ames
Bulbophyllum exquisitum Ames
Bulbophyllum facetum Garay, Hamer & Siegerist
Bulbophyllum fenixii Ames
Bulbophyllum filicoides Ames
Bulbophyllum flavescens (Blume) Lindl.
Bulbophyllum gibbsiae Rolfe
Bulbophyllum gimagaanense Ames
Bulbophyllum glandulosum Ames
Bulbophyllum gnomoniferum Ames
Bulbophyllum halconense Ames
Bulbophyllum invisum Ames
Bulbophyllum kettridgei (Garay, Hamer & Siegerist) J. J. Verm.
Bulbophyllum lancifolium Ames
Bulbophyllum lancilabium Ames
Bulbophyllum lancipetalum Ames
Bulbophyllum lasianthum Lindl.
Bulbophyllum lasioglossum Rolfe ex Ames
Bulbophyllum lasiopetalum Kraenzl.
Bulbophyllum latisepalum Ames & C. Schweinf.
Bulbophyllum laxiflorum (Blume) Lindl.
Bulbophyllum leibergii Ames & Rolfe
Bulbophyllum lepantense Ames
Bulbophyllum leptocaulon Kraenzl.
Bulbophyllum levanae Ames
Bulbophyllum levanae Ames var. **giganteum** Quis. & C. Schweinf.
Bulbophyllum leytense Ames
Bulbophyllum lipense Ames
Bulbophyllum lobbii Lindl.
Bulbophyllum loherianum (Kraenzl.) Ames
Bulbophyllum longiflorum Thouars
Bulbophyllum longimucronatum Ames & C. Schweinf.

Bulbophyllum longipetiolatum Ames
Bulbophyllum macranthum Lindl.
Bulbophyllum maculatum Boxall ex Naves
Bulbophyllum maculosum Ames
Bulbophyllum makoyanum (Rchb. f.) Ridl.
Bulbophyllum maquilinguense Ames & Quisumb.
Bulbophyllum marivelense Ames
Bulbophyllum masaganapense Ames
Bulbophyllum maxillare (Lindl.) Rchb. f.
Bulbophyllum mearnsii Ames
Bulbophyllum membranifolium Hook. f.
Bulbophyllum merrittii Ames
Bulbophyllum mindanaense Ames
Bulbophyllum mindorense Ames
Bulbophyllum mona-lisae Sieder & Kiehn †
Bulbophyllum monstrabile Ames
Bulbophyllum montense Ridl.
Bulbophyllum mutabile (Blume) Lindl.
Bulbophyllum nasseri Garay
Bulbophyllum negrosianum Ames
Bulbophyllum nemorale L. O. Williams
Bulbophyllum nymphopolitanum Kraenzl.
Bulbophyllum odoratum (Blume) Lindl.
Bulbophyllum ornatissimum (Rchb. f.) J. J. Sm.
Bulbophyllum orthoglossum Kraenzl.
Bulbophyllum othonis (Kuntze) J. J. Sm.
Bulbophyllum pampangense Ames
Bulbophyllum papillipetalum Ames
Bulbophyllum papulosum Garay
Bulbophyllum pardalotum Garay, Hamer & Siegerist
Bulbophyllum penduliscapum J. J. Sm.
Bulbophyllum peramoenum Ames
Bulbophyllum philippinense Ames
Bulbophyllum piestoglossum J. J. Verm.
Bulbophyllum pleurothalloides Ames
Bulbophyllum plumatum Ames
Bulbophyllum profusum Ames
Bulbophyllum puguahaanense Ames
Bulbophyllum putidum (Teijsm. & Binn.) J. J. Sm.
Bulbophyllum recurvilabre Garay
Bulbophyllum reflexum Ames & C. Schweinf.
Bulbophyllum reilloi Ames
Bulbophyllum rhizomatosum Ames & C. Schweinf.
Bulbophyllum santosii Ames
Bulbophyllum sapphirinum Ames
Bulbophyllum saurocephalum Rchb. f.
Bulbophyllum savaiense Schltr. subsp. **subcubicum** (J. J. Sm.) J. J. Verm.
Bulbophyllum schefferi (Kuntze) Schltr.
Bulbophyllum sempiternum Ames
Bulbophyllum sensile Ames
Bulbophyllum sibuyanense Ames
Bulbophyllum sigmoideum Ames & C. Schweinf.
Bulbophyllum stellatum Ames
Bulbophyllum subaequale Ames
Bulbophyllum superfluum Kraenzl.

Bulbophyllum surigaense Ames & Quisumb.
Bulbophyllum toppingii Ames
Bulbophyllum tortuosum (Blume) Lindl.
Bulbophyllum trigonosepalum Kraenzl.
Bulbophyllum unguiculatum Rchb. f.
Bulbophyllum uniflorum (Blume) Hassk.
Bulbophyllum vagans Ames & Rolfe
Bulbophyllum vaginatum (Lindl.) Rchb. f.
Bulbophyllum vermiculare Hook. f.
Bulbophyllum vinaceum Ames & C. Schweinf.
Bulbophyllum weberi Ames
Bulbophyllum wenzelii Ames
Bulbophyllum whitfordii Rolfe ex Ames
Bulbophyllum williamsii A. D. Hawkes
Bulbophyllum woelfliae Garay, Senghas & Lemcke
Bulbophyllum zambalense Ames
Bulbophyllum zamboangense Ames

RÉUNION (FRENCH) / RÉUNION (FRANÇAIS) / REUNIÓN (FRANCESCA)

Bulbophyllum caespitosum Thouars
Bulbophyllum clavatum Thouars
Bulbophyllum commersonii Thouars
Bulbophyllum cordemoyi Frapp. ex Cordem.
Bulbophyllum curvibulbum Frapp. ex Cordem.
Bulbophyllum cylindrocarpum Frapp ex Cordem.
Bulbophyllum cylindrocarpum Frapp ex Cordem. var. **aurantiacum** Frapp. ex Cordem.
Bulbophyllum cylindrocarpum Frapp ex Cordem. var. **olivacea** Frapp. ex Cordem.
Bulbophyllum densum Thouars
Bulbophyllum frappieri Schltr.
Bulbophyllum herbula Frapp. ex Cordem.
Bulbophyllum incurvum Thouars
Bulbophyllum lineare Frapp. ex Cordem.
Bulbophyllum longiflorum Thouars
Bulbophyllum macrocarpum Frapp. ex Cordem.
Bulbophyllum nervulosum Frapp. ex Cordem.
Bulbophyllum nutans (Thouars) Thouars
Bulbophyllum occlusum Ridl.
Bulbophyllum occultum Thouars
Bulbophyllum pendulum Thouars
Bulbophyllum prismaticum Thouars
Bulbophyllum variegatum Thouars

RWANDA / RWANDA (LE) / RWANDA

Bulbophyllum burttii Summerh.
Bulbophyllum cochleatum Lindl.
Bulbophyllum cochleatum Lindl. var. **bequaertii** (De Wild.) J. J. Verm.
Bulbophyllum cochleatum Lindl. var. **brachyanthum** (Summerh.) J. J. Verm.
Bulbophyllum cochleatum Lindl. var. **tenuicaule** (Lindl.) J. J. Verm.
Bulbophyllum comatum Lindl. var. **inflatum** (Rolfe) J. J. Verm.

Bulbophyllum expallidum J. J. Verm.
Bulbophyllum josephi (Kuntze) Summerh.
Bulbophyllum kivuense J. J. Verm.
Bulbophyllum oreonastes Rchb. f.
Bulbophyllum prorepens Summerh.
Bulbophyllum sandersonii (Hook. f.) Rchb. f.
Bulbophyllum unifoliatum De Wild.
Bulbophyllum vulcanicum Kraenzl.

SAMOA / SAMOA (LE) / SAMOA

Bulbophyllum atrorubens Schltr.
Bulbophyllum betchei F. Muell.
Bulbophyllum distichobulbum P. J. Cribb
Bulbophyllum ebulbe Schltr.
Bulbophyllum longiflorum Thouars
Bulbophyllum longiscapum Rolfe
Bulbophyllum membranaceum Teijsm. & Binn.
Bulbophyllum pachyanthum Schltr.
Bulbophyllum samoanum Schltr.
Bulbophyllum savaiense Schltr.
Bulbophyllum trachyanthum Kraenzl.

SAO TOME AND PRINCIPE / SAO TOMÉ-ET-PRINCIPE / SANTO TOMÉ Y PRÍNCIPE

Bulbophyllum cochleatum Lindl.
Bulbophyllum cochleatum Lindl. var. **tenuicaule** (Lindl.) J. J. Verm.
Bulbophyllum curvimentatum J. J. Verm.
Bulbophyllum falcatum (Lindl.) Rchb. f. var. **velutinum** (Lindl.) J. J. Verm.
Bulbophyllum imbricatum Lindl.
Bulbophyllum intertextum Lindl.
Bulbophyllum lizae J. J. Verm.
Bulbophyllum luciphilum Stevart
Bulbophyllum maximum (Lindl.) Rchb. f.
Bulbophyllum mediocre Summerh.
Bulbophyllum resupinatum Ridl.

SEYCHELLES / SEYCHELLES (LES) / SEYCHELLES

Bulbophyllum humblottii Rolfe
Bulbophyllum intertextum Lindl.
Bulbophyllum longiflorum Thouars

SIERRA LEONE / SIERRA LEONE (LA) / SIERRA LEONA

Bulbophyllum acutebracteatum De Wild.
Bulbophyllum barbigerum Lindl.
Bulbophyllum bidenticulatum J. J. Verm.
Bulbophyllum calyptratum Kraenzl.
Bulbophyllum calyptratum Kraenzl. var. **graminifolium** (Summerh.) J. J. Verm.
Bulbophyllum calyptratum Kraenzl. var. **lucifugum** (Summerh.) J. J. Verm.

Bulbophyllum cochleatum Lindl.
Bulbophyllum cocoinum Bateman ex Lindl.
Bulbophyllum colubrinum (Rchb. f.) Rchb. f.
Bulbophyllum comatum Lindl. var. **inflatum** (Rolfe) J. J. Verm.
Bulbophyllum denticulatum Rolfe
Bulbophyllum falcatum (Lindl.) Rchb. f.
Bulbophyllum falcatum (Lindl.) Rchb. f. var. **bufo** (Lindl.) J. J. Verm.
Bulbophyllum falcatum (Lindl.) Rchb. f. var. **velutinum** (Lindl.) J. J. Verm.
Bulbophyllum fuscum Lindl.
Bulbophyllum fuscum Lindl. var. **melinostachyum** (Schltr) J. J. Verm.
Bulbophyllum imbricatum Lindl.
Bulbophyllum intertextum Lindl.
Bulbophyllum lupulinum Lindl.
Bulbophyllum maximum (Lindl.) Rchb. f.
Bulbophyllum nigritianum Rendle
Bulbophyllum oreonastes Rchb. f.
Bulbophyllum parvum Summerh.
Bulbophyllum pipio Rchb. f.
Bulbophyllum pumilum (Sw.) Lindl.
Bulbophyllum resupinatum Ridl. var. **filiforme** (Kraenzl.) J. J. Verm.
Bulbophyllum saltatorium Lindl.
Bulbophyllum saltatorium Lindl. var. **calamarium** (Lindl.) J. J. Verm.
Bulbophyllum scaberulum (Rolfe) Bolus
Bulbophyllum scariosum Summerh.
Bulbophyllum tetragonum Lindl.

SINGAPORE / SINGAPOUR / SINGAPUR

Bulbophyllum acuminatum (Ridl.) Ridl.
Bulbophyllum botryophorum Ridl.
Bulbophyllum cleistogamum Ridl.
Bulbophyllum limbatum Lindl.
Bulbophyllum macrochilum Rolfe
Bulbophyllum makoyanum (Rchb. f.) Ridl.
Bulbophyllum mirum J. J. Sm.
Bulbophyllum nigropurpureum Carr
Bulbophyllum pileatum Lindl.
Bulbophyllum restrepia (Ridl.) Ridl.
Bulbophyllum ruficaudatum Ridl.
Bulbophyllum rugosum Ridl.
Bulbophyllum singaporeanum Schltr.
Bulbophyllum striatellum Ridl.
Bulbophyllum trifolium Ridl.

SOLOMON ISLANDS / ILES SALOMON / ISLAS SALOMÓN

Bulbophyllum antenniferum (Lindl.) Rchb. f.
Bulbophyllum betchei F. Muell.
Bulbophyllum blumei (Lindl.) J. J. Sm.
Bulbophyllum bulliferum J. J. Sm.
Bulbophyllum cerinum Schltr.
Bulbophyllum chloranthum Schltr.
Bulbophyllum citrinilabre J. J. Sm.

Bulbophyllum cylindrobulbum Schltr.
Bulbophyllum dennisii J. J. Wood
Bulbophyllum dichotomum J. J. Sm.
Bulbophyllum ebulbe Schltr.
Bulbophyllum epibulbon Schltr.
Bulbophyllum fractiflexum J. J. Sm.
Bulbophyllum fractiflexum J. J. Sm. subsp. **salomonense** J. J. Verm. & A. L. Lamb
Bulbophyllum graciliscapum Schltr.
Bulbophyllum gracillimum (Rolfe) Rolfe
Bulbophyllum grandiflorum Blume
Bulbophyllum hahlianum Schltr.
Bulbophyllum hassalli Kores
Bulbophyllum levatii Kraenzl.
Bulbophyllum longiflorum Thouars
Bulbophyllum longiscapum Rolfe
Bulbophyllum luckraftii F. Muell.
Bulbophyllum macranthum Lindl.
Bulbophyllum manobulbum Schltr.
Bulbophyllum medusae (Lindl.) Rchb. f.
Bulbophyllum melanoxanthum J. J. Verm. & B. A. Lewis
Bulbophyllum membranaceum Teijsm. & Binn.
Bulbophyllum microrhombos Schltr.
Bulbophyllum novae-hiberniae Schltr.
Bulbophyllum orbiculare J. J. Sm.
Bulbophyllum pachyanthum Schltr.
Bulbophyllum pachyglossum Schltr.
Bulbophyllum piestobulbon Schltr.
Bulbophyllum samoanum Schltr.
Bulbophyllum stenochilum Schltr.
Bulbophyllum stenophyllum Schltr.
Bulbophyllum trachyanthum Kraenzl.
Bulbophyllum trachyglossum Schltr.
Bulbophyllum ulcerosum J. J. Sm.
Pedilochilus ciliolatum Schltr.

SOUTH AFRICA / AFRIQUE DU SUD (L.) / SUDAFRICA

Bulbophyllum acutebracteatum De Wild. var. **rubrobrunneopapillosum** (De Wild.) J. J. Verm.
Bulbophyllum cochleatum Lindl.
Bulbophyllum elliottii Rolfe
Bulbophyllum sandersonii (Hook. f.) Rchb. f.
Bulbophyllum scaberulum (Rolfe) Bolus

SRI LANKA / SRI LANKA / SRI LANKA

Bulbophyllum crassifolium Thwaites ex Trimen
Bulbophyllum elegans Gardner ex Thwaites
Bulbophyllum elliae Rchb. f.
Bulbophyllum fischeri Seidenf.
Bulbophyllum macraei (Lindl.) Rchb. f.
Bulbophyllum maskeliyense Livera
Bulbophyllum petiolare Thwaites

Bulbophyllum purpureum Thwaites
Bulbophyllum thwaitesii Rchb. f.
Bulbophyllum tricarinatum Petch
Bulbophyllum trimeni (Hook. f. ex Trim.) J. J. Sm.
Bulbophyllum wightii Rchb. f.

SUDAN / SOUDAN (LE) / SUDÁN (EL)

Bulbophyllum cochleatum Lindl.
Bulbophyllum scaberulum (Rolfe) Bolus

SURINAME (REPUBLIC OF) / RÉBUBLIQUE DU SURINAME (LA) / REPÚBLICA DE SURINAME (LA)

Bulbophyllum bracteolatum Lindl.
Bulbophyllum kegelii Hamer & Garay

TAIWAN, PROVINCE OF CHINA

Bulbophyllum albociliatum (Liu & Su) Nakajima
Bulbophyllum albociliatum (Liu & Su) Nakajima var. **Weiminianum** T. P. Lin & Kuo Huang
Bulbophyllum aureolabellum T. P. Lin
Bulbophyllum drymoglossum Maxim.
Bulbophyllum flaviflorum (Liu & Su) Seidenf.
Bulbophyllum formosanum (Rolfe) Nakajima
Bulbophyllum hirundinis (Gagnep.) Seidenf.
Bulbophyllum japonicum (Makino) Makino
Bulbophyllum macraei (Lindl.) Rchb. f.
Bulbophyllum macraei (Lindl.) Rchb. f. var. **autumnale** (Fukuyama) S. S. Ying
Bulbophyllum melanoglossum Hayata
Bulbophyllum omerandrum Hayata
Bulbophyllum pectenveneris (Gagnep.) Seidenf.
Bulbophyllum pectinatum Finet
Bulbophyllum retusiusculum Rchb. f.
Bulbophyllum riyanum Fukuyama
Bulbophyllum rubrolabellum T. P. Lin
Bulbophyllum setaceum T. P. Lin
Bulbophyllum taiwanense (Fukuyama) Nakajima
Bulbophyllum tokioi Fukuyama
Bulbophyllum umbellatum Lindl.
Sunipia andersonii (King & Pantl.) P. F. Hunt

TANZANIA, (UNITED REPUBLIC OF) / RÉPUBLIQUE-UNIE DE TANZANIE (LA) / REPÚBLICA UNIDA DE TANZANÍA (LA)

Bulbophyllum bavonis J. J. Verm.
Bulbophyllum cochleatum Lindl.
Bulbophyllum cochleatum Lindl. var. **bequaertii** (De Wild.) J. J. Verm.
Bulbophyllum cochleatum Lindl. var. **brachyanthum** (Summerh.) J. J. Verm.
Bulbophyllum concatenatum P. J. Cribb & P. Taylor

Bulbophyllum elliottii Rolfe
Bulbophyllum encephalodes Summerh.
Bulbophyllum expallidum J. J. Verm.
Bulbophyllum fuscum Lindl. var. **melinostachyum** (Schltr) J. J. Verm.
Bulbophyllum gilgianum Kraenzl.
Bulbophyllum humblottii Rolfe
Bulbophyllum imbricatum Lindl.
Bulbophyllum inornatum J. J. Verm.
Bulbophyllum intertextum Lindl.
Bulbophyllum josephi (Kuntze) Summerh.
Bulbophyllum longiflorum Thouars
Bulbophyllum maximum (Lindl.) Rchb. f.
Bulbophyllum sandersonii (Hook. f.) Rchb. f.
Bulbophyllum scaberulum (Rolfe) Bolus
Bulbophyllum scaberulum (Rolfe) Bolus var. **crotalicaudatum** J. J. Verm.
Bulbophyllum stolzii Schltr.
Bulbophyllum unifoliatum De Wild.
Bulbophyllum unifoliatum De Wild. subsp. **flectens** (P. J. Cribb & Taylor) J. J. Verm.

THAILAND / THAÏLANDE (LA) / TAILANDIA

Acrochaene punctata Lindl.
Bulbophyllum abbrevilabium Carr
Bulbophyllum acuminatum (Ridl.) Ridl.
Bulbophyllum adangense Seidenf.
Bulbophyllum adjungens Seidenf.
Bulbophyllum affine Lindl.
Bulbophyllum albibracteum Seidenf.
Bulbophyllum albidostylidium Seidenf.
Bulbophyllum alcicorne C. S. P. Parish & Rchb. f.
Bulbophyllum allenkerrii Seidenf.
Bulbophyllum angusteovatum Seidenf.
Bulbophyllum annandalei Ridl.
Bulbophyllum apodum Hook. f.
Bulbophyllum auratum (Lindl.) Rchb. f.
Bulbophyllum auricomum Lindl.
Bulbophyllum biflorum Teijsm. & Binn.
Bulbophyllum biseriale Carr
Bulbophyllum bisetoides Seidenf.
Bulbophyllum bisetum Lindl.
Bulbophyllum bittnerianum Schltr.
Bulbophyllum blepharistes Rchb. f.
Bulbophyllum bractescens Rolfe ex Kerr
Bulbophyllum brevistylidium Seidenf.
Bulbophyllum capillipes C. S. P. Parish & Rchb. f.
Bulbophyllum careyanum (Hook. f.) Sprengel
Bulbophyllum cariniflorum Rchb. f.
Bulbophyllum comosum Collett & Hemsl.
Bulbophyllum concinnum Hook. f.
Bulbophyllum corallinum Tixier & Guillaumin
Bulbophyllum corolliferum J. J. Sm.
Bulbophyllum crassipes Hook. f.
Bulbophyllum cupreum Lindl.
Bulbophyllum dayanum Rchb. f.
Bulbophyllum dentiferum Ridl.

Bulbophyllum depressum King & Pantl.
Bulbophyllum dhaninivatii Seidenf.
Bulbophyllum dickasonii Seidenf.
Bulbophyllum didymotropis Seidenf.
Bulbophyllum dissitiflorum Seidenf.
Bulbophyllum echinulus Seidenf.
Bulbophyllum ecornutum (J. J. Sm.) J. J. Sm.
Bulbophyllum elassonotum Summerh.
Bulbophyllum farinulentum J. J. Sm.
Bulbophyllum forresti Seidenf.
Bulbophyllum gibbolabium Seidenf.
Bulbophyllum gracillimum (Rolfe) Rolfe
Bulbophyllum guttifilum Seidenf.
Bulbophyllum gymnopus Hook. f.
Bulbophyllum gyrochilum Seidenf.
Bulbophyllum haniffii Carr
Bulbophyllum helenae (Kuntze) J. J. Sm.
Bulbophyllum hirtulum Ridl.
Bulbophyllum hirtum (J. E. Sm.) Lindl.
Bulbophyllum hymenanthum Hook. f.
Bulbophyllum intricatum Seidenf.
Bulbophyllum kanburiense Seidenf.
Bulbophyllum khaoyaiense Seidenf.
Bulbophyllum khasyanum Griff.
Bulbophyllum lanuginosum J. J. Verm.
Bulbophyllum lasiochilum C. S. P. Parish & Rchb. f.
Bulbophyllum laxiflorum (Blume) Lindl.
Bulbophyllum lemniscatoides Rolfe
Bulbophyllum lemniscatum C. S. P. Parish ex Hook. f.
Bulbophyllum leopardinum (Wall.) Lindl.
Bulbophyllum liliacinum Ridl.
Bulbophyllum limbatum Lindl.
Bulbophyllum lindleyanum Griff.
Bulbophyllum longebracteatum Seidenf.
Bulbophyllum longibracteatum Seidenf.
Bulbophyllum longissimum (Ridl.) Ridl.
Bulbophyllum luanii Tixier
Bulbophyllum macranthum Lindl.
Bulbophyllum macrocoleum Seidenf.
Bulbophyllum medusae (Lindl.) Rchb. f.
Bulbophyllum membranaceum Teijsm. & Binn.
Bulbophyllum microglossum Ridl.
Bulbophyllum micropetalum Rchb. f.
Bulbophyllum microtepalum Rchb. f.
Bulbophyllum monanthos Ridl.
Bulbophyllum moniliforme C. S. P. Parish & Rchb. f.
Bulbophyllum morphologorum Kraenzl.
Bulbophyllum muscarirubrum Seidenf.
Bulbophyllum mutabile (Blume) Lindl.
Bulbophyllum nanopetalum Seidenf.
Bulbophyllum nesiotes Seidenf.
Bulbophyllum nigrescens Rolfe
Bulbophyllum nigripetalum Rolfe
Bulbophyllum nipondhii Seidenf.

Bulbophyllum notabilipetalum Seidenf.
Bulbophyllum odoratissimum (J. E. Sm.) Lindl.
Bulbophyllum orectopetalum Garay, Hamer & Siegerist
Bulbophyllum orientale Seidenf.
Bulbophyllum ovatilabellum Seidenf.
Bulbophyllum ovatum Seidenf.
Bulbophyllum pallidum Seidenf.
Bulbophyllum parviflorum C. S. P. Parish & Rchb. f.
Bulbophyllum patens King ex Hook. f.
Bulbophyllum pectinatum Finet
Bulbophyllum peninsulare Seidenf.
Bulbophyllum pentaneurum Seidenf.
Bulbophyllum phayamense Seidenf.
Bulbophyllum physocoryphum Seidenf.
Bulbophyllum picturatum (Lodd.) Rchb.f.
Bulbophyllum planibulbe (Ridl.) Ridl.
Bulbophyllum polliculosum Seidenf.
Bulbophyllum polyrhizum Lindl.
Bulbophyllum proboscideum (Gagnep.) Seidenf. & Smitinand
Bulbophyllum propinquum Kraenzl.
Bulbophyllum protractum Hook. f.
Bulbophyllum pseudopicturatum (Garay) Sieder & Kiehn †
Bulbophyllum psittacoglossum Rchb. f.
Bulbophyllum pulchellum Ridl.
Bulbophyllum pumilio C. S. P. Parish & Rchb. f.
Bulbophyllum purpurascens Teijsm. & Binn.
Bulbophyllum putidum (Teijsm. & Binn.) J. J. Sm.
Bulbophyllum putii Seidenf.
Bulbophyllum reclusum Seidenf.
Bulbophyllum reichenbachii (Kuntze) Schltr.
Bulbophyllum repens Griff.
Bulbophyllum reptans (Lindl.) Lindl.
Bulbophyllum retusiusculum Rchb. f.
Bulbophyllum rolfeanum Seidenf. & Smitinand
Bulbophyllum rubroguttatum Seidenf.
Bulbophyllum rufilabrum C. S. P. Parish ex Hook. f.
Bulbophyllum rufinum Rchb. f.
Bulbophyllum rugosisepalum Seidenf.
Bulbophyllum sanitii Seidenf.
Bulbophyllum scaphiforme J. J. Verm.
Bulbophyllum secundum Hook. f.
Bulbophyllum shanicum King & Pantl.
Bulbophyllum shweliense W. W. Sm.
Bulbophyllum siamense Rchb. f.
Bulbophyllum sicyobulbon C. S. P. Parish & Rchb. f.
Bulbophyllum sillemianum Rchb. f.
Bulbophyllum simplicilabellum Seidenf.
Bulbophyllum smitinandii Seidenf. & Thorut
Bulbophyllum spathulatum (Rolfe ex E. Cooper) Seidenf.
Bulbophyllum stenobulbon C. S. P. Parish & Rchb. f.
Bulbophyllum striatum (Griff.) Rchb. f.
Bulbophyllum suavissimum Rolfe
Bulbophyllum subtenellum Seidenf.
Bulbophyllum sukhakulii Seidenf.
Bulbophyllum sutepense (Rolfe ex Downie) Seidenf. & Smitinand

Bulbophyllum taeniophyllum C. S. P. Parish & Rchb. f.
Bulbophyllum tenuifolium (Blume) Lindl.
Bulbophyllum thaiorum J. J. Sm.
Bulbophyllum tortuosum (Blume) Lindl.
Bulbophyllum trichocephalum (Schltr.) T. Tang & F. T. Wang
Bulbophyllum tricorne Seidenf. & Smitinand
Bulbophyllum tricornoides Seidenf.
Bulbophyllum tridentatum Kraenzl.
Bulbophyllum tripaleum Seidenf.
Bulbophyllum tripudians C. S. P. Parish & Rchb. f.
Bulbophyllum triviale Seidenf.
Bulbophyllum umbellatum Lindl.
Bulbophyllum unciniferum Seidenf.
Bulbophyllum vaginatum (Lindl.) Rchb. f.
Bulbophyllum violaceolabellum Seidenf.
Bulbophyllum wallichi Rchb. f.
Bulbophyllum wangkaense Seidenf.
Bulbophyllum wendlandianum (Kraenzl.) Dammer
Bulbophyllum xylophyllum C. S. P. Parish & Rchb. f.
Drymoda picta Lindl.
Drymoda siamensis Schltr.
Monomeria barbata Lindl.
Monomeria longipes (Rchb. f.) Aver.
Sunipia andersonii (King & Pantl.) P. F. Hunt
Sunipia angustipetala Seidenf.
Sunipia annamensis (Ridl.) P. F. Hunt
Sunipia australis (Seidenf) P. F. Hunt
Sunipia bicolor Lindl.
Sunipia cumberlegei (Seidenf) P. F. Hunt
Sunipia grandiflora (Rolfe) P. F. Hunt
Sunipia kachinensis Seidenf.
Sunipia minor (Seidenf) P. F. Hunt
Sunipia racemosa (Sm.) T. Tang & F. T. Wang
Sunipia rimannii (Rchb. f.) Seidenf.
Sunipia soidaoensis (Seidenf) P. F. Hunt
Sunipia thailandica (Seidenf & Smit.) P. Hunt
Sunipia virens (Lindl.) P. F. Hunt
Sunipia viridis (Seidenf) P. F. Hunt
Trias intermedia Seidenf. & Smitinand
Trias mollis Seidenf.
Trias nana Seidenf.
Trias nasuta (Rchb. f.) Stapf
Trias oblonga Lindl.
Trias picta (C. S. P. Parish & Rchb. f.) C. S. P. Parish ex Hemsl.
Trias rosea (Ridl.) Seidenf.

TOGO / TOGO (LE) / TOGO (EL)

Bulbophyllum falcatum (Lindl.) Rchb. f.

TONGA / TONGA (LES) / TONGA

Bulbophyllum longiscapum Rolfe

Bulbophyllum membranaceum Teijsm. & Binn.
Bulbophyllum pachyanthum Schltr.

TRINIDAD AND TOBAGO / TRINITÉ-ET-TOBAGO (LA) / TRINIDAD Y TOBAGO

Bulbophyllum kegelii Hamer & Garay

UGANDA / OUGANDA (L.) / UGANDA

Bulbophyllum carnosisepalum J. J. Verm.
Bulbophyllum cochleatum Lindl.
Bulbophyllum cochleatum Lindl. var. **bequaertii** (De Wild.) J. J. Verm.
Bulbophyllum cochleatum Lindl. var. **brachyanthum** (Summerh.) J. J. Verm.
Bulbophyllum cocoinum Bateman ex Lindl.
Bulbophyllum encephalodes Summerh.
Bulbophyllum falcatum (Lindl.) Rchb. f.
Bulbophyllum fuscum Lindl. var. **melinostachyum** (Schltr) J. J. Verm.
Bulbophyllum josephi (Kuntze) Summerh.
Bulbophyllum longiflorum Thouars
Bulbophyllum maximum (Lindl.) Rchb. f.
Bulbophyllum oreonastes Rchb. f.
Bulbophyllum oxychilum Schltr.
Bulbophyllum saltatorium Lindl. var. **albociliatum** (Finet) J. J. Verm.
Bulbophyllum sandersonii (Hook. f.) Rchb. f.
Bulbophyllum schimperianum Kraenzl.
Bulbophyllum vulcanicum Kraenzl.

UNITED STATES OF AMERICA (THE) / ETATS-UNIS D.AMÉRIQUE (LES) / ESTADOS UNIDOS DE AMÉRICA (LOS

Bulbophyllum pachyrachis (A. Rich.) Griseb.

VANUATU / VANUATU / VANUATU

Bulbophyllum absconditum J. J. Sm.
Bulbophyllum atrorubens Schltr.
Bulbophyllum betchei F. Muell.
Bulbophyllum dichotomum J. J. Sm.
Bulbophyllum ebulbe Schltr.
Bulbophyllum graciliscapum Schltr.
Bulbophyllum levatii Kraenzl.
Bulbophyllum longiflorum Thouars
Bulbophyllum longiscapum Rolfe
Bulbophyllum membranaceum Teijsm. & Binn.
Bulbophyllum microrhombos Schltr.
Bulbophyllum minutipetalum Schltr.
Bulbophyllum neo-caledonicum Schltr.
Bulbophyllum neoebudicus (Garay, Hamer & Siegerist) Sieder †
Bulbophyllum rhomboglossum Schltr.

Bulbophyllum samoanum Schltr.
Bulbophyllum santoense J. J. Verm.
Bulbophyllum savaiense Schltr.
Bulbophyllum stenophyllum Schltr.
Bulbophyllum vutimenaense B. A. Lewis
Pedilochilus hermonii P. J. Cribb & B. Lewis

VENEZUELA / VENEZUELA (LE) / VENEZUELA

Bulbophyllum aristatum (Rchb. f.) Hemsl.
Bulbophyllum bracteolatum Lindl.
Bulbophyllum dunstervillei Garay
Bulbophyllum exaltatum Lindl.
Bulbophyllum geraense Rchb. f.
Bulbophyllum lehmannianum Kraenzl.
Bulbophyllum manarae Foldats
Bulbophyllum meridense Rchb. f.
Bulbophyllum meristorhachis Garay & Dunst.
Bulbophyllum morenoi Dodson & Vasquez
Bulbophyllum oerstedii (Rchb. f.) Hemsl.
Bulbophyllum pachyrachis (A. Rich.) Griseb.
Bulbophyllum roraimense Rolfe
Bulbophyllum steyermarkii Foldats
Bulbophyllum vareschii Foldats

VIET NAM / VIET NAM (LE) / VIET NAM

Bulbophyllum abbrevilabium Carr
Bulbophyllum affine Lindl.
Bulbophyllum ambrosia (Hance) Schltr.
Bulbophyllum andersonii (Hook. f.) J. J. Sm.
Bulbophyllum annamense (Garay) Sieder & Kiehn †
Bulbophyllum apodum Hook. f.
Bulbophyllum arcutilabium Aver.
Bulbophyllum astelidum Aver.
Bulbophyllum atrosanguineum Aver.
Bulbophyllum averyanovii Seidenf.
Bulbophyllum bariense Gagnep.
Bulbophyllum bisetoides Seidenf.
Bulbophyllum blepharistes Rchb. f.
Bulbophyllum boulbetii Tixier
Bulbophyllum bryoides Guillaumin
Bulbophyllum careyanum (Hook. f.) Sprengel
Bulbophyllum catenarium Ridl.
Bulbophyllum clandestinum Lindl.
Bulbophyllum clipeibulbum J. J. Verm.
Bulbophyllum concinnum Hook. f.
Bulbophyllum corallinum Tixier & Guillaumin
Bulbophyllum crassiusculifolium Aver.
Bulbophyllum dalatense Gagnep.
Bulbophyllum dayanum Rchb. f.
Bulbophyllum delitescens Hance
Bulbophyllum elassonotum Summerh.

Bulbophyllum elatum (Hook. f.) J. J. Sm.
Bulbophyllum emarginatum (Finet) J. J. Sm.
Bulbophyllum evrardii Gagnep.
Bulbophyllum farreri (W. W. Sm.) Seidenf.
Bulbophyllum fibratum (Gagnep.) Seidenf.
Bulbophyllum fischeri Seidenf.
Bulbophyllum flabellum-veneris (J. Koenig) Seidenf. & Ormerod ex Aver.
Bulbophyllum flaviflorum (Liu & Su) Seidenf.
Bulbophyllum frostii Summerh.
Bulbophyllum funingense Z. H. Tsi & H. C. Chen
Bulbophyllum furcatum Aver.
Bulbophyllum guttulatum (Hook. f.) Balakr.
Bulbophyllum hiepii Aver.
Bulbophyllum hirtulum Ridl.
Bulbophyllum hirtum (J. E. Sm.) Lindl.
Bulbophyllum hirundinis (Gagnep.) Seidenf.
Bulbophyllum hymenanthum Hook. f.
Bulbophyllum ignevenosum Carr
Bulbophyllum insulsum (Gagnep.) Seidenf.
Bulbophyllum kanburiense Seidenf.
Bulbophyllum khasyanum Griff.
Bulbophyllum kontumense Gagnep.
Bulbophyllum laxiflorum (Blume) Lindl.
Bulbophyllum lemniscatoides Rolfe
Bulbophyllum lockii Aver & Averyanova
Bulbophyllum longibrachiatum Z. H. Tsi
Bulbophyllum longiflorum Thouars
Bulbophyllum luanii Tixier
Bulbophyllum macraei (Lindl.) Rchb. f.
Bulbophyllum macranthum Lindl.
Bulbophyllum macrocoleum Seidenf.
Bulbophyllum mastersianum (Rolfe) J. J. Sm.
Bulbophyllum monanthum (Kuntze) J. J. Sm.
Bulbophyllum moniliforme C. S. P. Parish & Rchb. f.
Bulbophyllum morphologorum Kraenzl.
Bulbophyllum ngoclinhensis Aver.
Bulbophyllum nigrescens Rolfe
Bulbophyllum odoratissimum (J. E. Sm.) Lindl.
Bulbophyllum orectopetalum Garay, Hamer & Siegerist
Bulbophyllum orientale Seidenf.
Bulbophyllum parviflorum C. S. P. Parish & Rchb. f.
Bulbophyllum pectenveneris (Gagnep.) Seidenf.
Bulbophyllum pectinatum Finet
Bulbophyllum penicillium C. S. P. Parish & Rchb. f.
Bulbophyllum pinicolum Gagnep.
Bulbophyllum poilanei Gagnep.
Bulbophyllum protractum Hook. f.
Bulbophyllum psittacoglossum Rchb. f.
Bulbophyllum pumilio C. S. P. Parish & Rchb. f.
Bulbophyllum purpureifolium Aver.
Bulbophyllum putidum (Teijsm. & Binn.) J. J. Sm.
Bulbophyllum refractum (Zoll.) Rchb. f.
Bulbophyllum reptans (Lindl.) Lindl.
Bulbophyllum retusiusculum Rchb. f.
Bulbophyllum rufinum Rchb. f.

Bulbophyllum scaphiforme J. J. Verm.
Bulbophyllum schwarzii Sieder
Bulbophyllum secundum Hook. f.
Bulbophyllum semiteretifolium Gagnep.
Bulbophyllum shweliense W. W. Sm.
Bulbophyllum sigaldiae Guillaumin
Bulbophyllum simondii Gagnep.
Bulbophyllum smitinandii Seidenf. & Thorut
Bulbophyllum spadiciflorum Tixier
Bulbophyllum spathulatum (Rolfe ex E. Cooper) Seidenf.
Bulbophyllum stenobulbon C. S. P. Parish & Rchb. f.
Bulbophyllum striatum (Griff.) Rchb. f.
Bulbophyllum strigosum (Garay) Sieder & Kiehn †
Bulbophyllum subebulbum Gagnep.
Bulbophyllum taeniophyllum C. S. P. Parish & Rchb. f.
Bulbophyllum thaiorum J. J. Sm.
Bulbophyllum tixieri Seidenf.
Bulbophyllum tortuosum (Blume) Lindl.
Bulbophyllum tripudians C. S. P. Parish & Rchb. f.
Bulbophyllum umbellatum Lindl.
Bulbophyllum vietnamense Seidenf.
Bulbophyllum wallichi Rchb. f.
Bulbophyllum xylophyllum C. S. P. Parish & Rchb. f.
Monomeria barbata Lindl.
Sunipia andersonii (King & Pantl.) P. F. Hunt
Sunipia annamensis (Ridl.) P. F. Hunt
Sunipia cirrhata (Lindl.) P. F. Hunt
Sunipia dichroma (Rolfe) Bân & D. H. Duong
Sunipia pallida (Aver.) Aver.
Sunipia racemosa (Sm.) T. Tang & F. T. Wang
Sunipia scariosa Lindl.
Trias nasuta (Rchb. f.) Stapf

WALLIS AND FUTUNA ISLANDS (DEPENDENT TERRITORY OF FRANCE) / ISLAS WALLIS Y FUTUNA (TERRITORIO DEPENDIENTE DE FRANCIA) / ILES WALLIS-ET-FUTUNA (TERRITOIRE DÉPENDANT DE LA FRANCE

Bulbophyllum longiscapum Rolfe

ZAMBIA / ZAMBIE (LA) / ZAMBIA

Bulbophyllum cochleatum Lindl.
Bulbophyllum elliottii Rolfe
Bulbophyllum encephalodes Summerh.
Bulbophyllum expallidum J. J. Verm.
Bulbophyllum fuscum Lindl. var. **melinostachyum** (Schltr) J. J. Verm.
Bulbophyllum gravidum Lindl.
Bulbophyllum injoloense De Wild. subsp. **pseudoxypterum** (J. J. Verm.) J. J. Verm.
Bulbophyllum intertextum Lindl.
Bulbophyllum josephi (Kuntze) Summerh. var. **mahonii** (Rolfe) J. J. Verm.
Bulbophyllum lupulinum Lindl.

Bulbophyllum maximum (Lindl.) Rchb. f.
Bulbophyllum oreonastes Rchb. f.
Bulbophyllum sandersonii (Hook. f.) Rchb. f.
Bulbophyllum scaberulum (Rolfe) Bolus
Bulbophyllum unifoliatum De Wild.

ZIMBABWE / ZIMBABWE (LE) / ZIMBABWE

Bulbophyllum ballii P. J. Cribb
Bulbophyllum elliottii Rolfe
Bulbophyllum encephalodes Summerh.
Bulbophyllum fuscum Lindl. var. **melinostachyum** (Schltr) J. J. Verm.
Bulbophyllum humblottii Rolfe
Bulbophyllum intertextum Lindl.
Bulbophyllum josephi (Kuntze) Summerh.
Bulbophyllum longiflorum Thouars
Bulbophyllum maximum (Lindl.) Rchb. f.
Bulbophyllum oreonastes Rchb. f.
Bulbophyllum rugosibulbum Summerh.
Bulbophyllum sandersonii (Hook. f.) Rchb. f.
Bulbophyllum scaberulum (Rolfe) Bolus
Bulbophyllum unifoliatum De Wild. subsp. **infracarinatum** (G. Will.) J. J. Verm.
Chaseella pseudohydra Summerh.

New names, combinations and synonyms in *Bulbophyllum* Thouars

A. Sieder and M. Kiehn

Note: While this checklist was in preparation for publication, the 2nd edition of the "Orchids of Madagascar" (Hermans, J., Hermans, C., Du Puy, D., Cribb, P. & Bosser J. 2007) appeared, taking up three names originally included in this paper as taxa in need for being renamed. In order to avoid unnecessary synonyms, the three names of Hermans et al. (2007) are used here and marked with an asterix (*). All other new names and descriptions contained in Hermans et al. (2007) will be included in the follow-up to this checklist.

a) New names:

Bulbophyllum arminii **Sieder & Kiehn nom. nov.**
New name for *Hapalochilus bandischii* Garay, Hamer & Siegerist
　　　　Garay, L. A., Hamer, F. & Siegerist, E. S. (1995): Inquilia Orchidacea.
　　　　Orchidaceae Plaerumque Levyanae. Lindleyana 10: 179.
　　　　Type: W. H. Bandisch s. n., holo AMES
　　　　Origin: Papua New Guinea
　　　　Fig.: Garay, L. A., Hamer, F. & Siegerist, E. S. (1995): Inquilia
　　　　Orchidacea. Orchidaceae Plerumque Levyanae. Lindleyana 10: 180,
　　　　Fig. 3, A.
　　　　Non *Bulbophyllum bandischii* Garay, Hamer & Siegerist
　　　　Garay, L. A., Hamer, F. & Siegerist, E. S. (1992): Bemerkung zu
　　　　Bulbophyllum Section *Hyalosema*: *Bulbophyllum bandischii* Garay,
　　　　Hamer & Siegerist spec. nov., *Bulbophyllum ornithorhynchum* (J. J.
　　　　Sm.) Garay, Hamer & Siegerist, *B. unitubum* J. J. Sm.. Orchidee
　　　　(Hamburg) 43: 262.
　　　　Type: E. S. Siegerist 90-7-3, W. Bandisch 28, AMES
　　　　Origin: Papua New Guinea
　　　　Fig.: Garay, L. A., Hamer, F. & Siegerist, E. S. (1992): Bemerkung zu
　　　　Bulbophyllum Section *Hyalosema*: *Bulbophyllum bandischii* Garay,
　　　　Hamer & Siegerist spec. nov., *Bulbophyllum ornithorhynchum* (J. J.
　　　　Sm.) Garay, Hamer & Siegerist, *B. unitubum* J. J. Sm.. Orchidee
　　　　(Hamburg) 43: 263.
Named after Prof. DI Dr. tech. Armin Szilvinyi (1895-1966), first president of the
Austrian Orchid Association (1949-1957).

Bulbophyllum bulhartii **Sieder & Kiehn nom. nov.**
New name for *Bulbophyllum pulchrum* Schltr.
　　　　Schlechter, R. (1912): Die Orchidaceen von Deutsch-Neu-Guinea.
　　　　Repert. Spec. Nov. Regni Veg. Beih. 1: 710 (publ. Dec. 1912)
　　　　Type: Schlechter 20186, holo B, destroyed
　　　　Origin: New Guinea
　　　　Fig.: Schlechter, R. (1923-1928): Figuren-Atlas zu den Orchidaceen
　　　　von Deutsch-Neu-Guinea. Feddes Repert. Spec. Nov. Regni Veg. XXI:
　　　　Tafel CCXXXI, Nr. 876.
　　　　Syn.: *Hapalochilus striatus* Garay & W. Kittr.
　　　　Garay, L. A. & Kittredge, W. (1986): Notes from the Ames Orchid
　　　　Herbarium. Bot. Mus. Leafl. 30 (3) (1985, publ. 1986): 57.
　　　　Non *Bulbophyllum striatum* (Griff.) Rchb. f.

Reichenbach, H. G. f. (1861): Synopsis plantarum phanerogamicarum. *Bulbophyllum*. In Walpers, W. G., Ann. Bot. Syst. 6: 257.
Syn.: *Dendrobium striatum* Griff.
Griffith, W. (1851): Notulae ad plantas Asiaticas. Part III : Monocotyledonous plants: 318.
Types: Hooker & Thompson s.n., Griffith s.n. (not traced)
Origin: India
Non *Bulbophyllum pulchrum* (N. E. Br.) J. J. Sm.
Smith, J. J. (1912): *Bulbophyllum* Thou. Sect. *Cirrhopetalum*. Bull. Jard. Bot. Buitenzorg 2 (8): 27.
Syn.: *Cirrhopetalum pulchrum* N. E. Br.,
Brown, N. E. (1886): Illustr. Hortic. 33: 139.
Type: Teysmann s.n., holo BO
Origin: Papua New Guinea
Named after Prof. Dr. Vinzenz Bulhart (1885–1965), second president of the Austrian Orchid Society (1957-1965).

Bulbophyllum vakonae Hermans*
New name for *Bulbophyllum ochrochlamys* Schltr.

Schlechter, R. (1924): Orchidaceae Perrierianae. Ein Beitrag zur Orchideenkunde der Insel Madagascar. Repert. Spec. Nov. Regni Veg. Beih. 33: 190.
Type: H. Perrier 11852, iso P
Origin: Madagascar
Fig.: Perrier, H. (1941): Orchidees. Fl. Madagasc. 49 fam. (1)
Non *Bulbophyllum ochrochlamys* Schltr.
Schlechter, R. (1913): Die Orchidaceen von Deutsch-Neu-Guinea. Repert. Spec. Nov. Regni Veg. Beih. 1: 856.
Type: Schlechter R. 17300
Origin: New Guinea
Fig.: Schlechter, R. (1923-1928): Figuren-Atlas zu den Orchidaceen von Deutsch-Neu-Guinea. Feddes Repert. Spec. Nov. Regni Veg. XXI: 1137.
= *Bulbophyllum absconditum* J. J. Sm.

Bulbophyllum mattesii Sieder & Kiehn nom. nov.
New name for *Bulbophyllum oblanceolatum* Schltr.

Schlechter, R. (1913): Die Orchidaceen von Deutsch-Neu-Guinea. Repert. Spec. Nov. Regni Veg. Beih.
1: 817; as „oblanceotum".
Type: Schlechter 17829, holo B, destroyed
Origin: New Guinea
Fig.: Schlechter, R. (1923-1928): Figuren-Atlas zu den Orchidaceen von Deutsch-Neu-Guinea. Feddes Repert. Spec. Nov. Regni Veg. XXI: Tafel CCLXXVIII, Nr. 1064.
Non *Bulbophyllum oblanceolatum* King & Pantl.
King, G. & Pantling, R. (1897): Some new Indo-Malayan Orchids. J. Asiat. Soc. Bengal, Pt. 2, Nat. Hist. LXVI (II/3): 586.
Type: Wray 980 (not traced)
Origin: Malaysia
Named after Paul Mattes, MBA, (1926-2001), fourth president of the Austrian Orchid Association (1982-1988), highly qualified orchid photographer and excellent specialist of the genus *Paphiopedilum*.

Bulbophyllum mona-lisae **Sieder & Kiehn nom. nov.**
New name for *Bulbophyllum melanoglossum* Kraenzl.
   Kraenzlin, F. W. L. (1916): Orchidaceae Novae. Ann. K. K. Naturhist.
   Hofmus. XXX: 60.
   Type: Loher s.n., holo B (?lost); iso AMES
   Origin: Philippines
   Non *Bulbophyllum melanoglossum* Hayata
   Hayata, B. (1914): Icones Plantarum Formosanarum 4. Orchids. Icon.
   Pl. Formosan.: 49
   Type: Taihoku s. n., TOKYO
   Origin: Taiwan
Named after Mona Lisa Steiner (1915-2000). Born in Vienna; studied botany and
zoology at the University of Vienna until 1938. Emigrated to the Philippines in
1938, where she was assistant at the Botany Department of the University of the
Philippines in Manila. PhD at Vienna University in 1954. Returned to Vienna in
1965. Focus of her research was the orchid flora of the Philippines and the Pacific
Islands.

Bulbophyllum perseverans **Hermans***
New name for *Bulbophyllum graciliscapum* H. Perrier
   Perrier, H. (1937): Les Bulbophyllums de Madagascar. Not. Syst.
   (Paris) 6 (2): 107.
   Type: H. Perrier 16513, holo P
   Origin: Madagascar
   Non *Bulbophyllum graciliscapum* Schltr.
   Schlechter, R. (1905): Nachträge zur Flora der Deutschen
   Schutzgebiete in der Südsee (mit Ausschluß Samoa`s und der
   Karolinen) 203.
   Type: Schlechter 13924, holo B destroyed, iso BM, K
   Origin: New Guinea
   Non *Bulbophyllum graciliscapum* Ames & Rolfe
   Ames, O. (1915): The genera and species of Philippine orchids.
   Orchidaceae (Ames) 5: 175-176.
   Type: E. B. Copeland 1127, holo AMES, iso K
   Origin: Philippines
   = *Bulbophyllum apoense* **Schuit. & De Vogel**
   Non *Bulbophyllum graciliscapum* Summerh.
   Summerhayes, V. S. (1954): African Orchids XXII Kew Bull. 8: 579.
   Type: Le Testu 5787, holo K, iso P
   Origin: Africa
   = *Bulbophyllum saltatorium* **Lindl. var. *albociliatum* (Finet) J. J.**
    Vermeulen

Bulbophyllum reifii **Sieder & Kiehn nom. nov.**
New name for *Bulbophyllum rhizomatosum* Schltr.
   Schlechter, R. (1923): Neue Orchidaceen Papuasiens. Bot. Jahrb. Syst.
   58: 131.
   Type: C. Ledermann 7979, K
   Origin: New Guinea
   Non *Bulbophyllum rhizomatosum* Ames & C. Schweinf.
   Ames, O. & Schweinfurth, C. (1920): The orchids of Mt. Kinabalu,
   British North Borneo. Orchidaceae (Ames) 6: 194 f.
   Type: Clemens 106, AMES
   Origin: Borneo
   Non *Bulbophyllum rhizomatosum* Schltr.

Schlechter, R. (1924): Orchidaceae Perrierianae. Ein Beitrag zur
Orchideenkunde der Insel Madagascar. Repert. Spec. Nov. Regni Veg.
Beih. 33: 171-240.
Type: H. Perrier 15735, iso K.
Origin: Madagascar
= *Bulbophyllum obtusilabium* **W. Kittr.**
Named after Reg. Rat. Ing. Kurt Reif (1920-1999), fifth president of the Austrian
Orchid Society (1994-1999).

Bulbophyllum ormerodianum Hermans*
New name for *Bulbophyllum abbreviatum* Schltr.

Schlechter, R. (1924): Orchidaceae Perrierianae. Ein Beitrag zur
Orchideenkunde der Insel Madagascar. Repert. Spec. Nov. Regni Veg.
Beih. 33: 198.
Type: H. Perrier 8054, iso. P
Origin: Madagascar
Non *Bulbophyllum abbreviatum* Rchb. f.
Reichenbach, H. G. f. (1881): Gard. Chron. 2: 70.
Type: Reichenbach Herb. 49447, W
Syn.: *Cirrhopetalum abbreviatum* Rchb. f.
Reichenbach, H. G. f. (1881): Gard. Chron. 2: 70.
Type: Reichenbach Herbarium 49447, W
Syn.: *Bulbophyllum trigonopus* Rchb. f.,
Reichenbach, H. G. f. (1881): Gard. Chron. 2: 71.
Type: Reichenbach Herbarium 49447, W
Syn.: *Cirrhopetalum trigonopus* Rchb. f.
Reichenbach, H. G. f. (1881): Gard. Chron. 2: 71.
Type: Reichenbach Herbarium 49447, W
Origin: Philippines
Descriptions of *B. trigonopus* and *B. abbreviatum* are based on the same type
collection (Reichenbach Herb. 49447).

Bulbophyllum wechsbergii Sieder & Kiehn nom. nov.
New name for *Bulbophyllum rhynchoglossum* Schltr.

Schlechter, R. (1912): Die Orchidaceen von Deutsch-Neu-Guinea
Repert. Spec. Nov. Regni Veg. Beih. 1: 716.
Type: R. Schlechter 19665, holo B, destroyed
Origin: Papua New Guinea
Fig.: Schlechter, R. (1923-1928): Figuren-Atlas zu den Orchidaceen
von Deutsch-Neu-Guinea. Feddes Repert. Spec. Nov. Regni Veg. XXI:
Tafel CCXXXIII, Nr. 885.
Syn.: *Hapalochilus reflexus* Garay & W. Kittr.
Garay, L. A. & Kittredge, W. (1986): Notes from the Ames Orchid
Herbarium. Bot. Mus. Leafl. 30 (3) (1985, publ. 1986): 56.
Non *Bulbophyllum reflexum* Ames & Schweinf.
Ames, O. & Schweinfurth, C. (1920): The orchids of Mt. Kinabalu,
British North Borneo. Orchidaceae (Ames) 6: 192-194.
Type: Clemens 384, AMES
Origin: Borneo
Non *Bulbophyllum rhynchoglossum* Schltr.
Schlechter, R. (1910): Orchidaceae novae et criticae CLII, Decas XIV -
XV. Feddes Repert. Spec. Nov. Regni Veg. 8: 569.
Type: Schlechter 15851, holotypus B, destroyed, iso K (only known
from the type collection)
Origin: Borneo

Fig.: Blütenanalysen neuer Orchideen. IV. Indische und malesische
Orchideen. Feddes Repert. Spec. Nov. Regni Veg. LXXIV, Tafel 63,
Nr. 250.

Named after Walter Wechsberg (1940-2004), orchid gardener at the Austrian
Federal Gardens Schönbrunn until his retirement in 2001; outstanding cultivator
and breeder of orchids.

b) New combinations:

Bulbophyllum annamense (Garay) Sieder & Kiehn comb. nov.

Syn.: *Cirrhopetalum annamense* Garay
Garay, L. A. (1999): Orchid species currently in cultivation. Harvard
Pap. Bot. 4 (1): 306.
Type: Sigaldi 144 (C.R.S.T.151), holo P
Origin: Viet Nam
Fig.: Garay, L. A. (1999): Orchid species currently in cultivation.
Harvard Pap. Bot. 4: 316, Figure 4, C.

Bulbophyllum cercanthum (Garay, Hamer & Siegerist) Sieder & Kiehn comb. nov.

Syn.: *Cirrhopetalum cercanthum* Garay, Hamer & Siegerist
Garay, L. A., Hamer, F. & Siegerist, E. (1996): Inquilina Orchidacea II.
Lindleyana 11: 231.
Type: Charles Nishihira s. n., AMES
Origin: Borneo
Fig.: Garay, L. A., Hamer, F. & Siegerist, E. (1996): Inquilina
Orchidacea II. Lindleyana 11: 232, Fig. 3, A.

Bulbophyllum confusum (Garay, Hamer & Siegerist) Sieder & Kiehn comb. nov.

Syn: *Cirrhopetalum confusum* Garay, Hamer & Siegerist
Garay, L. A., Hamer, F. & Siegerist, E. S. (1994): The Genus
Cirrhopetalum and the genera of the
Bulbophyllum alliance. Nordic J. Bot. 14: 622.
Type: Pl. 121 of King, G. & Pantling, R. (1898): Orchids of the
Sikkim-Himalaya. Ann. Bot. Gard. Calcutta 8: 88.
Origin: W-Himalaya

Bulbophyllum neoebudicus (Garay, Hamer & Siegerist) Sieder & Kiehn comb. nov.

Syn.: *Hapalochilus neoebudicus* Garay, Hamer & Siegerist
Garay, L. A., Hamer, F. & Siegerist, E. S. (1995): Inquilia Orchidacea.
Orchidaceae Plaerumque Levyanae. Lindleyana 10: 182.
Type: Missouri Botanical Garden s.n., holo AMES
Origin: Vanuatu
Fig.: Garay, L. A., Hamer, F. & Siegerist, E. S. (1995): Inquilia
Orchidacea. Orchidaceae Plaerumque Levyanae. Lindleyana 10: 178,
Fig. 2, C.

Bulbophyllum pseudopicturatum (Garay) Sieder & Kiehn comb. nov.

Syn.: *Cirrhopetalum pseudopicturatum* Garay
Garay, L. A. (1999): Orchid species currently in cultivation. Harvard
Pap. Bot. 4 (1): 308.

Type: C. Parisch 23, Holotype K; Isotype Reichenbach Herbar No. 49424, W.
Origin: Thailand
Fig.: Garay, L. A. (1999): Orchid species currently in cultivation. Harvard Pap. Bot. 4: 316, Figure 4, D.

Bulbophyllum strigosum **(Garay) Sieder & Kiehn comb. nov.**
Syn.: *Rhytionanthos strigosum* Garay
Garay, L. A. (1999): Orchid species currently in cultivation. Harvard Pap. Bot. 4 (1): 311.
Type: Without specific locality, ex Hort. Botanical Garden of the University of Vienna, Anton Sieder 250/92, holo WU, iso AMES
Origin: Viet Nam
Fig.: Garay, L. A. (1999): Orchid species currently in cultivation. Harvard Pap. Bot. 4: 315, Fig. 3, D.

c) New synonyms:

Bulbophyllum macrochilum **Rolfe**
Rolfe, R. A. (1896): Bull. Misc. Inform. Kew 16: 45.
Type: Haviland s.n., holo K
Origin: Borneo
Syn. nov.: *Bulbophyllum vanessa* **King & Pantl.**
King, G. & Pantling, R. (1897): Some new Indo-Malayan Orchids. J. Asiat. Soc. Bengal, Pt. 2, Nat. Hist. LXVI (II/3): 587.
Type: Ayre s.n., Scortecchini drawing no. 434 (K)
Origin: Malaysia

Bulbophyllum elegans **Gardner ex Thwaites**
Gardner, G. (1861). Enum. Pl. Zeyl.: 298.
Type: s.coll. C. P. 2350 holo PDA, K
Origin: Sri Lanka
Syn. nov.: *Bulbophyllum balaeniceps* **Rchb. f.**
Reichenbach, H. G. (1863) Hamburger Garten-Blumenzeitung 19: 208.
Type: could not be traced
Remark: this taxon is not a synonym to *Bulbophyllum napellii* **Lindl**.

d) Species to be excluded from *Bulbophyllum* s. lat.:

Bulbophyllum urceolatum **A. D. Hawkes**
Type: Adams, P. A. 10, Herb. Univ. California
Origin: Micronesia
Fig. Hawkes, A. D. (1952): Notes on a Collection of orchids from Ponape, Caroline Islands. Pacific Sci. 6: 4.
According to Schuiteman, Orchid Monogr. 8: 57 (1997), and personal communication of A. Schuiteman via J. J. Vermeulen (Leiden), this taxon seems to be a synonym to *Mediocalcar paradoxum* (Kraenzl.) Schltr. var. *robustum* (Schltr) Schuiteman. Definitely, it is not a member of the genus *Bulbophyllum*.

ANNEX I

NOTES